Annual Student Conference 2019

Held at the 2019 AIChE Annual Meeting

Orlando, Florida, USA
10 - 15 November 2019

ISBN: 978-1-7138-0548-9

Printed from e-media with permission by:

Curran Associates, Inc.
57 Morehouse Lane
Red Hook, NY 12571

AIChE
120 Wall Street, FL 23
New York, NY 10005-4020

Phone: (800) 242-4363
Fax: (203) 775-5177

www.aiche.org

Additional copies of this publication are available from:

Curran Associates, Inc.
57 Morehouse Lane
Red Hook, NY 12571 USA
Phone: 845-758-0400
Fax: 845-758-2633
Email: curran@proceedings.com
Web: www.proceedings.com

TABLE OF CONTENTS

583126 A Day in the Life of a Process Engineer

What does a Process Engineer do at an energy company? What are some of the daily tasks that Process Engineers work on and challenges they face? What skills are needed beyond engineering skills to excel? How do you transition from the academic environment to the professional world? Join Process Engineers at Chevron as they describe what they do on a daily basis and how they integrate a variety of professional skills including teamwork, communication and networking, to be successful.

583110 Making a Lasting Impact: The Lifetime Engineering Applications Project

The Lifetime Engineering Applications Project (LEAP) is a technical project started by AIChE at UCLA. It teaches students how to use process design software to model real-life industrial processes while also considering economic and ethical factors. In only two years, several unique projects have been developed, such as alcohol distillation, ammonia production, crude oil processing, and biofuels synthesis. Since only simulation software is required for the course, the project is low cost yet greatly impactful on the students, as they leave with a deeper understanding of chemical engineering fundamentals and are better prepared for both their senior design projects and challenges they may face in industry. The leads of the project gain experience with preparing and presenting lesson plans while having fun. During this presentation, the current leads of LEAP will be discussing their experiences and how the project has grown.

583111 Mentorship

Mentorship is a great way to involve underclassmen in your student chapter and create a tight-knit community within your department. Come learn about the mentorship program at the University of Michigan AIChE Chapter, including what types of events are successful and how to start your own mentorship program.

583122 Finances and Your First Job – Financial Planning for ChemE Careers

How do I start saving money? What benefits should I enroll in? How do I start planning for retirement? Join the Young Professionals Committee (YPC) for a discussion on the benefits and rewards of financial planning when starting your first job. This session will feature YPC members in a variety of careers (graduate school, industry, government, etc.) and discuss tips for money management and planning for retirement. YPC members will discuss topics such as investing in 401Ks, budgeting, benefits packages, student debt, and planning for retirement. YPC members will also discuss topics that they wish they knew when starting their first jobs.

586427 Process Simulation in the Real World

Most universities only use process simulators in the senior design course. What happens when you graduate? Where would you use these types of tools in the "real world"? Join our interactive workshop & win some fun ChemE prizes!

583112 How to Create a Diverse, Encouraging and Socially Responsible Chapter

AIChE CIMATEC through the "Social Responsibility Nucleus" has developed activities with the proposal to develop concepts of social responsibility in our members and community. The Project has encourage awareness and recognition around women in engineering, minority affairs, LGBTQ+ inclusion, and environmental topics. The presentation of the Workshop will aim to present the ways we can develop social activities with the Chapter and how we can create a social responsibility team, as well as how to involve companies and industrial companies in the activities carried out. The presentation will take as an example the AIChE LGBTQ + and Allies Student Ice Cream Social conducted by our Chapter fully supported by AIChE's LGBTQ + & Allies Initiative and the Foundation's #AllforGood - Engineering for Inclusion program. The event brought together more than 70 people, including students, professors, industry professionals, and university collaborators.

583113 Marketing Your Chapter

This workshop is intended to share best approaches when it comes to communicating student chapters on social media and through email. It is meant to discuss the do's and don'ts of student chapter instagrams, twitters, YouTube accounts, and Facebook pages. Furthermore, it will discuss ideas on how to get more outreach to freshmen and sophomores through email lists, events, class announcements, and more! We can also go through how to structure an efficient student chapter website, and how to run a successful marketing committee. Our marketing committee uses great tools such as Canva and Hootsuite to create lovely, professional flyers and plan our posts, respectively. We will discuss how to utilize tools and the marketing structure we currently have: a marketing committee head, social media director, graphic designer, and newsletter chief. Our online footprint has increased exponentially these past few years and it has helped with attendance, sponsorship, and outreach.

583136 Effective Interviewing for Industrial Positions

Successful interviewing isn't a matter of tricks and techniques - it's about principles and preparation. It's about knowing the company you're speaking to, knowing what you have to offer and being able engage in a professional discussion with a potential colleague. Greg has interviewed hundreds of chemical engineering students for jobs in industry and will share insights and approaches you can use to make your job interviews more effective.

583145 Elsevier Data Analytics Tools for Undergraduate Research

Research as an undergraduate student can be a challenge, requiring a significant commitment outside of the classroom and oftentimes eating up a large portion of an undergraduate student's free time. Although undergraduate research experience is often well worth the extra effort, maximizing the value obtained from the undergraduate-graduate student/PI relationship is a goal which all who partake in these efforts share. In order to help maximize the understanding of the underlying research literature, as well as the goal of the individual research project, Elsevier has a suite of research tools which can be used by both student and mentor to help improve the undergraduate research experience. In this session, we will discuss how to use these tools to promote overall understanding of the existing literature and spaces where gaps exist, as well as improving knowledge of the underlying science required for valuable contribution in the lab. Whether you are a new professor hoping to foster a strong undergraduate research team within your lab, a freshman student looking to get into research, or a senior writing your NSF GRFP application this session will provide helpful tools and valuable time saving search strategies.

583139 Green Collar Engineering Opportunities for Chemical Engineers

This workshop will engage students in an interactive discussion about the challenges, advantages, and career opportunities for chemical engineers in the environmental field. A brief overview presentation of opportunities in industry, academia, consulting and government will be followed by an extensive question and answer session to help the students better understand their potential career options.

583123 Washington Internships for Students of Engineering

Since 1980, the Washington Internships for Students of Engineering program has prepared future leaders of the engineering profession in the United States for the increasingly important issues at the intersection of science, technology, and public policy. Every year AIChE's WISE program selects outstanding engineering students to conduct research on a public policy issue and learn how engineers can contribute to public policy decisions in complex technological matters. This session will feature AIChE WISE program alumni, who will be sharing stories about their experience and will talk about career opportunities for chemical engineers in public policy. The deadline to apply for the 2020 AIChE WISE Program is December 31, 2019. All students who are interested in applying are encouraged to attend.

583137 Applying to Graduate School and Research Fellowships

Why should you pursue a graduate degree in engineering? What should you consider when applying to graduate school? How do you ace your fellowship and graduate program applications? How can AIChE help you prepare for graduate school and excel as a graduate student as you pursue your degree? Members of AIChE's Young Professionals Committee (YPC) will be discussing answers to all of these questions and more. This session will feature a presentation follow by a panel of current graduate students and recent graduates in YPC. Join us for a great session discussing everything you need to prepare for your graduate school career.

583114 Developing and Maintaining Connections to Industry Partners

Maintaining partnerships with industry contacts can be one of the most valuable and rewarding elements of a Student Chapter's operations. From supplying organization funding to improving job prospects for chapter members, maintaining close, high-value ties with corporate partners yields many benefits for a student chapter. However, it can often be difficult to initiate or maintain these relationships; in this workshop, Purdue AIChE will discuss its best practices and past experiences on how to initiate contact with a company representative, how to maintain continuity as students graduate and company representatives change roles, how to maximize a chapter's value proposition to a company, and how to get the most out of corporate relationships.

583115 University of Michigan and Universidad Nacional de Colombia - Medellín: How to Develop Your Sister Chapter Program

La Universidad Nacional de Colombia - Medellín and the University of Michigan - Ann Arbor have been sister chapters since 2015. The workshop will cover how to become involved in the sister chapter program as well as how to improve sister chapter relations. Students from each chapter will discuss the benefits to having a sister chapter and the fulfilling relationship that we have developed over the past several years. Both chapters will share their experience as a sister chapter, experiences in meeting in person, events that we hold together, and how we have maintained close and permanent contact.

583140 Nuclear Workforce for Chemical Engineers

The main objective of this workshop is to highlight a rise in nuclear workforce and to point out the sheer number of positions that will become available to graduates from chemical engineering programs. Students will have an opportunity to learn about how to become a part of nuclear science, engineering and technology and how to ensure that young professionals in chemical engineering major to relate themselves for the foreseeable future of the nuclear national laboratories, suppliers/industries, and universities.

583128 The Professional Engineering Career Path

This presentation discusses how becoming a Professional Engineer (PE) can enhance a chemical engineer's career. Students learn about the PE licensure process and the two exams required to become licensed. The Computer Based Testing (CBT) exam content for both the Fundamentals of Engineering (FE) Exam and Practice of Engineering (PE) exams are described. The presentation utilizes embedded videos of young PEs and stresses the value of taking the FE exam before, or shortly after, graduation. It aims to demystify the FE/PE process. The workshop leaders are Joe Cramer and Bill Parrish who are chemical engineering PE's and active members of the AIChE's Licensing and Professional Development Committee as well as the National Council of Examiners for Engineering and Surveying's Chemical PE Exam Committee. Both have also served on AIChE's Career and Educational Operating Council which oversees licensing for AIChE. The workshop encourages Q&A and interaction with attendees.

583143 The Role of Mixing in Fast, Competitive Chemical Reactions

Many valuable products are made by reactions that are both fast and competitive. In a perfectly mixed reactor the concentrations of the reacting species are uniform throughout the volume and the outcome of the reaction is determined by kinetics. In order to achieve this condition, the reactants must be mixed at a rate that is at least equal to the rate of reaction. The ratio of these rates is a Damkohler number.

When scaling-up the Damkohler number must be kept constant to endure that the reaction environment is the same at the two scales. If it is not, the yield of the desired product will reduce and more by-products will be formed. A classic example of this is the nitration of benzene where the desired product is nitrobenzene and the by products are di- and trinitrobenzene.

In this seminar a qualitative approach to characterizing the performance of impellers operating in the turbulent regime will be discussed. Impellers in stirred tanks can be considered pumps (as they are machines that move fluid) and can be characterized in terms of their power input, flow and shear generation and efficiency. This understanding can help to determine which impeller type is best suited to a mixing process.

Next the research that has been carried to study the role of mixing on competitive chemical reactions will be briefly reviewed and the recommendations regarding design and operation of these processes will be summarized. Finally, real examples demonstrating the application of this technique will be presented.

583146 WIN an internship with AVEVA Software

Enter the 2020 competition with AVEVA and get the opportunity to apply your process design skills to solve a real-world design problem using the new generation Simulation Platform SIMCENTRAL.

Workshop panelists: Dr Richard Turton the lead author of two textbooks, "Analysis, Synthesis, and Design of Chemical Processes" (5th edition) and "Chemical Process Equipment Design", both books are published by Prentice-Hall (Pearson). He has extensive experience in the process systems engineering area and routinely teaches courses in the design and simulation areas and Mike Donahue, Senior Consulting Manager, Aveva. He received his MS in Engineering and BS in Chemical Engineering from Cal Poly where he also teaches Process Design as an Adjunct Lecturer.

583141 A Hitchhiker's Guide through Nanoscale Science and Engineering Forum

The world of nanoscale materials has become vast and overly complex. Numerous avenues for research, development, and application have made understanding the impacts of the world of nano more complicated. Whether it is nanobiotechnology or energy-related fields, nanoscale materials have become an important thrust for future growth of science and technology. In this workshop, we will delve deep into the world of nanoscale science and engineering. Examples from nanobiotechnology, carbon nanomaterials, and nanomaterials for energy-related applications will showcase how you can be a part of this ever-growing world. An interactive discussion panel will help audience members understand the different career paths available for pursuing work involving nanoscale materials. It is expected participants will leave with a broader and fuller understanding of nanoscale science and engineering and the importance of their contributions to its continual evolution.

583147 Electrolyte Modeling Basics

The Electrolyte Modeling Basics workshop is designed for students interested in learning (or to re-learn) about electrolyte chemistry concepts and the fundamentals of modeling aqueous solutions. A lecture using OLI Studio and OLI Flowsheet software will be used to demonstrate how complex electrolyte systems can be modeled. Students will leave with a qualitative picture of how ions and molecules behave in water and a better understanding of solution properties such as pH, activity coefficients and alkalinity. This course is appropriate for those new to OLI studio, professionals working with this software and those looking for a refresher course in its underlying theory.

583125 Executing a Chapter's Garbage Idea Using Entrepreneurship Principles

In November 2017, the Texas Tech University Student Chapter had an unsophisticated idea: to pick up trash. By May 2019, they picked up 4,500+ pounds of trash in Lubbock neighborhoods. While coming up great ideas is important, executing garbage ideas well has tremendous value. An AIChE Student Chapter, a startup, and a large innovative company all reside in environments of high uncertainty. For this reason, the Lean Startup method's entrepreneurship principles apply: (1) entrepreneurs are everywhere, (2) entrepreneurship is management, (3) validated learning, (4) innovation accounting, (5) build-measure-learn. Learning objectives: - understand what the 5 principles of Lean Startup method are - see how they apply to an AIChE Student Chapter event, MyCleanLBK - empower chapters with tools to execute events well - put chapter leadership at ease by illustrating that although coming up with great ideas is hard/important, executing "garbage ideas" well is easy and has tremendous value.

583129 Inflection Points: a Panel on Critical Career Decisions in Chemical Engineering

Study hard, graduate, get your first chemical engineering job: these are the typical career steps outlined to any student looking at their career choices ahead after graduation. But what's that like in real life? This panel will be a facilitated discussion of chemical engineers at different stages of their careers focusing on the critical career decisions they made along the way. This mix of personal stories and audience interaction will show that there is no one answer to building a chemical engineering career. Panelists include speakers from Google, Wacker Chemical, Shell, and Lubrizol.

583116 Hosting a Successful Regional Student Conference

After hosting about 400 students and professionals at the 2019 North Central Regional Student Conference, the University of Toledo planning committee will share their experience and offer tips on how to host a successful regional student conference. Topics will range from big items (e.g., money and space requirements) to little things (e.g., swag items and those special diet meals). This workshop will include discussions on successful activities, problems encountered and their potential solutions, and a basic suggested timeline to keep the Planning Committee on track. Engaging students, faculty, alumni, local professionals, and others helped make our meeting successful. We will also answer your questions and give advice on how to enjoy the planning process while making the most out of the opportunity to host a student conference in your region!

583117 How to Organize a Successful Conference? PROCESA 2019: 1st AIChE Colombia Student Regional Conference

This year members of Universidad Nacional de Colombia Sede Medellín Student chapter had the opportunity to host the 1st AIChE Student Regional Conference in Colombia but also in Latin America. This was the opportunity to show all the abilities and experience the chapter has gained all these years. This event meant a lot to Chapter members but also for all the AIChE Colombia Network of chapter that is formed. Be host of the conference in Colombia was the scenario to boost networking between students, professors and industry members, but mostly as a reference of the strength of AIChE in the country with image of teamwork of its members. The experience left lessons and tools, this presentation will cover general points to have in mind when you're planning a conference, tips to get closer to students and professionals, budget planification and new ideas to get more impact with the conference.

583144 World of Particle Technology

More than 80% of gasoline, 70% of polyolefins and a plethora of other products are made using fluidized bed (FB) technology. From gasification to drying, FBs and circulating FBs provide the distinct advantage of high heat transfer and solids mobility. These features have resulted in several breakthrough technologies with better temperature control and the ability to move solids from a reduction to an oxidation environment. This workshop will focus on some of these breakthrough technologies. Billions of pounds of bulk solids are processed and handled every year by the US process industries, yet most chemical engineers are ill-equipped to deal with the complexities of engineering science of solids processing. Hence, plants and products suffer with lost production, inability to achieve design production rates, off grade or off specification products. During this session, we will take a look at the fun and exciting world of solids processing.

583767 Mechanistic Studies of the Decomposition of Various Copper Salt

Copper (II) sulfate pentahydrate and copper (II) nitrate trihydrate are two common chemicals which have a variety of applications in in art, agriculture, and chemistry. Particularly, they are used widely as copper precursor for catalysts synthesis. However, their decomposition mechanisms are still under debate. Prior research has focused on determining the decomposition mechanisms of these copper salts by using only one or two characterizations methods. In this study, we used three different characterization methods: mass spectroscopy, thermal gravimetric analyzation (TGA), and Fourier-transform infrared spectroscopy (FTIR), to determine the decomposition mechanisms of copper sulfate and copper nitrate. By combining three different characterization methods, the decomposition mechanisms of these copper salts were determined with a high level of level certainty. Using this method, it was determined that copper sulfate first lost two molecules of water around 75 ^{o}C and then another two molecules of water around 100 ^{o}C. After this, copper sulfate lost one more molecule of water around 225 ^{o}C. Then one molecule of SO_x was produced around 650 ^{o}C. The decomposition of copper nitrate started with the loss of one water molecule around 100 ^{o}C. Then, at 175 ^{o}C, a molecule of water, nitric oxide, and oxygen gas were lost. At 250 ^{o}C, one molecule of both nitric oxide and oxygen gas were given off. After this, hydrogen gas was lost at around 600 ^{o}C.

587958 The Synthesis and Reaction Kinetics of Reversible Epoxies in a Diels Alder Reaction

Epoxies are an important class of thermosetting polymers for many long-term applications such as adhesives, structural materials, paints, and coatings. Due to their thermosetting nature, epoxies are hard and robust, unfortunately they also are hard to remove, recycle and reuse. Epoxies capable of reverse polymerization are being studied in order to address these issues. In this research, reversible epoxies were synthesized by introducing the Diels-Alder reaction groups to epoxy monomers. A Diels-Alder cycloadduct was formed from the reaction of 1,1'-(Methylenedi-4,1-phenylene) bismaleimide and furfuryl glycidyl ether. Formation of the Diels-Alder cycloadduct was confirmed using Fourier Transform-Infrared (FTIR) Spectrometry. Next, the forward and reverse Diels-Alder reaction – corresponding to re-polymerization and depolymerization, respectively – was monitored by FTIR measurements at 90 oC and 110 oC as a function of exposure time. FT-IR measurements were taken for the sample at various temperatures from 90-120°C. FT-IR absorption peaks relevant to the reverse Diels-Alder reaction were gradually stronger as longer exposure time at 90 oC and 110 oC. The equilibrium shift was observed toward the reverse reaction dominant side at higher temperature by comparison of FTIR spectra at 90 oC and 110 oC. The reversible epoxy resins will be cured with various curing agents and their reaction kinetics will be studied.

584752 Surface Reconstructions in Perovskite Oxide Catalysts for Oxygen Evolution Reaction Using Ab Initio Grand Canonical Monte Carlo Simulation

Strontium titanate ($SrTiO_3$) is a promising active material for catalysis of the oxygen evolution reaction (OER) which is an important area of research for alternative energy solutions through fuel cell technologies. The effectiveness and abundance of component materials of STO relative to traditional platinum electrodes suggest perovskite catalysts as a replacement material for traditional platinum electrodes. Greater understanding of the structure of STO under different conditions, such as those employed in the OER, will aid this direction.

Reconstructing the surfaces of perovskite catalysts can circumvent factors that limit the OER. Theoretical and experimental studies on SrTiO3 have shown that a TiO2 rich surface can reduce the overpotential of the catalyst for this reaction.

These reconstructions are expressed under different chemical environments of the material surfaces and influence the surface chemistry and effectiveness of the ability to catalyze the reaction. Surface reconstructions of the perovskite catalysts was explored using automated ab initio Grand Canonical Monte Carlo (aiGCMC) methods to sample across thermodynamically relevant surface configurations. The method allows for the generation of surface phase diagrams of the STO systems which is important in predicting and understanding the surface chemistry behavior of the materials under differing conditions. This work employs the aiGCMC method for the STO (001) terminated system to develop an understanding of the surface configurations in phase space which can be related to experimental parameters.

587646 Ab Initio Study into the Thermodynamic Stability and Intrinsic Activity of $La_{0.75}Sr_{0.25}MnO_3$ Vacancies As Bifunctional Oer/ORR Catalyst in Alkaline Media

Density Functional Theory (DFT + U) study of stability and electrochemical activity for $(La_{0.75}Sr_{0.25})_{1+x}MnO_{3-d}$ perovskite oxides was conducted. For investigation of thermodynamic stability, phase diagrams were constructed using total energy values calculated from DFT + U. Additionally, vacancy formation energies were determined by means of chemical potential for both oxygen rich and poor conditions. As a measure of activity in the Oxygen Evolution Reaction (OER) and Oxygen Reduction Reaction (ORR), the overpotential of various perovskite oxides were calculated. The free energies of the intermediates in the OER/ORR mechanisms were required to determine the overpotentials; as a result, free energies were determined by combining the total energies of the intermediates with zero-point energies (within the harmonic approximation) and finite temperature terms. Through comparison of the overpotentials, DFT + U predicts that some of the vacancies have higher intrinsic activity than pristine $La_{0.75}Sr_{0.25}MnO_3$. Furthermore, the enhanced reactivity of $(La_{0.75}Sr_{0.25})_{1+x}MnO_{3-d}$ oxides is correlated with the oxygen p-band center distance from the Fermi level.

Acknowledgements: The work is supported by ACS-PRF [58740-UR6] and used the Extreme Science and Engineering Discovery Environment (XSEDE) TACC at the stampede2 through allocation [TGDMR140131].

587886 Phosphorus-Mediated Reductive C--N Bond Coupling: A Path to Functionalized Azaheterocycles

Nitrogen heterocycles are of significant industrial interest due to their abundance in natural products and other organic molecules with great biological potential. As such, methods which can readily access C–N bonds are of great synthetic interest. Current prevailing strategies, such as the Ullman, Buchwald-Hartwig, and Chan-Lam coupling reactions, are transition-metal dominated, requiring the usage of either copper or noble metals. Researchers in the Radosevich group have previously synthesized a biphilic phosphetane catalyst which catalyzes bond formation between nitroarenes and phenylboronic acids. This new coupling strategy is fundamentally different from previous palladium-based methods in that it operates via a PIII/PV=O redox cycle.

While this phosphetane catalyst is easily synthesized and can be used at reasonable loadings (15%), its major drawback is that it is not commercially available. Phosphines, as O-atom acceptors, were of special interest as alternative reagents, although their nucleophilic nature ensures that they cannot be regenerated. In my research, we showed that stoichiometric amounts of triethylphosphine, an inexpensive commercial agent, can drive the deoxygenative coupling of nitroarenes and boronic acids to generate di(hetero)arylamine products in a direct synthetic operation without the need for any transition metal additives. A variety of diarylamine products were prepared in good yield using this method, revealing versatility with regards to functional groups on both coupling partners. Additionally, this method offers access to traditionally difficult substrates such as nitroarenes containing electron-withdrawing groups.

These C--N coupling techniques can then be used to access difficult azaheterocyclic frameworks with multiple sites for reaction. If one were to add groups to a benzimidazole, the presence of two nitrogen atoms which can tautomerize ensures two possible sites for addition. When the molecule is symmetric—as is the parent nucleus—this is not a problem, but in asymmetric instances, it is nontrivial to ensure functional groups end up on the correct side of the molecule. Here, we use the phosphetane catalyst described earlier, which can also catalyze an intramolecular Cadogan cyclization, to synthesize various azaheterocycles. This method allows us to approach pharmaceutically relevant molecules, and has been successfully applied to the synthesis of oxindoles, indoles, quinoxalines, and benzimidazoles.

588172 Novel Energy Transfer Mechanism for Dielectric Cu_2o Nanostructures in Photocatalysis.

Most Plasmonic Metal Nanocatalysts (PMN) are plagued by inherent joule loss, band-gap limitations and the inability to showcase strong magnetic fields. In addition, several of the metal catalysts currently being used in industrial and pharmaceutical applications are toxic in nature and require expensive downstream operations. Fortunately, in the past few years, copper-based photocatalysts have been identified as an alternative to conventional catalysts. These copper-based catalysts are appealing due to their relative abundance, low cost, low toxicity and heterogenous nature. Previous reports suggest that researchers believe copper-based photocatalysts undergo energy transfers via plasmonic mechanisms as a result of Localized Surface Plasmon Resonance effects (LSPR) and standard semiconductor mechanisms that come into play once the semiconductor material's bandgap energy barrier is exceeded. A new energy transfer theory was developed. Through novel experimental modeling and computational methods (FDTD simulations), mechanism types were isolated and then tested for degradation under various conditions. The concept demonstrated in our study unveils a new way to conduct solar light harvesting and photocatalysis. Although we demonstrated the concept using Cu_2O particles, this concept can be broadly applied to other metal oxide semiconductor particles including titanium dioxide. The successful degradations narrowed down the mechanism possibilities and provided statistical support for this theory in application in not only photocatalysis, but an abundance of industries such as but not limited to: pharmaceutical, energy fuels, thin film solar cells, CO_2 reduction, specialty chemicals, agrochemicals, as well as water decontamination among other related industries.

585975 Sustainable Plastic Precursors from Hydrothermal Catalytic Cracking of Palmitic Acid: A Reaction Pathway Study

Petroleum-derived naphtha cracking is currently used to produce important plastics precursors such as xylene and toluene. Due to growing concerns of climate change and efforts to produce clean energy, a sustainable, petroleum-free method of producing these important chemical process precursors is necessary. Research has found promise in the hydrothermal catalytic cracking of naturally occurring fatty acids, specifically palmitic acid found in palm oil, to form aromatics. A variety of catalysts have been used for this process, the most notable being HZSM-5 which has been proven to be effective at selectively producing aromatics and converting biomass [1]. While prior work shows promise, more work is necessary to investigate palmitic acid cracking due to the lack of focus on reaction pathway and less-than-desirable aromatic yields. We hypothesized that decreasing water and catalyst loading as well as using nano-HZSM-5 could increase the yield of aromatics, specifically xylene and toluene, compared to prior work. Batch reactions of palmitic acid catalytic cracking were conducted in supercritical conditions (400°C and 3200 psi) with water and nano-HZSM-5 catalyst, and products were characterized using gas chromatography with mass spectrometry (GC/MS), with flame ionization detection (GC/FID), and with thermal conductivity detection (GC/TCD). It was found that water loading of 15 weight percent produced higher yields of xylene and toluene than reported literature values. We also set out to understand the reaction pathway by varying the length of the reaction time (0 to 45 minutes) and the temperature (350, 400) as well as running hydrothermal cracking of the fatty acid without catalyst. From our results, we speculate that the reaction occurs through intermediates at the catalyst surface and is preceded primarily by non-limiting decarboxylation and chain-cracking of the palmitic acid. It has been found that palmitic acid catalytic cracking can produce promising yields of aromatics, and our investigation of the reaction pathway can aid in potential scale-up design decisions for the process. Work on this project is being continued to thoroughly investigate the reaction pathway of intermediates and the stability of products during the reaction.

<u>References:</u>

[1] Na Mo and Phillip E. Savage

ACS Sustainable Chemistry & Engineering **2014** *2* (1), 88-94

DOI: 10.1021/sc400368n

585427 Understanding Cation Effects in the Hydrogen Evolution Reaction on Cu(100) Surfaces

The hydrogen evolution reaction (HER) is a half-cell electrochemical process in which H_2 evolves from water electrolysis. This reaction competes with the electroreduction of CO_2 into hydrocarbon fuels, decreasing electrocatalyst selectivity towards desirable hydrocarbon products. In the present theoretical work, we studied cation effects in the HER on Cu (100) surfaces. Copper is the only metal electrocatalyst that can convert CO_2 through electroreduction into hydrocarbon fuels. Being able to prevent the competing HER reaction could help make CO_2 electroreduction into a practical process. We seek to elucidate the mechanism through which alkali metal cations hinder the HER from a fundamental-molecular perspective. Although several researchers have reported that the HER is not significantly affected by the presence of alkali cations relative to other products, preliminary experimental results from collaborators suggested that there is a cation-size dependence effect on the HER. Our hypothesis is that an electric field results from cation presence near the electrode surface, affecting energy barriers for the H* adsorption Volmer step by interacting with the partially charged transition state. To test our hypothesis, computational techniques based on the Density Functional Theory (DFT) were used to model the Volmer step reaction mechanism. The cation effects were implicitly taken into account by applying electric fields to our model. Our results are in qualitative agreement with experiments that show that increasing the cation size increases the HER rate for same charge cations. The cation effect mechanism still to be elucidated from further calculations and experimentation.

585673 Role of GSSG in the Metal-Mediated Oxidative DNA Damage with Iron

Glutathione Disulfide (GSSG) plays an important role as an anti-oxidant, including metal-mediated oxidative DNA damage. Iron can generate harmful reactive oxygen species (ROS), such as hydroxyl radical (•OH) by the Fenton reaction, and the presence of GSSG may mitigate •OH production by coordinating iron ions. Oxidative damage by ROS numerous diseases and conditions such as diabetes and aging. On the other hand, ROS are essential for metabolic processes such as homeostasis, and selective and site-specific ROS generation can be used in chemotherapeutics . Thus, understanding ROS production and monitoring the amount of ROS in the human body is of great importance, which makes understanding the interaction between iron ions and GSSG beneficial. The present study attempts to understanding the ability of GSSG to coordinate Fe(III) using Isothermal Titration Calorimetry (ITC). The titration curve shows the stoichiometry of the reaction and can provide thermodynamic quantities, such as enthalpy and entropy for this reaction. These quantities can be ultimately used to find the equilibrium constant for the coordination reaction between iron ions and GSSG.

583401 Understanding V2+/V3+ Kinetics in Acidic Electrolytes for Vanadium Redox Flow Batteries

Vanadium redox flow batteries (VRFBs) are a promising technology for grid energy storage because of their scalability, long lifetimes, and low maintenance needs. However, they suffer from low energy density, which limits their widespread usage. To help improve the efficiency of VRFBs, we investigated the mechanism of the kinetically limiting V^{2+}/V^{3+} redox reaction on glassy carbon. We first study vanadium complexation with anions in common HCl and H_2SO_4 electrolytes by comparing experimental UV-Vis spectra to Time-Dependent Density Functional Theory (TDDFT) calculations, as well as DFT-predicted free energies of V^{3+}-anion complexes in solution. DFT predictions using the hybrid B3LYP-D3 functional and an explicit/implicit solvent model indicate that the free energies of the complexes follow the order of $[V(H_2O)_5SO_4]^+ < [V(H_2O)_4Cl_2]^+ < [V(H_2O)_6]^{3+}$, in agreement with experimental observations. The V^{2+}/V^{3+} kinetics are ~2.5 times faster in HCl than in H_2SO_4 or HCl/H_2SO_4, attributed to the presence of $[V(H_2O)_4Cl_2]^+$. We also confirm that the V^{2+}/V^{3+} reaction proceeds via an adsorbed intermediate and propose a bridging mechanism through adsorbed *Cl (in HCl) and *OH (in H_2SO_4 or HCl/H_2SO_4). We hypothesize the *Cl bridge enhances the electron transfer compared with the *OH bridge for metal ions because of the higher polarizability of *Cl compared with *OH. A bridging mechanism through *Cl is supported by even faster redox kinetics in HBr than HCl, and is further studied by comparing our calculated halide adsorption energies and polarizabilities on a glassy carbon surface to the observed experimental kinetics. These key findings open avenues for electrolyte engineering to ensure the presence of more effective bridges to increase V^{2+}/V^{3+} redox kinetics in VRFBs.

586299 Towards Efficient and Stable Nifeox Alkaline Oxygen Evolution Electrocatalysts

Alkaline anion exchange membrane electrolyzers have high potential to support renewable energy systems through energy storage in the form of hydrogen. However, the efficiency of these electrolyzers is significantly hindered by the sluggish kinetics of the oxygen evolution reaction (OER). High overpotentials are needed to overcome this kinetic hinderance, which translates to higher energy requirements for water electrolysis. Effective electrocatalysts can lower the overpotential for OER and water electrolysis, but these catalysts need to be highly active, stable, and inexpensive. The most active catalysts for OER are IrO_x and RuO_x; however, these catalysts contain precious transition metals that are prohibitively costly to deploy on a large scale. Nickel-Iron oxide catalysts ($NiFeO_x$) are non-precious alternatives for OER catalysis, which have shown higher activity than IrO_x in alkaline conditions. However, the long-term stability of these $NiFeO_x$ has not been adequately demonstrated to date.

We have synthesized carbon-supported $NiFeO_x$ nanoparticles using a straightforward wet impregnation method. This catalyst can achieve 10 mA/cm^2 at 300 mV overpotential at room temperature and at 240 mV at 70°C in alkaline aqueous solution. The stability of this catalyst, however, is poor in concentrated alkaline electrolytes. Thus, we are working to determine the cause of instability in this catalyst. Our working hypothesis is that the poor stability results primarily from oxidative degradation of the carbon support. We have evaluated this hypothesis by integrating accelerated degradation electrochemistry experiments with detailed materials analysis techniques, including SEM, XRD, and Raman Microscopy.

586000 Ni-Based Catalysts for Electrocatalysis in Proton Exchange Membrane Fuel Cells

Ni-based Catalysts for Electrocatalysis in Proton Exchange Membrane Fuel Cells

Matt DiBiase[1], Thomas Gascoigne[1], Enoch A. Nagelli[1], Jiangtian Li[2], Rongzhong Jiang[2], Deryn Chu[2*]*

[1]Department of Chemistry & Life Science United States Military Academy, West Point, New York 10996

[2]U.S. Army Combat Capabilities Development Command, Army Research Laboratory

2800 Powder Mill Rd., Adelphi, MD 20783-1107

Proton Exchange Membrane (PEM) Fuel Cells are a field of great interest for their efficiency advantages over combustion technology. However, conventional methods of electrolysis to produce H_2 and O_2 gas necessary for PEM fuel cells rely on expensive catalysts, Pt and IrO_2. Despite exceptional efficiency of these catalysts, their high costs prevent industry scale up and production. We present non-precious Ni-based catalysts as alternatives. Of the Ni catalysts characterized, NiS and NiFe layered double hydroxide (LDH) together provided the smallest total potential of 1.7 V for overall water splitting (hydrogen evolution reaction/HER and oxygen evolution reaction/OER). However, Linear Sweep Voltammetry illustrated that NiFe LDH had the lowest overpotential of the two, contributing only 0.3 V to the total potential. Nevertheless, the total potential of 1.7 V is comparable to the industrial standard (1.8-2.0V), and only 0.2 V above the best value of 1.5 V from a combination of the state-of-the-art OER and HER catalysts.

KEYWORDS: Proton Exchange Membrane, Overpotential, Oxygen Evolution Reactions, Hydrogen Evolution Reactions

CONTACT: Dr. Deryn Chu, Email: deryn.d.chu.civ@mail.mil , Dr. Enoch Nagelli, Email: enoch.nagelli@westpoint.edu

583492 Machine Learning with a Genetic Algorithm for Enhanced Catalysis Research

Many catalysts involve metal nanoclusters dispersed on metal oxide supports because of their stability and ability to catalyze a broad range of important reactions. Multiple factors including the size, chemical composition, and metal-support interaction can impact the performance of nanocluster catalysts, thus there is great opportunity for their design. However, nanoclusters adopt a variety of sizes and shapes that are difficult to know a priori. Consequently, a major challenge for understanding nanoclusters is properly modeling their structures. Although pioneering work has introduced optimization algorithms to find nanocluster structures, a major barrier between routine study of nanoclusters on supports and nanoclusters of greater than ~15 atoms is the computational intensity of Density Functional Theory (DFT) employed for electronic calculations. To address this challenge, we use machine learning (ML) combined with DFT and a Genetic Algorithm (GA), a type of global optimization algorithm, to accelerate the structure search supported nanoclusters. This modular software is dubbed “GAxML”. We also show how Active Learning (AL) can be used to reduce the computational expense of the GA. The incorporation of both ML and AL into a DFT-based GA can greatly accelerate the ability to accurately model the structure of nanocluster catalysts, opening up possibilities for exploring catalytic mechanisms and comparing with experimental measurements with greater fidelity.

585291 Characterization of Carbon Based Catalyst Supports Using Cyclic Voltammetry and Electrochemical Impedance Spectroscopy for Carbon Dioxide Reduction Applications

Catalysts are instrumental in making otherwise expensive chemical processes time and cost efficient, and their effectiveness can be maximized by increasing their active surface area. By depositing catalytic material onto a support with maximized surface area, catalyst activity can be increased without a substantial volume of catalytic material. For electrochemical processes, certain carbon materials are ideal for this purpose because they exhibit high surface areas, electrical conductivity, and resistance to thermal and chemical degradation. The determination of the electrochemically active surface area within any given carbon support is therefore of great interest for material selection and as a baseline for performance characterization. The electroactive surface area can be determined on a conductive material by exploiting the effect of the electrochemical double layer, which can be modeled as a capacitor. A capacitance from this double layer can be calculated by measuring the charge transferred when a given potential is applied. This capacitance is proportional to, and therefore can be used to calculate, the electrochemically active surface area of a conductive material. By using cyclic voltammetry, electrochemical impedance spectroscopy, and specific capacitance values for carbon from literature, specific electrochemically active surface areas and resistances of various carbon based supports were calculated. From these, candidates were chosen for catalyst deposition for the reduction of supercritical carbon dioxide into chemical fuels.

587415 New Materials for CO_2 Reduction Electrocatalysis

Carbon dioxide upgrading is motivated by the issues humanity is facing because of global warming. This is due to the anthropogenic CO_2 emissions such as the burning of fossil fuels, deforestation, and fertilization etc. Numerous methods such as reusing, recycling, use of alternative transportation methods, reducing meat intake, and planting more trees are suggested in order to reduce the global carbon footprint. A possible remedy would be to use the emitted CO_2 in order to produce value-added products, such as liquid fuels, to meet the world's ever-increasing energy demand. Carbon dioxide is an extremely stable molecule; thus, its reduction is an energy-intensive process. Efficient and robust catalysts are needed in order to lower the high activation energy barrier of CO_2 conversion. For this project, monometallic and bimetallic basic copper chloride nanomaterials were prepared by pulsed laser in liquids synthesis by my faculty advisor Astrid M. Müller as electrocatalyst precursors; ad-metals were Sn, Bi, or In. Pulsed laser in liquid synthesis was used for its ability to control the size and composition of the catalyst and for faster synthesis. We employed scanning electron microscopy (SEM) with elemental mapping by energy-dispersive X-ray analysis to image these catalysts and determine the metal ratios.

587720 Electrocatalytic Hydrogenation of 4-Propylguaiacol, a Bio-Oil Model Compound

Conversion of lignocellulosic biomass to renewable hydrocarbon fuels involves biomass deconstruction, which follows either a biochemical or thermochemical pathway. Fast pyrolysis is a thermochemical process that involves high temperature heating of lignocellulosic biomass for short times in the absence of oxygen, yielding approximately 70% bio-oil by weight. The physical and chemical properties of bio-oil hinder its use as a drop-in replacement for fossil fuels. Bio-oil contains reactive oxygenates and it is corrosive, which impede long-term storage and transportation. Processing it in refineries to make transportation fuels using classical petroleum-style hydro-treatment is possible provided it is first upgraded. Electrocatalytic Hydrogenation, **ECH** is proposed as an upgrading technology that operates under mild conditions (temperatures less than 80°C and atmospheric pressure). It involves *in-situ* hydrogen production by water electrolysis, which is reacted at the cathode to hydrogenate bio-oil and increase its stability and energy density. In this process, intermittent renewable electricity from solar and wind energy is stored in chemical bonds formed with hydrogen atoms. In this study, the ECH of 4-propylguaiacol (2-methoxy-4-propylphenol), a lignin derived bio-oil model compound, was performed using a ruthenium-based electrocatalyst. The effects of temperature and cell voltage were investigated to observe differences in product speciation and yields. Products were analyzed using gas chromatography–mass spectrometry **(GC/MS**). Using relatively high temperatures of 80°C and cell voltages of ~50 volts resulted in increased conversion of 4-propylguaiacol to propylcyclohexane.

583951 Nickel Phosphides As Electrocatalysts for Hydrogen Evolution

Optimizing catalysts for the hydrogen evolution reaction (HER) is a critical step toward the efficient production of H_2(g) fuel from water or low-carbon (C2, C3, C4) fuels from the reduction of carbon dioxide. It has been demonstrated experimentally that certain transition-metal phosphides, specifically the nickel phosphides Ni_2P and Ni_5P_4, efficiently catalyze the HER at a smaller cost than common Pt-based electrocatalysts. However, the HER mechanism on nickel phosphides remains unclear. Using density functional theory and thermodynamics, we explore both the vacuum and aqueous reconstructions of low-index surfaces of Ni_2P, Ni_5P_4, Ni_3P, $Ni_{12}P_5$ for the HER as a precursor to exploring the reconstructions' role in the carbon dioxide reduction reactions. These reconstructions reveal which active-site surfaces are most likely to appear on a nickel phosphide nanoparticle at various overpotentials and which hydrogen and phosphorous surface reconstructions are most likely at various overpotentials. This study emphasizes the importance of considering applied bias and aqueous surface stability in predicting the HER-active sites in the electrocatalytic system, mechanism, and overpotential, and highlights the primary role of P reconstructions in HER catalysis on transition-metal phosphides.

587297 Design Strategies for Efficient Non-Stoichiometric Mixed Metal Oxide Electrocatalysts

Design Strategies for Efficient Non-Stoichiometric Mixed Metal Oxide Electrocatalysts

Krishna Patel, Samji Samira, Xiang-Kui Gu and Eranda Nikolla

Department of Chemical Engineering and Materials Science, Wayne State University, Detroit

Non-precious mixed metal oxides belonging to the pervosite family with a general composition of $A_{n+1}B_nO_{3n+1}$ (n = 1, 2, 3 ... ; A = alkaline earth (or) rare metal; B = transition metal) are ideal candidates for cost-effective catalysts for oxygen electrocatalysis [1]. When n = 1, oxides belong to the first-series Ruddlesden-Popper (R-P) phase, while as $n \rightarrow 8$, they exhibit a perovskite structure. These non-stoichiometric oxides exhibit mixed ionic and electronic conductivities as well as high surface oxygen exchange rates which are crucial properties for electrocatalysts [2]. Further, these oxides can also accommodate >90% of the metals in the periodic table giving rise to a vast phase space of material compositions [2]. However, a wide range of phase space makes it difficult to identify optimal oxides for targeted chemical reactions. Also, the complexity of these oxides has resulted in using bulk descriptors as there is no surface related experimentally measurable descriptors [2]. Therefore, it is imperative to develop structure-performance relationships that can guide the identification of active oxide electrocatalysts.

In this presentation, a combinatorial approach involving theoretical studies and precisely-controlled electrochemical measurements is used to identify measurable activity descriptors that can guide the identification of active nonstoichiometric oxides. Oxygen reduction reaction (ORR) plays a major role in efficient and sustainable electrochemical energy storage and conversion technologies [1, 2]. ORR was used as a probe reaction to describe the activity and stability of these oxides. The effect of composition and crystal structure for a vast phase space was studied. A variety of oxides were synthesized using a citric acid based sol-gel method followed by surface characterization using TEM imaging. We show that the experimentally measurable parameter, surface reducibility, can be used as a descriptor to predict the electrochemical activity and stability of oxides, thus leading to the identification of highly active and stable catalysts for oxygen electrocatalysis.

References

1. Samira, S., Gu, X., Nikolla, E., Design Strategies for Efficient Nonstoichiometric Mixed Metal Oxide Electrocatalysts: Correlating Measurable Oxide Properties to Electrocatalytic Performance. *ACS Catal.* **2019,** *Accepted.*
2. Gu, X. K.; Samira, S.; Nikolla, E., Oxygen Sponges for Electrocatalysis: Oxygen Reduction/Evolution on Nonstoichiometric, Mixed Metal Oxides. *Chem. Mater.* **2018,** *30*, 2860-2872.

586613 Oxygen Reduction Reaction on Gold Electrode Surfaces in Acidic Medium

The oxygen reduction reaction (ORR) attains a major role in energy conversion complexes given it is catalyzed by fuel cells for the efficient and sustainable production of electricity. In acidic medium, the reduction of oxygen is favored in a two-electron transfer pathway which involves two reduction products: hydrogen peroxide and water. The reduction performance of three different gold electrodes with varying surface areas was tested as a sustainable and more earth-abundant alternative to the reduction of molecular oxygen on platinum polycrystalline electrodes. Such comparison would lead to the maximization of hydrogen peroxide production in acidic medium. For this study, the electrochemical reduction of oxygen was examined in a 0.1M $HClO_4$ solution by identifying the reduction potentials of oxygen in the distinct gold electrodes utilized as electrocatalysts. With each individual potential, electric current present in solution was measured as a function of time via amperometry. ^1H NMR spectroscopy was performed in order to characterize the product and determine the efficiency of the production of hydrogen peroxide in each gold electrode. The theoretical yield of hydrogen peroxide when compared within the three gold electrodes presented no direct proportion to surface area as expected. Such incongruence could be due to difference in crystalline structure within the electrodes. The sufficient production of hydrogen peroxide would give way to oxidation of compounds of interest such as ethylene glycol in order to further explore their performance as fuel alternatives.

585119 Mesoporous Silica Supported Perovskite Oxide for Low Temperature Thermochemical CO_2 Conversion

Today, carbon dioxide (CO_2) emission levels are higher than at any other point in human history, and chemical reactions can be used to establish a synthetic carbon cycle, which would reduce emissions and produce liquid fuel. The reverse water-gas shift chemical looping process (RWGS-CL) is a reaction of interest because it is effective for converting large amounts of CO_2 and the resulting carbon monoxide (CO) can be used to produce hydrocarbons that are liquid fuel or chemical intermediates. Perovskite oxides are an effective catalyst for the RWGS-CL based on converting CO_2 to CO at efficient conversion temperatures, ability to form oxygen vacancies, and structural stability. The perovskite oxide used is $La_{0.75}Sr_{0.25}FeO_3$ (LSF). Recent studies have shown silica (SiO_2) as the most efficient support material for the perovskite oxide catalyst used in RWGS-CL. The form of SiO_2 that has been experimented with to provide support for LSF is SBA-15, because of its large pore volume and surface area, coupled with potential to support smaller particles of the perovskite increasing the number of active sites available, which would enhance conversion. Here we have analyzed adaptations to the synthesis procedure of perovskite oxides and SBA-15 to develop a single synthesis procedure for a composite material. These materials are characterized by X-ray Diffraction, N_2-physisorption, CO_2-Chemisoprtion, and temperature-programmed reduction, oxidation, and RWGS-CL experiments.

586980 Hydrodeoxygenation of Guaiacol with Ru Catalysts on Various Activated Carbon Supports

Bio-oil is a promising sustainable fuel alternative to traditional sources such as crude oil and natural gas. However, due to its high oxygen content, low heating value, and high viscosity, bio-oil cannot be used directly as a transportation fuel. Therefore, bio-oil requires additional upgrading to turn it into more valuable fuels. The hydrodeoxygenation (HDO) of bio-oil removes the oxygen groups in the bio-oil, turning it into lighter products. Activated carbon is known as a typical catalyst support for HDO reactions. In this study, various activated carbons were used as catalyst supports to load Ru. Commercial activated carbon (derived from fossil fuels) was first investigated for the HDO reaction. For comparison, activated carbon derived from food waste (FWAC) and miscanthus (MISAC) were also examined. Food waste was collected from UConn dining services, and then pyrolyzed into bio-char and physically activated by steam. MISAC was also prepared by pyrolysis and steam activation. All the supports were impregnated with 5% of Ru. The carbon supports were characterized by elemental analysis, N2 sorption-desorption, scanning electron microscope (SEM) and X-ray powder diffraction (XRD). In this study, guaiacol was used as a model compound in the HDO reaction. Decane was used as a solvent to dilute the reactant and octane was used as an internal standard. The reaction pathway was derived from the experimental data and kinetic parameters were calculated. 5% Ru-FWAC showed the highest reaction rate, probably due to its increased hydrophilicity. These results suggest that Ru metal on an activated carbon support is a highly active catalyst in the organic phase for HDO. There is also a great potential for using carbon supports derived from sustainable biomass sources like food waste.

583894 In-Situ Production of Hydrogen Peroxide Via Electrochemical Reduction of Anthraquinone Electrodes

Hydrogen peroxide, H_2O_2, is leveraged as an oxidizing agent in several industrial applications, including the bleaching of cotton and wood pulp and the treatment of wastewater. While demand for the chemical compound has steadily increased, commercial methods of producing hydrogen peroxide have remained costly and fundamentally unchanged: sequential hydrogenation, filtration, oxidation and extraction stages require a significant amount of capital to support both process infrastructure and energy consumption.

Here we present the novel use of an electrochemical cell with a solid-state polyanthraquinone (PAQ) cathode to mediate the electrochemical generation of hydrogen peroxide via the temporal separation of its reduction and oxidation. This is achieved by a sequence of electrochemical reduction of the immobilized quinones in an acidic electrolyte, draining of the cathodic half-cell in the electrochemical cell, and then producing hydrogen peroxide with a flow of alternating water and oxygen slugs through the cell. As such, the quinone serves as a mass (protons) and energy (electrons) transfer agent. This allows the for the temporal separation of electrolyte and product streams, so that we produce the hydrogen peroxide in an environment free of contaminants and effectively eliminate the need for precious metal catalysts and energy-intensive purifications.

Further, we have developed a mechanical pumping system operating in tandem with the electrochemical cell to sequentially fill/drain catholyte, oxidize the intermediate, and capture the hydrogen peroxide in a recycled stream, thereby operating continuously and increasing the concentration of the peroxide solution with each pass through the electrochemical cell. As such, hydrogen peroxide concentration exceeds 800 PPM within three hours of system operation. The electrochemical cell operates with a faradaic efficiency of over 30% and the PAQ cathodes have been observed to resist degradation for over 35 hours of system operation.

586314 Metal Mixing of Copper Oxide Sorbents for Sulfur Contaminant Removal from Hydrocarbon Streams

Sulfur contaminants in natural gas and hydrocarbons at even trace amounts (1-1000 ppm-mol) poison catalysts, corrode pipelines and are hazardous to the environment due their conversion to sulfur dioxide (SO_2) during combustion, which contributes to acid rain. In industry, sulfur contaminants, such as hydrogen sulfide (H_2S), are commonly removed through metal-oxide based chemisorption, where H_2S reacts over a metal oxide solid surface at elevated process temperatures (473-1073K). These processes are attractive because of their high removal capacity and highly pure effluent streams. Recently, there has been an interest in low-temperature H_2S metal-oxide based chemisorption due to safety and economic reasons. Thus, copper oxide (CuO) is an attractive sorbent because of the highly favorable thermodynamics of the reaction that not only allow for low-temperature removal, but also allow for the decrease of H_2S concentrations from several thousand ppm to sub-ppm levels. The mixing and doping of oxides have been shown to increase the reactivity of sorbents with contaminants. Little work however has been conducted to understand the effects of the structural and chemical changes resulting from the introduction metals to CuO sorbents on the sorption capacity.

In this work, CuO sorbents were mixed with various metals (Ba, Mn, Bi, Mg, Al, Zn, Ni, Co and La) to investigate the effect of introducing these metals on the sorbent removal capacity. The doped CuO spherical nanoparticles were synthesized via co-precipitation. The synthesized nanomaterials were then characterized using a variety of techniques and tested for low-temperature H_2S removal by running packed bed sulfidation experiments at 1000 ppm-vol inlet H_2S concentration, temperature of 323K and pressure of 1 atm. The experiments showed that doped CuO-sorbents with crystallite sizes larger than 12 nm perform poorly. It was found that all metal dopants with the exception of La and Zn resulted in decreased capacities compared to those expected from pure CuO samples with similar crystallite sizes. La doped CuO sorbents resulted in about a 75% increase in sorption capacity compared to pure CuO samples with similar crystallite sizes.

585784 Molybdenum Disulfide Thin Films As a Catalyst for Hydrogen Evolution

In the search for alternative energy sources, hydrogen fuel cells have appeared as a promising clean source of energy, particularly in automobiles. The fuel cell requires hydrogen gas, which is produced through the hydrogen evolution reaction. Researchers have searched for a cheap, efficient catalyst for this reduction reaction, which may make implementation of this technology on an industrial scale more realistic. Molybdenum disulfide thin films are relatively inexpensive and easy to make through electrodeposition, so they make promising candidates for catalysts. Doping the films with nickel has been shown to change the properties of the film, so may improve its catalytic activity. Catalysis testing for the hydrogen evolution reaction demonstrates that while hydrogen begins to reduce on a non-catalytic indium tin oxide (ITO) film at -0.6V, the reduction will occur at around -0.4 and -0.3V for MoS_2 and Ni-doped MoS_2, respectively. This indicates that less power will be required to start the hydrogen evolution reaction when Ni-doped MoS_2 is used as a catalyst. Cyclic voltammetry of a ferrocyanide/ferricyanide redox reaction using Ni-doped MoS_2 demonstrates that the shifted hydrogen reduction peak is not due to an increase in surface area or an increase in electrical conductivity. Thus, doping the MoS_2 thin films with Ni significantly increases their efficiency in this critical reaction.

588160 Synthesis of Molybdenum Trioxide and Characteristics of H_xMoO_3 Bronze Formation

Many modern chemical processes rely on highly polluting petroleum as a hydrogen source. Replacing petroleum with a cleaner alternative, such as water, could greatly enhance the sustainability of the chemical industry. Electrochemical methods allow the conversion of water into hydrogen; however, there is currently not a good understanding of how to integrate such electrochemical techniques into existing industrial reactions. With the goal of improving the sustainability of industrial chemical conversions, the gap between thermal and electrochemical catalysis needs to be bridged. Reducible metal oxides, such as tungsten trioxide (WO_3) and molybdenum trioxide (MoO_3), are useful models for these studies, as they can take up hydrogen both thermally (hydrogen spillover) and electrochemically (hydrogen intercalation). Our group recently reported on these processes in WO_3,[1] but MoO_3 is known to have a higher hydrogen capacity.[2] This suggests that MoO_3 might exhibit enhanced reactivity, and so this work focuses on hydrogen uptake in MoO_3.

Two MoO_3 synthesis methods were investigated in order to produce a stable, reducible oxide. The selected method produced MoO_3 with an orthorhombic crystal structure, confirmed using XRD. Cyclic voltammetry (CV) was performed to probe intercalation dynamics, where multiple discrete intercalation events were observed. Comparison to CVs of WO_3 reveal the MoO_3 has a lower overpotential for intercalation suggesting hydrogen insertion into MoO_3 is faster than into WO_3. This study provides a foundation for further work on hydrogen uptake in H_xMoO_3 and provides evidence that MoO_3 might be more useful than WO_3 for thermo-electrochemical hydrogenation chemistries.

[1] Miu, E. V., & McKone, J. R. (2019). *Journal of Materials Chemistry A*, 7(41), 23756-23761.
[2] Genin, C. Driouiche, A., Gerand, B., &Figlarz, M. (1992). *Solid State Ionics*, 53,315-323.

585088 Enhancing Electrolytic Ozone Selectivity and Activity with SnO2- Sb-Ni-Gd Electrocatalysts

Electrolytic synthesis of ozone is unique from other synthesis techniques due to its ability to generate ozone directly in water with generally lower equipment costs relative to competing methods. The selectivity and stability of the catalyst material used in electrolytic synthesis is, however, a major disadvantage when compared to long term operation of other methods. Many investigators have contributed to finding new, highly selective, and minimally hazardous catalysts for ozone synthesis, such as Ni and Sb doped SnO_2, but as of yet none have proven adequately energy efficient nor stable. In this investigation, we perform electrochemical testing and ozone quantification on a new material combination: Gd, Ni, and Sb doped SnO_2 films. SnO_2-Sb-Ni-Gd films exhibit substantially higher electrochemical activity and selectivity for electrolytic ozone generation than films doped with just Sb and Ni or just Sb and Gd. Faradaic efficiencies exceeding 60% were exhibited by this catalyst, giving an optimistic nod towards the viability of this material and meriting further testing into its stability.

585864 Mixed Metal Oxides for Thermochemical Energy Storage

As more non-dispatchable renewable generation resources are integrated to the power grid, pressure is exerted on the existing fossil fleet to provide flexible power output to accommodate their intermittent nature. This will lead to fossil plants operating at lower efficiencies and capacity factors due to rapid cycling. Bulk energy storage has the potential to relieve the strain on these plants by allowing fossil plants to fire during periods of low power prices or demand. Although electricity storage technologies can be used to collect the electric power generated by a fossil plant, alternatively, thermal energy can potentially be captured from ultra-supercritical steam (580-620°C) via thermochemical energy storage. This study sought to evaluate different mixed-metal oxide (MMO) compounds with a pyrochlore crystal structure ($A_2B_2O_7$) to be used as thermochemical energy storage media. Specifically, both Fe-based lanthanum zirconate pyrochlores ($La_2Fe_xZr_{2-x}O_{7-d}$) and manganese oxides ($Mn_{1-x}Fe_xO_{3-d}$) were synthesized and characterized at various temperatures and red-ox cycling experiments. The impact of transition metal substitution (3, 6, and 9 wt.% Fe) into the B-site of the pyrochlore on the heats of reduction and oxidation was measured using microreactors. Fundamental material characterization such as BET surface area and XRD powder diffraction was conducted to supplement the reaction experiments. Temperature programmed reduction (TPR) experiments revealed that the pyrochlore reduction temperature was lowered with an increase in Fe weight loading. The lanthanum zirconate pyrochlore with 6 wt.% Fe loading displayed more consistent stability over 10 red-ox cycles as compared to the manganese oxide with 12 wt.% Fe loading. Differential scanning calorimetry (DSC) showed that the energy storage capacity of manganese oxide was found to be significantly more than lanthanum zirconate. Future work involves additional red-ox cycling experiments (e.g. 100+ cycles, varying temperatures, and cycle times) to discern the long-term stability of future TCES materials.

584516 Phosphonic Acid Precoatings on Supported Metal Catalysts for CO2 Reduction

Phosphonic acid (PA) self-assembled monolayers have been shown to modify catalytic behavior at the metal-support interface of supported-metal catalysts. Here, aminomethyl phosphonic acid (NH_2MPA) was used on supported Ni catalysts to promote CO selectivity in CO_2 hydrogenation. The NH_2MPA may improve CO selectivity by withdrawing electron density from the Ni. It has been demonstrated that electron-deficient nickel sites bind CO more weakly than electronically neutral ones, which allows CO to desorb before further hydrogenation to CH_4 occurs. Typically, PAs have been deposited on the support after the metal catalyst is already present. However, deposition of PA onto supported Ni catalysts leads to blockage of Ni sites, which reduces catalytic activity. "Reverse deposition," where the support material is precoated with PAs prior to metal deposition, would hypothetically allow for PA-modification at the metal-support interface without active site blockage. NH_2MPA precoated Ni on Al_2O_3 showed increased selectivity to CO in CO_2 reduction reaction studies compared to uncoated Ni. This result was supported by infrared spectroscopy studies showing an apparent weakening of CO binding on the catalyst when the support is coated with NH_2MPA. Catalytic activity still decreased for the precoated sample, but the reduction was much less than that of the sample that underwent normal deposition. To assess the generality of the reverse deposition method, n-butyl phosphonic acid precoated Au and Pt supported catalysts were also investigated. Electron microscope images of these catalysts revealed larger particle size for samples, which may be the cause of lowered catalytic activity for precoated Ni. Future work could look to alleviate this loss in metal dispersion from PA precoating in order to optimize CO selectivity without blocking active sites.

586423 A Flow Cell-Based Approach to Characterizing the Kinetics of Redox Flow Battery Electrolytes

Wind and solar energy installations will be deployed at an increasing rate in the coming decades due to increasing demand for carbon-free electricity. However, energy from these sources is intermittent and peak production fails to coincide with peak demand. In order for the grid to become completely reliant on renewable energy, energy storage systems must be adjacently deployed to support the grid during low production periods. The redox flow battery (RFB) is one of the most promising new technologies to store and use very large quantities of renewable energy on a continuous basis.

The energy efficiencies of flow batteries are often limited in part by slow electron-transfer reactions, and this challenge is driving new research to develop electrodes and redox couples that exhibit fast electron transfer kinetics. In this context, we are working to understand the influence of electrocatalysis on the practical performance of established and emerging flow battery chemistries. Our prior work focused on critically assessing rotating disk electrode (RDE) voltammetry techniques—which are widely used for kinetics studies in fuel cell research—for measuring electron transfer rates in RFBs. Based on the results, we concluded that the unique characteristics of flow batteries necessitate the development of new experimental techniques that better approximate their operating conditions.

Our current approach employs microelectrodes in a flow cell to eliminate the confounding effects of mass transport and electrical resistance in RFB electrokinetics experiments. Results to date show that microelectrode methods can resolve electron-transfer kinetics at least an order of magnitude faster than conventional RDE experiments. We also found that flowing the electrolyte can mitigate electrode fouling effects in the $Pt|FeCl_2||FeCl_3|Pt$ material system, which yields better estimates of the true electron-transfer kinetics. Ongoing work in our lab is now focused on characterizing the kinetics of several RFB couples at carbon fiber microelectrodes in a channel-flow configuration, since carbon is desirable for use in RFBs due to its low cost.

584023 High Temperature Alkene Oligomerization on Single Site Cobalt Catalyst

Nonrenewable resources make up 80% of worldwide energy consumption and are being depleted faster than they can regenerate themselves. It is vital that new energy resources are discovered to provide for society's current and future needs. Shale gas, which contains short-length alkanes (methane, ethane, propane), has steadily become a more prominent resource over the past decade because of advancements in drilling technology. Ethane and propane can be catalytically converted into longer chains through a two-step process of dehydrogenation, to convert alkanes to olefins, followed by oligomerization, to convert olefins to higher molecular weight hydrocarbons that can be used in fuels and petrochemicals. While these processes are traditionally performed under different conditions, they share common intermediates. A stable single-site cobalt catalyst (Co^{2+}/SiO_2) was synthesized to perform high-temperature ethylene oligomerization. The Co^{2+} single site structure was confirmed using X-ray absorption spectroscopy (XAS) at the Advanced Photon Source (APS), Argonne National Laboratory. The C-H bond activation capability of Co^{2+}/SiO_2 was benchmarked against literature using propane dehydrogenation at 550°C and propylene hydrogenation at 200°C with conversions up to 25% and selectivities up to 99%. Ethylene oligomerization performed at atmospheric pressure and temperatures greater than 350°C resulted in a large selectivity (>50%) towards butenes as well as the formation of hexenes (>30% selectivity) and trace amounts of other higher molecular weight hydrocarbons. At higher pressures, these reactions are predicted to condense to liquid fuel products that are more compatible with the United States transportation methods.

587661 Role of Proton Source in Lithium-Mediated Ammonia Synthesis

Electrochemical ammonia synthesis has attracted a lot of attention as a possible alternative to the Haber-Bosch process, as it allows for supplanting high reaction pressures and temperatures with electric potential. This may allow for a more modular and lower capital cost method of ammonia fertilizer production. Many chemistries have been reported to produce ammonia at ambient conditions in aqueous solutions. However, many of them suffer from low rates and low selectivities for ammonia due to competition from the hydrogen evolution reaction. Nonaqueous methods hold promise due to the higher nitrogen solubility in nonaqueous solvents, as well as the ability to control proton source activity. One nonaqueous chemistry that has shown promise in the literature is the lithium metal-mediated one. Lithium-mediated ammonia synthesis uses metallic lithium's reactivity toward nitrogen to fix it, forming lithium nitride. The lithium nitride then reacts in situ with a proton source to produce ammonia and lithium ions. The lithium ions and an electric current are used to regenerate metallic lithium.

This work studies the role of proton sources in the process, which are important for protonating the lithium nitride, as well as for activating the lithium for nitrogen fixation. A wide array of proton sources, with different proton acidities and levels of steric hindrance, was examined systematically to determine their performance. Performance at high currents density was correlated to proton donor properties, creating a predictive model of proton source activity. The concentrations at which ammonia Faradaic efficiency is maximized was determined for each proton source at low current density, in order to minimize transport limitations. Faradaic efficiencies well above previously reported values were observed.

588025 TiO_2 Catalysts: Micropowders and Nanofibers in Sustainable Reactions

TiO_2 Catalysts: Micropowders and Nanofibers in Sustainable Reactions

A. Jernigan[1], S. Nealy[2], W. Maniukiewicz[3] , M. Modelska[3], M. Binczarski[3], T. Maniecki[3], I. Witonska[3], and A. Stanishevsky[2], C. West[1]

[1] *Department of Chemical Engineering, University of South Alabama, USA*

[2] *Department of Physics, University of Alabama at Birmingham, USA*

[3] *Institute of General and Ecological Chemistry, Faculty of Chemistry, Lódz University of Technology, Poland*

Catalytic technologies are vital in the advancement and development of many industries, including chemical, biological, and environmental. One area of focus in catalysis for several decades has been trying to find an efficient means of generating clean hydrogen, as this remains a barrier to widespread implementation of environmentally friendly hydrogen fuel cell technologies. An abundant source of hydrogen is water but liberating hydrogen from water is limited by both reaction kinetics and thermodynamics. Using photocatalysis with readily available biofuels as sacrificial reagents addresses those issues in part and presents the opportunity to harness solar energy to drive the reaction. The most studied photocatalyst is titania, and this work investigates the potential of titania nanofibers for those applications. Some of the inherent disadvantages are addressed using small amounts of other metals, such as its relative inactivity in the visible light range and rapid reversal of electron excitation. One metal incorporated is platinum, which is known to improve hydrogen generation rates, but other, less expensive metals are to be to reduce the amount of platinum needed and thus reduce the catalyst cost.

In catalyst optimization, there is always a focus on finding ways to increase the available surface area and thus the efficiency and rate of reaction. Amid this push for smaller technologies, attention has been recently drawn to the potential of nanomaterials as an alternative to micromaterials in catalysis. Nanofibers are gaining significant interest as an alternative to micropowders with the advantage of greatly increased active surface area without the issue of agglomeration as is seen in nanopowders. In this work, TiO_2 nanofibers synthesized using needleless AC electrospinning were prepared for comparison to TiO_2 micropowders through the screening reaction of photocatalytic reduction of CO_2 to produce methanol. Future plans are to compare the sustainable production of hydrogen by photoreforming of ethanol on TiO2 nanofibers and micropowders doped with CuO and Pt.

Funding:

NSF-IRES grants 1558268, 1708600, and 1852207

USA Office of Undergraduate Research

USA Honors College

USA College of Engineering

USA Study Abroad Office

582968 Advanced Oxidation of Hydrothermal Liquefaction Aqueous Effluent Using Microscale Based Reactors

Wet oxidation is a process that allows for removal of organic pollutants from contaminated water effluents. In this process, oxygen is flowed through the effluent to oxidize the organic compounds into carbon dioxide and water. Such purification methods are crucial in processes such as hydrothermal liquefaction, where large volumes of dilute effluent are generated from biocrude oil production. Current research is focused on the catalysts available to improve the efficiency of oxidizing these organic molecules. However, at optimal oxidation rates with catalyst, most modern reactor designs are still limited by the diffusion rate of oxygen through the effluent. At small enough scales (<100 mm), diffusion limitations faced by traditional tube reactors can be minimized, improving the rate of oxidation by making the process reaction-rate limited instead of diffusion-rate limited. For this purpose, a modular microscale-based reactor was designed for removal of phenol from aqueous solution. A UV-C lamp was used to catalyze photolysis; solutions were reacted at 20°C, 1 atm and with five, ten, and fifteen-minute residence times. A 1000 ppm phenol solution composed of phenol dissolved in water acted as a model contaminated effluent. In the second phase of this experiment, solutions treated with biochar at various concentrations were reacted to determined its photocatalytic abilities High Pressure Liquid Chromatography (HPLC) analysis was used to quantify the concentration of both phenol and oxidation intermediates, such as hydroquinone and catechol. The extent of oxidation was determined to be greatest for biochar-treated solutions, with little or no major change in oxidation rates with variation of distance from UV light or treatment with additional oxygen.

585843 Advanced Reactor Design for Carbon-Free Ammonia SOFCs: An Automated Profiling Approach

There is a growing need for alternative low emission energy sources. Solid Oxide Fuel Cells (SOFCs) are a high efficiency, relatively durable, fuel flexible and low emission alternative to common energy sources. Ammonia is an attractive fuel for SOFCs because it contains no carbon, is easy to store, and is highly efficient due to endothermic cracking. This coupled with the hydrogen oxidation reaction allows for efficient heat utilization. In this study, a method for automated temperature control of the SOFC is developed in conjunction with axial temperature and concentration profiles of tubular SOFCs (*t*-SOFCs) in order to investigate the coupled ammonia decomposition and hydrogen oxidation reactions. Automated control of the *t*-SOFC provides for easy mass data recording allowing for hours of temperature and current data to be stored. Fast ammonia decomposition over the Ni catalyst is evident at the inlet of *t*-SOFC and complete ammonia conversion is confirmed above 600°C. It is found that direct ammonia-fueled *t*-SOFC and an equivalent hydrogen-nitrogen fueled *t*-SOFC provide identical performances. The temperature and concentration profiles validate that the efficient heat integration inside ammonia-fueled *t*-SOFC is feasible if operated at 700°C and below. The results confirm that the overall reactions proceed in a two-step process: the endothermic decomposition of ammonia followed by the exothermic oxidation of hydrogen. These results support foundational background for the development of autothermal ammonia SOFCs.

587445 Homogeneous Zirconocene Catalyzed α-Olefin Polymerization in a Microreactor with in-Situ IR Thermography

A 3D printed microreactor and IR camera are used to study a series of homogeneous zirconocene catalyzed α-olefin polymerization reactions. Microreactors have some advantages over batch reactors used in the study of chemical reactions. They use smaller volumes reactants and help save costs and minimize waste. The concentrations of the reactants and the temperature vary in the axial direction of the micro-channel. This allows the study of the full extent of a reaction by taking a snapshot of the reaction channel. Experiments are performed which demonstrate the practicality of studying exothermic reactions using IR thermography. An IR camera is placed facing the IR transparent top face of the microreactor. The reactor maintained a uniform temperature using a Peltier heating element. The setup is isolated inside of a vacuum chamber to minimize convection and thermal ‘noise’ recorded by the IR camera. Simulating hot fluid flow highlighted the channels in the thermal image and showed negligible heat conduction by the channel walls. The α-olefin, 1-Hexene and activator (B(C6F5)3) solution is reacted with a homogeneous zirconocene catalyst (SBI-Zr-Me2) solution by feeding the two streams into the microreactor. A thermal signature produced by the reaction was observed down the length the mixing region. Using an energy balance the total conversion can be determined at each point down the length of the channel. This ability to study the reaction with in-situ IR thermography opens the door to automated research of reaction chemistry and kinetics.

588061 Catalytic Methane Dehydroaromatization with Microwave Heating

In industry, methane is converted into aromatics through an indirect route: methane to syn-gas and then syn-gas to aromatics. Methane dehydroaromatization (DHA) is a direct route that would be more economical. The DHA process we use is a catalytic reaction involving a zeolite, a microporous aluminosilicate mineral, for higher selectivity of desired aromatics. Due to the thermodynamic unfavorability of the reaction, high temperature operating conditions are required. At high temperatures, coking (solid carbon depositions) of the catalyst occurs and inhibits the reaction. Microwave (MW) heating of the catalyst is currently under investigation. The main advantage is selective heating which can reduce coking. By incorporating metals such as gallium, into the structure, the hypothesis is that the metals induce dipoles which can increase MW sensitivity. Current work involves synthesis of zeolites with different metal compositions and silicon to aluminum ratios in the framework to study how MW sensitivity is affected.

586697 Phenol Degradation By Photo Fenton Reaction Using Fly Ash As Catalyst

Industrial wastewater containing phenolic compounds represents a serious discharge problem due to its low biodegradability, high toxicity and ecological aspects, as well as causing considerable damage to ecosystems and human health. That is why the need arises to look for efficient processes that allow the elimination of toxic pollutants from these waters, generally resistant to conventional treatments, respecting the guidelines present in Resolution 0631 of 2005, where the permissible limits for dumping this waste into bodies of water in Colombia.

Advanced oxidation processes have proven highly effective and the coupling of these systems as a pretreatment to conventional biological processes is positioned as an economically viable alternative for the elimination of toxic and persistent pollutants, such as petroleum derivatives. One of the advanced oxidation processes under study is the Fenton process with ultraviolet light, this allows mineralization and not only the removal of contaminants.

The use of a solid waste from the thermoelectric industries, called fly ash, is proposed. The annual global generation of these is approximately 750 million tons per MW of installed power, which means an estimated total production of fly ash in Spain of 7.5 x 103 t / year. This residue has properties that allow them to have a range of possibilities for use on an industrial scale, its chemical composition is very attractive to be used in the Photo-Fenton process due to the presence of iron oxide and $TiO2$ in the structure. In this sense, this research aims to carry out the degradation of phenol using fly ash from the thermoelectric company Termotasajero S.A., to evaluate its catalytic effectiveness in the Photo Fenton process.

586616 Micro Heat Exchanger for Direct Methane to Methanol Oxidation

This project aims to develop an integrated mixer-reactor-heat exchanger micro tube scalable module for the non-catalytic direct partial oxidation of methane in natural gas (NG) to methanol – what we will call "direct methane to methanol" (DMTM).

We hypothesize that the required residence times can be reduced to <20 s by increasing reactor zone temperature to near 450°C. This is possible using an integrated micro tube-based mixer and preheater, a short, sharp reaction zone and a rapid quench. Advances in micro tube fabrication for high pressure (laser welding allows pressures >1500 psi or 100 bar) and temperature control (quench zones <1 cm length), combined with passivation of reactor surfaces, will make possible economical DMTM with air and without added radical initiators.

This project's experimentation to date consists of studying the passivation. We sulfided metal coupons using either elemental sulfur or liquid sulfiding agents such as dimethyl disulfide and dimethyl sulfoxide. The liquid agents are broken down into a mixture of elemental sulfur, hydrogen sulfide and light hydrocarbons (mostly methane) at 350-500°C, with N_2 or N_2/H_2 at ~130 kPa, reacting with the coupons. Coupons tested included the steels I625, 904L, 304L and C1010, and Ni foil.

These experiments were done over 3-20 day periods. The liquid agents proved better in forming coherent films on the coupons and the added H_2 was useful in breaking down the liquid agents.

The sulfilded coupons were weighed and the films analyzed by X-ray fluorescence (XRF), X-ray photoelectron spectroscopy (XPS), atomic emission spectroscopy (AES), FTIR, and thermogravimetric analysis-differential scanning calorimetry (TGA-DSC).

Once all data were analyzed, it was decided to use I625 for future work. Alloy I625 is a high Ni-Cr steel (58 wt% Ni min., 20-23% Cr); the Ni and Cr form surface sulfides, which from past work can be inert to oxidizing conditions. Suflided coupons are being tested under high-pressure air.

The microreactor system, made entirely of I625 alloy, is being fabricated by Mezzo Technologies. Mezzo used its proprietary software to design the shell and tube sections.

In the meantime, oxidation stability tests on previously coated coupons are taking place in the reactor system that was previously used in the sublimed sulfur agent reactions. These tests are also looking at the chemical inertness of the coupons.

584025 Catalytic Transfer Hydrogenolysis of Bio-Polyols to Renewable Chemicals over Bimetallic PtPd/C Catalysts: Size-Dependent Activity and Selectivity

Owning to the lack of fossil fuel reserves and environmental problems, considerable attention has been paid to the production of biofuels and valuable chemicals from renewable biomass resources. In the traditional hydrogenation process, the addition of hydrogen and high temperatures and pressures are usually needed. The catalytic transfer hydrogenation/hydrogenation without external hydrogenation is considered to be much more atomic efficient and more economical. However, in the process of hydrogen production, hydrogen transfer and hydrogenation, the multifunctional of catalyst structure has not been studied extensively. In this study, bimetallic PtPd/C catalysts with different particle sizes (1.9~6.0 nm) were prepared by different methods. Our study with several kinds of surface characterization confirms the formation of alloyed PtPd nanoparticles, and PtPd/C catalyst has strong size-dependent activity and selectivity trend for sorbitol transformation. The important finding in this work is that, although the kinetics of partial dehydrogenation (C-H cleavage) shows size insensitivity, the reforming (C-C cleavage) and hydrolyzing (C-O cleavage) reactions show strong size determining activity and selectivity in the range of 1.9~6.0 nm. The selectivity of value-added glycols and derivatives over the synergistic PtPd/C catalysts can reach as high as 87% even under inert atmosphere. The experimental results with various polyols provide the basis for the rational sorbitol catalytic transfer hydrogenation reaction network, and verify that PtPd/C nanocatalysts show excellent performance in catalytic transfer hydrogenation of sugar polyols to prepare sugar alcohols and alcohols with high added value. And the results of this basic research is of reference value for the rational design of the next generation of solid catalysts for efficient hydrogenation.

584672 Selective Cleavage of Lignin β-O-4 Ether Linkage By Pd/CeO_2 Catalyst with Unique Metal-Support Interaction

The catalytic activity of heterogeneous catalysts based on noble metals is investigated for the conversion of lignin to precursors of various commodity chemicals. Here we use a model compound containing a β-O-4 ether linkage to understand its cleavage mechanism and kinetics as analogous to that in lignin. Various parameters including temperature, pressure, solvent, and choice of metals and supports have been optimized to obtain the most practical and efficient conditions for industrial applications. It has been found that Pd is the most capable noble metal catalyst tested for β-O-4 cleavage compared to Pt and Ru. The support CeO_2 gives the highest selectivity towards acetophenone (rather than hydrogenated products) when compared to other supports including both ''reducible'' (TiO_2) and ''irreducible'' (Al_2O_3 and SiO_2) metal oxides. The unique support effects of CeO_2 when paired with Pd make it the most effective support. To optimize process temperatures, conversion of the model compound, 2-phenoxy-1-phenylethan-1-ol (PPE) was investigated at 50, 70, 90, and 110 °C under 5 bar H_2. While the conversion of PPE was found to increase with temperature, we have found that increasing the pressure does not follow the same trend. When varying the hydrogen pressure between 5 and 50 bar at a constant 80 °C, the highest conversion (59%) after 4 hours is achieved at only 10 bar. Here, we propose a pressure (H_2) dependent mechanism that involves balancing hydride and PPE on the catalyst surface to cleave the β-O-4 linkage in PPE to selectively form acetophenone, phenol, cyclohexanol, and cyclohexanone. Our results show interesting insights into the various interactions hydrogen has with aromatic substrates on the surface of metal nanoparticles and lead to the design and optimization of a catalytic reaction that can selectively produce commodity chemical precursors from renewable feedstock, such as polystyrene and Nylon-6.

585641 Electrochemical Synthesis of Urea Derivatives Via Ionic Liquid Electrolytes, Primary Amines, and Carbon Dioxide.

Urea is the most synthesized organic compound by mass in the world and 80% of urea available is used to make fertilizer. Since the need for fertilizer is growing every year, the need for urea is increasing as a consequence. Current industry methods for synthesizing urea are expensive, not energy efficient, and have low yields. We are developing an electrochemical reaction which can be performed assisted with ionic liquid as a catalyst to synthesize urea in higher efficiency and at a lower cost. The reaction was delivered successfully in various primary amines in an attempt to synthesize the ureal derivatives. Different potentials and reaction times were tested on all of the amines used in the reaction but had no effect. The reaction was then improved by purging CO2 into the ionic liquid before addition of the primary amine. This gave significantly higher yields with all amines and a 50% reduction in reaction time. Reactions were then performed with ionic liquid and without ionic liquid and the reactions containing ionic liquid were formed product while the reactions without ionic liquid formed only salt. As a result of future findings, the use of Urea in current industries and laboratories could be significantly enhanced in regard to the amount of time, cost, and resources needed.

587884 Ceria Supports for Reforming Catalysts Used in Short Contact Time Catalytic Partial Oxidation

Various catalysts were observed in a quartz tube reactor and their performance was analyzed and compared. The catalysts were evaluated on their ability to convert air, CH_4 and CO_2 into CO and H_2. Multiple support structures are being tested using the same catalyst composition, a series of tests was run using a 3-D printed support composed of CeO_2 and Al_2O_3, and another set of tests utilized a foam support, composed of Al_2O_3. The tests were run using a Pt coated support followed by a Ni-coated support. It has been determined that the addition of Co on the Ni catalysts had no effect on the conversion rate. Previous testing was done using Rh instead of Pt, the values were compared to observe the effectiveness of Pt against a more expensive material. The goal is to scale up the size of the reactor and compare results to determine if this process is industrially viable.

588305 Raman-Spectrokinetics for Gaining Insights on Support Effects of Supported Vanadium Oxide Catalysts

Understanding the link between observed structure and measured reactivity of supported metal oxide catalysts remains a challenging goal for the complex catalytic systems used in oxidative and non-oxidative dehydrogenation of light hydrocarbons.[1]Supported ternary metal oxide catalysts are one such system; their reactivity is influenced by catalyst coverage, relative amounts of species on the support, and the properties the of support. In the study presented, we used a combination of operando Raman spectroscopy and conventional time-resolved reaction kinetics to determine experimentally the redox reaction rates of surface species directly from the Raman spectra. This approach, coined Raman-spectrokinetic, overcomes a major limitation associated with elucidating kinetic parameters by conventional methods, via mass spectrometry (MS) and gas chromatography (GC). The Raman-spectrokinetic approach allows for observation of reaction rates at a specific catalytic site whereas conventional MS and GC analyses only provide the sample's overall reaction rates (average rates from the entire samples' sites). The study presented is a continuation of an earlier study which explored the oxidation of $(VO_x)_m/(NbO_y)_n/SiO_2$catalysts.[2]Herein, we aim to elaborate on earlier work by reporting on the effects of catalyst support material by utilizing the Raman-spectrokinetic approach to gain new insights on the oxidation kinetics of supported vanadium oxide catalysts, following the evolution/depletion of the vandyl (V=O) vibration as a function of O_2/H_2concentration. With the knowledge that turnover frequency in catalysts remains constant as a function of coverage, our observation of the same phenomena via Raman-spectrokinetics renders support for the validity of the method. The goal of this study was to broaden our use of the Raman-spectrokinetic approach into reactivity of metal oxides at different MO_xcoverage (2D and 3D species), on different supports (SiO_2, Al_2O_3, TiO_2, ZrO_2), and with different surface structures. The insights gained in investigating the effects of these variables on reactivity and stability will further ongoing efforts to rationally design catalysts and achieve unprecedented catalytic performance with supported metal oxide catalysts.

[1]Carrero, C. A.; Schloegl, R.; Wachs, I. E.; Schomaecker, I. E. *ACS Catalysis*, **2014**, 4 (10), 3357-3380

[2]Moncada, J.; Adams, W.R.; Thakur, R.; Julin, M.; Carrero, C.A. *ACS Catalysis*, **2018**, 8 (10), 8976-8986

584194 Synthesis of Novel Two-Dimensional Graphene-like Nanomaterials for Catalysis and Energy Applications

Modern catalytic technologies are costly and inefficient at targeting the harmful byproducts of refining processes. However, recent studies show that contemporary catalysts stand to improve through doping and colloidal synthetic methods to produce low-cost, highly active hybrid materials. The key to unlocking the full potential of these hybrid materials lies in maximizing the available active surface area to influence hydrodenitrogenation, hydrodeoxygenation and hydrodesulfurization. Here, our work shows promise in working at the nanoscale to tune the properties of these catalysts and to produce them at a fraction of the cost of comparable, high-efficacy catalysts. This project targets the synthetic study of single nanosheets of Molybdenum disulfide (MoS_2), a highly investigated catalyst for electrochemical studies and energy applications, produced via colloidal synthesis using Cobalt and Nickel dopants. Our results demonstrated an increased catalytic activity for the hydrogen evolution reaction – signaling an expectation of promising results for hydrodesulfurization, which has a similar chemical pathway - when utilizing Cobalt as a dopant in conjunction with specific synthetic parameters including nucleation time, reaction temperature, dopant concentration, and the Sulfur precursor. After determining viable candidates for further reaction-based studies through Linear Sweep Voltammetry, we confirmed the structure of our materials through X-Ray Diffraction and examined the incorporated dopant percentage through Inductively Coupled Plasma Mass Spectrometry. Thus, by assurance of our characterizations, we have developed a method of improving the catalytic activity of Molybdenum disulfide by producing Cobalt-doped MoS_2 nanosheets with the goal of innovating cost-effective, consistent, and alternative catalysts that can optimize current processing methods in chemical manufacturing.

588129 Synthesis of an Alloy-Based Ni-Fe/C Catalyst for the Alkaline Oxygen Evolution Reaction

Hydrogen gas is industrially relevant as a raw material for numerous chemical processes, including high-volume industrial chemicals such as ammonia and methanol. Currently, hydrogen is produced using reforming processes that involve fossil fuels and result in large releases of pollutants like CO_2. Hydrogen production using electrolysis is a benign alternative. Alkaline anion-exchange membrane electrolyzers are promising for clean hydrogen production because the noncorrosive environment of the alkaline electrolyte enables the use of non-precious metal catalysts. However, practical implementation of these electrolyzers is hindered by high overpotentials for the oxygen evolution reaction (OER). The oxidizing conditions of OER also affect the stability and conductivity of catalysts, leading to poor performance. Ni-based catalysts are widely studied for OER in alkaline conditions; furthermore, incorporation of Fe into Ni catalysts has shown promise by decreasing the onset potential compared to pure Ni. We are studying Ni–Fe composites derived from homogeneous metallic alloy precursors with the goal of creating alkaline OER catalysts that exhibit improved electron transport properties than those derived from pure oxides.

We synthesized Ni–Fe alloy nanoparticles on an oxidized Vulcan carbon support using a wet-impregnation technique. The as-synthesized catalyst was activated for OER using electrochemical cycling. Systematic variation of carbon and Fe content in the Ni–Fe/C catalysts showed that a catalyst with 70:30 ratio of Ni:Fe and 48 wt% carbon content gave the smallest onset overpotential of ~270 mV. This catalyst attained 10 mA/cm^2 at an overpotential of 300 mV in 0.1 M KOH. It also exhibited enhanced activity at higher electrolyte concentration and at elevated temperature, but its stability was poor in concentrated electrolyte. Work is now focused on understanding the origin of this stability limitation.

586799 Visible Light Induced Photocatalytic Hydrogen Evolution Using a CdS-Ni2P Hybrid Aerogel System

Due to the increasing global energy demand and the climate change impact of CO_2 from energy production, it is essential to construct clean energy production and storage systems. Photocatalytic evolution of hydrogen from water by visible light serves as a promising and appealing pathway. CdS nanoparticles are good visible light absorbers but not efficient hydrogen evolution catalysts, while Ni_2P is shown to be an exceptional electrocatalyst for the hydrogen evolution reaction (HER). Integration of Ni_2P nanoparticles with CdS nanoparticles using a sol-gel approach is expected to yield a hybrid system that will enable the efficient transfer of photo-generated carriers on CdS to catalytically active Ni_2P sites. In this work, the assembly of pre-formed nanoparticles to make CdS-Ni_2P hybrid aerogels will be described and photocatalytic HER data will be presented. The performance of these novel architectures will be compared to CdS nanoparticles and CdS aerogels. The role of interparticle coupling, porosity and surface passivation on catalytic efficiency will also be discussed. Research reported in this presentation was supported by the National Institutes of Health under Award Numbers GM118981, GM118982 and GM118983, the National Science Foundation under Grant Numbers CHE1367702 and DMR1361470, the Wayne State University President M. Roy Wilson BUILD Scholars Program and the Wayne State University Graduate School. The content is solely the responsibility of the authors and does not necessarily represent the official views of the National Institutes of Health, the National Science Foundation or Wayne State University.

588137 Solar Water Disinfection Using Visible Light

Abstract: Stable In Situ H_2O_2 Generation from an Au nanoparticle/SnO_2 System

Author: Sarah Glass

Metal/Semiconductor heterojunctions are an attractive solution for photodisinfection of water. These systems use sunlight to oxidize water and reduce dissolved O_2 to form reactive oxygen species (ROS) which can kill bacteria and viruses in water. Commonly used semiconductors for these systems are TiO_2 and ZnO both of which utilize only the ultraviolet (UV) portion of sunlight (~5% of the spectrum by energy). This is due to the bandgap of these semiconductors being greater than 3 eV. If the visible portion of sunlight was able to be used, a more effective system could be made. Visible active semiconductors like C_3N_4 have been widely pursued but suffer from instability in aqueous media. An alternative approach is to use metal nanoparticles to drive plasmonic and interband transitions to generate "hot" carriers (e.g. holes and electrons) that are separated at the metal/semiconductor junction. Gold nanoparticles are ultrastrong absorbers of visible light, which makes up 40% of the solar spectrum. This opens the opportunity for a more efficient system that uses ~8x more light provided by the sun.

Here we present the Au-nanoparticle/SnO_2 system that uses a wide-bandgap semiconductor, Tin Oxide, and plasmon resonance effects from gold nanoparticles. Gold nanoparticles strongly absorb visible light through their plasmon resonances and interband transitions with an absorption spectrum that is highly tunable based on the size and shape of the particle and the local environment they are in. SnO_2 is a highly stable, wide bandgap semiconductor that does not absorb visible light due to its wide bandgap (e.g. > 3.8 eV). It also has a favorable conduction band alignment with gold accepting the hot-electron and driving O_2 reduction at the SnO_2/water interface.

Our Au nanoparticle/SnO_2 systems results show that visible light can be used to generate H_2O_2. The test results support our initial hypothesis that using a stable wide bandgap semiconductor in conjunction with metal nanoparticles will generate ROS using visible light (> 450 nm).

587511 Homogeneous Hydrolysis of a Thermally Degrading Tracer

A major obstacle in commercializing geothermal resources is predicting the potential for cooling of production wells, particularly in the presence of cold water reinjection. Thermally degrading tracers such as phenyl acetate may enable advanced warning of "premature thermal breakthrough" if compounds with appropriate and known temperature-dependent reaction kinetics can be identified. This work investigates the potential for accelerated reaction rates in the presence of mineral surfaces (i.e., a heterogeneous reaction path). A homogeneous aqueous solution buffered with citric acid and sodium phosphate dibasic was spiked with phenyl acetate and the hydrolysis reaction was measured by spectrophotometry for temperatures ranging from 20 to 50 °C. The activation energy and pre-exponential factor were calculated from the Arrhenius equation and the hydrolysis of phenyl acetate displayed a low activation energy. This means that phenyl acetate could be a suitable chemical tracer for a wider range of temperatures and have useful applications as a well maintenance tool for low-temperature geothermal energy wells.

583252 High Efficiency Encapsulation of Fischer-Tropsch Catalyst in ZSM-5

Fischer-Tropsch Synthesis (FTS) is hydrogenation of carbon monoxide to hydrocarbons. The main challenge is that this reaction gives a wide range of products that follow the Anderson-Shultz-Flory distribution [1]. We are interested in designing an active catalyst for FTS with high selectivity toward C_2-C_4 olefins, which are high demand raw materials for other chemicals, plastics, and synthetic fuel [2].

Fe is widely used for FTS due to its high activity [1], and ZSM-5 has shown to promote selectivity toward light olefins [3]. We are attempting to use novel synthesis methods to encapsulate Fe nanoparticles within Na-ZSM-5 for a better performing catalyst. Our control catalyst was synthesized using incipient wetness impregnation of Fe on Na-ZSM-5.

Our catalysts were characterized using XRD, TEM, N_2 physisorption and CO chemisorption. The reactor studies were done under a differential packed bed reactor to study the performance of our catalysts for FTS. Our catalyst has shown to be a mesoporous material with high activity compared to our control catalyst.

1. Abelló, S. & Montané, D. Exploring Iron-based Multifunctional Catalysts for Fischer-Tropsch Synthesis: A Review. ChemSusChem 4, 1538–1556 (2011).
2. Zacharopoulou, V. & Lemonidou, A. Olefins from Biomass Intermediates: A Review. Catalysts 8, 2 (2017).
3. Ma, Z. & Porosoff, M. D. Development of Tandem Catalysts for CO2 Hydrogenation to Olefins. ACS Catalysis 9, 2639–2656 (2019).

584585 Electronic Descriptor of Single Metal-Oxo Species on Phthalocyanine and Porphyrin Functionalized Graphene Towards Methane Activation Process

Catalysts are needed to expedite the chemical reactions by lowering the energy required for the reactions to produce high-value-added chemicals. In this work, single-metal active sites based on 10 different phthalocyanine/porphyrin functionalized graphene materials have been screened and characterized for the C–H bond activation of methane to methanol conversion process using ab initio density functional theory (DFT) calculations. The results show that a radical mechanism is the predominant mechanism for the C–H bond activation process of methane for these type of materials. By recognizing the inverse correlation between the metal-oxo species formation energy and the C–H bond activation energy, both can be directly linked to a better electronic descriptor known as the metal charge state of the metal-oxo species. The optimal performance of the catalysts can be attributed to the metal charge state value of the metal-oxo species, with an optimum value of 1.30±0.03. The GCoPc system seems to be a potential candidate for optimum utilization of energy for complete methane to methanol conversion process. The catalytic activity of this system can also be tuned by controlling the metal charge state of Co *via* both substrate doping and ligand exchange.

587179 Insights on the Synthesis of Nanostructured Non-Stoichiometric Mixed Metal Oxide Electrocatalysts Via Reverse Microemulsion Method

Nonstoichiometric mixed metal oxides, such as perovskites (ABO_{3-d}), double perovskites (AA'B2O5+d), and Ruddelsden Popper oxides (An+1BnO(3n+1)+d) (where A is a lanthanide and/or an alkali earth metal and B is a transition metal cation), are recognized as promising electrocatalysts for oxygen reduction/evolution reactions - key reactions for a number of energy conversion and storage technologies. These materials are characterized by their excellent ionic-electronic conductivities, as well as thermal and redox stabilities. Recent findings have demonstrated a strong relationship between mixed metal oxide surface properties (i.e., structure and composition) and their electro/catalytic activity. However, the traditional synthetic methods employed for synthesizing these oxides offer little to no control over the surface structure or composition of these materials. Moreover, high calcination temperatures are traditionally needed to obtain pure oxide phases, which lead to agglomeration of the oxide nanoparticles, decreasing their final surface areas. In this direction, the reverse microemulsion approach has emerged as synthetic method for successfully producing complex oxides with controlled surface properties along with high surface areas. However, limited understanding of the synthesis parameters that govern the final particle properties and the universality of applying this synthesis method to a wide range of mixed metal oxides of different compositions/structures exists.

In this contribution, we investigated the effect of key synthesis parameters of the reverse microemulsion method on the properties of the resulting oxide nanoparticles. We found that the condensation reaction rate within the reverse micelle (intramicellar exchange) is an important parameter controlling the final particle morphology. Overall, higher concentrations of the precipitating agent resulted in fast reaction rates, leading to formation of short nanorods, while lower concentrations led to slower reaction rates with longer nanorods being produced. This is owed to the compromise between intramicellar and intermicellar exchanges, where the growth of rods is sustained by slowing the intramicellar condensation rate (precipitation of metal hydroxides within the reverse micelle), which allows for intermicellar exchange (fusion of metal containing reverse micelles) to take place. Furthermore, we provide insights on the universality of this synthetic approach to a wide range of mixed metal oxides with varying crystal structures and compositions.

588151 Learning on the Job: An Active Learning Calculator for Atomistic Simulations

Neural network potentials have been used to accurately predict the energy and forces on atomic structures, acting as suitable substitutes for traditional electronic structure calculators. However, these predictions often diverge from true values when the neural network encounters images that deviate considerably from those found in the training set. In these cases, it would be desirable to include this image in the training set and re-train the network so it can better understand this region of the potential energy surface, but the manual selection of these images can be tedious and impractical. We present a method whereby a bootstrap ensemble of neural network calculators is used as a hands-free calculator that can identify poorly understood regions on the potential energy surface and retrain the calculator as necessary: for every image, an assessment of prediction certainty is made by evaluating the ensemble width, used as a measure for uncertainty. For images where the ensemble is uncertain, we call the electronic structure calculator and retrain the ensemble calculator on that image. A scheme for accelerated retraining of the neural network is also presented within. We apply this calculator to a 400-image sample of a single water molecule where one of the bond lengths and the bond angle is varied. We show the ability for the ensemble calculator to train itself when it is uncertain about an image, and display how retraining on just a few images allows the calculator to understand the region much better. This active learning calculator should allow for the hands-free training of accurate neural networks to perform electronic structure calculations through sample regions not covered by the training data, considerably reducing computation time and cost for future calculations that retread previously investigated sections of the potential energy surface.

587830 Role of Pore Structure on the Sintering Effect of Pt Diesel Oxidation Catalysts

The USDRIVE Low Temperature Oxidation Catalyst Test Protocol recommends an accelerated aging test for Pt diesel oxidation catalysts (DOCs) to be performed at 800 °C in flowing air. However, at such high temperatures, Pt sinters readily to form large particles due to significant vapor phase transport of PtO2. One approach we used to slow the rate of Pt sintering was by alloying Pt with Pd [1]. A second approach was to trap the mobile Pt species with an oxide such as PdO that reacts to form Pt-Pd bimetallic particles [2], or using CeO2, which traps Pt in atomically dispersed form [3]. Recently, we have explored the role of pore confinement to determine if it was possible to slow the vapor phase transport of PtO2, and therefore, the rate of Pt sintering. Since PtO2 vapor has negligible binding to silica [4], it provides the ideal support to study the role of pore structure. We studied MCM-41 and SBA-15 as silica supports and found that, although they slowed the rate of Pt sintering after aging at 800 °C, the resulting Pt DOCs was not very reactive due to blocking of the pores by the Pt particles, which in turn blocked accessibility of the smaller Pt particles retained inside the pores. This inaccessibility is due to high pore diameter to length ratios of MCM-41 and SBA-15. Here we explore other silica with a more open pore structure, such as silica nanospheres with an open pore morphology exhibiting a lower aspect ratio than MCM-41 and SBA-15. We found that we could slow the rate of Pt sintering and improve the reactivity using a model reaction, such as CO oxidation. The results are then compared to Pt on a commercial Davisil silica support, which we use as the control for this study.

Acknowledgements

The research was supported by supported by NSF GOALI grant CBET-1707127 and the NSF REU site award CBET- 1560058

References:

[1] C. Carrillo *et al*, Applied Catalysis B: Environmental **218** (2017), p. 581.

[2] C. Carrillo *et al*, Journal of Physical Chemistry Letters **5** (2014), p. 2089.

[3] J. Jones *et al*, Science **353** (2016), p. 150.

[4] P. N. Plessow and F. Abild-Pedersen, ACS Catalysis **6** (2016), p. 7098.

587680 The Impact of Low Molar Doping of Low Valence Metals on Sm_2O_3 Catalysts for the Oxidative Coupling of Methane

Approximately 85% of natural gas is composed of methane. Converting methane to higher value molecules has been a major focus in the field of catalysis for decades. Ethylene is of particular interest as a target molecule as it has a variety of applications such as it's polymerization into polyethylene, the most used plastic in the world. Converting methane to ethylene through the oxidative coupling of methane (OCM) is a catalyzed process. An ideal catalyst requires the presence of basic sites and a high active surface area. While these properties can be quantified and understood easily, the interactions between a catalyst and various dopants are not fully clear. Synthesizing an effective catalyst for OCM requires an understanding of these interactions. Understanding the effects of certain dopants on a catalyst provides valuable information on how that dopant impacts catalyst performance for a variety of catalysts. In this study, the effects of molar doping at 0.1% and 1% of low valence metals onto samarium oxide (Sm_2O_3), were investigated. Sm_2O_3 was chosen as the base catalyst because it is one of the most active and stable single oxide catalysts for the methane coupling reaction, so understanding the behavior of the metals doped into Sm_2O_3 can help in the synthesis of more effective catalysts for OCM. It was found that several dopants such as lithium and silver improved the conversion and selectivity of the catalyst. At certain molar doping percentages, both the activity and stability of Sm_2O_3 was improved. Increasing the amount of dopant did not improve performance variables for all catalysts indicating the uniqueness of each dopant and the balance needed to find an optimal dopant concentration.

588348 Nickel Oxide Silica Based and Commercial Nickel Oxide Nanosorbcats for Adsorption of Cationic and Anionic Dyes Followed By Thermo-Oxidative Decomposition

Wastewater contains a variety of pollutants that are basically classified into organic such as dyes and inorganic substances. The contamination of surface and groundwater with synthetic dyes is a serious environmental problem and is a threat to human beings. These dyes could be cationic, anionic, or neutral. For instance, many industries use pigments to color their products such as textiles, plastics, paper, and so forth. The discharge of the produced wastewater effluents into the environment is worrying for toxicological and aesthetical reasons. Therefore, their removal from wastewater is of great interest. All processes for removing such pollutants from wastewater have their own limitations. However, the adsorption process is recognized to be an effective and economic process for cleaning wastewater. In addition, adsorptive remediation of wastewater on metal-oxide-based nanoparticles has been reported in the literature extensively. Recent studies showed significant awareness of metal oxides as nanoadsorbents and nanocatalysts (nanosorbcats) for cleaning wastewater, due to their individual chemical and physical properties, which make them ascendant to the conventional adsorbent/catalysts.

Herein, in this study, the employment of in-house prepared nickel oxide silica-based (SBN)s and commercial nickel oxide (NiO) nanoparticles as nanosorbcats for cleaning up synthetic wastewater was investigated. Different model molecule pollutants; mainly methylene blue (MB) and bromocresol green (BCG) exemplifying cationic and anionic pollutants, respectively, were removed from synthetic wastewater samples by SBNs and NiO nanoparticles. Batch isotherm adsorption studies in addition to kinetics and thermodynamics were performed for the removal of both dyes from wastewater. The experimental adsorption data has been modeled using the Sips adsorption isotherm model to understand the adsorption behavior of these molecules. Fast adsorption kinetics of both nanoparticles have been observed; NiO nanoparticles took 60 minutes to reach saturation of BCG while SBNs took only 2 minutes for the case of MB. Moreover, both nanoparticles have shown high uptakes of about 99% and 96% at low concentrations of MB and BCG. Furthermore, thermodynamic parameters like the changes in free Gibbs energy (ΔG), enthalpy (ΔH), and entropy (ΔS) of the system were calculated. The catalytic activity behavior of spent nanoparticles was investigated *via* thermogravimetric Analysis (TGA), to check out the applicability of the nanosorbcats concept.

Consequently, the regeneration, uptake saturation capacities and the performances of spent SBNs and commercial NiO nanoparticles after each adsorption cycle were studied and compared with the virgin ones. The detailed findings of this study will be discussed further during the poster presentation.

587869 Metal Screening of Ceria-Zirconia Based Catalyst for Low Temperature Dry Reforming of Methane

Low temperature reforming catalysts are desirable for intensification of the process for producing renewable liquid fuels from biogas. In this study, Ceria-Zirconia based catalysts with various metal loadings were synthesized and characterized for efficacy in the dry reforming of methane (CH_4+CO_2-->$2H_2$+2CO). Building upon optimal combinations and suggestions found in literature[1], various weight percentages of palladium (Pd), iron (Fe), ruthenium (Ru), or cobalt (Co) were loaded onto $Ce_{0.6}Zr_{0.4}O_2$ (CZO). Nickel and magnesium were also incorporated in predetermined weight percentages from published research to obtain a balance between catalyst activity and stability[2]. Characterization techniques included X-ray diffraction to identify bulk composition, hydrogen temperature programmed reductions (H to determine activation temperatures of catalysts, and nitrogen physisorption to define pore structures. Temperature programmed reactions were executed to verify 10% and 50% conversions of both CH_4 and CO_2 for palladium and ruthenium samples; these were prioritized due to the promising low reduction temperatures found via H_2-TPR. Ru samples gave the best performance in terms of lower reduction and conversion temperatures compared to the synthesized Pd samples and platinum (Pt) samples found in previous papers. 0.16Ru-1.4Ni-1Mg-CZO outperformed 0.16Pt-1.39Ni-1Mg-CZO and 0.16Pd-1.4Ni-1Mg-CZO by at least 20°C and 30°C, respectively. Considering the price of Ru has been, on average, 13 times cheaper than that of Pt over the past ten years[3], Ru can serve as a replacement for Pt in terms of cost and performance. Future work related to this project will involve time on stream studies to test the stability of the ruthenium catalysts and post reaction characterizations of tested catalysts.

Word Count: 253

References

(1) Elsayed, N. H.; Roberts, N. R. M.; Joseph, B.; Kuhn, J. N. Low Temperature Dry Reforming of Methane over Pt–Ni–Mg/Ceria–Zirconia Catalysts. *Applied Catalysis B: Environmental* **2015**, *179*, 213–219. https://doi.org/10.1016/j.apcatb.2015.05.021.

(2) Sokefun, Y. O.; Joseph, B.; Kuhn, J. N. Impact of Ni and Mg Loadings on Dry Reforming Performance of Pt/Ceria-Zirconia Catalysts. *Ind. Eng. Chem. Res.* **2019**. https://doi.org/10.1021/acs.iecr.9b01170.

(3) Xia, J.; Volokh, M.; Peng, G.; Fu, Y.; Wang, X.; Shalom, M. Low-Cost Porous Ruthenium Layer Deposited on Nickel Foam as a Highly Active Universal-PH Electrocatalyst for the Hydrogen Evolution Reaction. *ChemSusChem* **2019**, *12* (12), 2780–2787. https://doi.org/10.1002/cssc.201900472.

587915 *Oxidative Coupling of Methane Using Mn-Na_2WO_4/TiO_2 Catalysts: effect of the Support on Reaction Performance*

Ethylene is a vital chemical for the synthesis of most petrochemical products, and highly demanded in the industrial field. As the demand increases, various methods of converting abundant methane into useful ethylene are drawing more attention. The current commercial process of producing ethylene from methane is a two-stage process that first converts methane into syngas (steam reforming) and then synthesizes it with ethylene. However, this process suffers from low energy efficiency and high capital and operating costs. The OCM is energetically more efficient than endothermic steam reforming because it is exothermic, and this one-step process can reduce capital and operating costs. However, the OCM can proceed to deep oxidation and homogeneous reaction to form CO or CO_2. Thus, avoiding formation of CO or CO_2 is critical for improving yield of C_2 products. The Mn- Na_2WO_4/SiO_2 catalyst has shown promise for OCM. In this study, the Mn-Na_2WO_4 was supported with various types of TiO_2 (Mn-Na_2WO_4/TiO_2) to determine the most promising catalyst to enhance C_2 yield and methane conversion. These catalysts were also further compared to a previously studied Mn- Na_2WO_4/SiO_2 catalyst. Each catalyst was prepared using incipient wetness impregnation method followed by a specific calcination procedure. The tests were conducted over a range of temperatures from 725 to 800 ^{o}C in a fixed-bed reactor. The gases exiting from the reactor were analyzed using a gas chromatograph equipped with a flame ionization detector. The surface area and crystallinity of catalysts were analyzed using BET method and X-ray diffraction (XRD), respectively. In this study, the best catalyst was determined to be 1.9% Mn – 5% Na_2WO_4/TiO_2 (A), which has the largest surface area among the examined TiO_2 supports, suggesting that an increased surface area of the supporting metal may result in enhanced C_2 selectivity and methane conversion. In this context, future investigations are necessary to understand relationship or correlation between catalyst performance and surface area of the supporting catalysts.

588358 Metal Containing ZSM-5 Catalysts for the Dry Reforming of Methane: Exploring Synergistic Effects

The abundance of natural gas in the U.S. brought by the development of new fracking technologies for its extraction has triggered a notable renaissance of the U.S. chemical manufacturing industry. The selective activation of C-H bonds in the most abundant component in natural gas, methane, is one of the greatest challenges in catalysis. Methane can be converted into raw materials or commodity chemicals as follow: i) indirectly via syngas or ii) directly via, for instance, selective oxidation or dehydroaromatization. Along with methane, emphasis has been placed on the utilization of CO_2 to produce chemicals and fuels do to their abundance and negative impact to the environment. Both CH_4 and CO_2 are the main contributors of the greenhouse effect. As a result, Dry Reforming of Methane (DRM) is an attractive route to convert CH_4 and CO_2 into syngas with an H_2/CO ratio of 1:1, which is the optimal ratio to produce liquid hydrocarbons through the Fischer-Tropsch process. However, among other factors, the multiple undesired side reactions such as methane thermal degradation, carbon monoxide disproportionation, and water gas shift has, have limited the potential application of DRM at an industrial scale. These reactions adversely affect the H_2/CO ratio and the overall catalyst's performance. The major challenge to address, as a result of the mentioned side-reactions, is the rapid catalyst deactivation due to the formation of coke.

In this study, we aim to improve DRM using a series of Fe/Ni/ZSM-5 catalysts. First, various methodologies have been used to prepare these materials trying to determine the most important variables influencing the final bulk and surface structure of the catalyst. We use various characterization techniques such as N_2-physisorption, X-Ray diffraction (XRD), X-Ray photoelectron spectroscopy (XPS), Scanning Electron Microscopy (SEM), and Raman spectroscopy. Reactivity tests are performed in a home-made reaction setup equipped with various mass flow controllers, a stainless-steel reactor, a tubular furnace with a temperature controller, and a gas chromatograph (GC) for the analysis of the products. Steady-state kinetics has shown a stable catalytic activity and selectivity towards syngas for Fe/Ni/ZSM-5 compared to the Ni/ZSM-5 catalyst, indicating that Fe is suppressing the formation of coke during the reaction. We aim to determine the variables and the optimal catalyst composition to enhance the catalytic performance of conventional Ni-based catalysts. We will also perform in-situ/opernaro Raman studies to compliment the kinetic data and stablish a more conclusive structure-reactivity/selectivity/stability relationship.

585893 Computational Screening of Environmental Barrier Coating Materials for Nuclear Thermal Propulsion Engines

Although nuclear thermal propulsion (NTP) rocket engines have the potential to far outperform the state-of-the-art chemical combustion engines for interplanetary travel, corrosive interactions between the hydrogen propellant and the uranium fuel element acts as a serious barrier of entry for the application of NTP engines. Environmental Barrier Coatings (EBC) have been considered previously to protect the elements but coating techniques used did not enough control to be effective. This research investigated the use of Atomic Layer Deposition (ALD) as a technique to coat fuel spheres with a nanoscale EBC to prevent hydrogen attack on the fuel. ALD allows a conformal, uniform, and nanometer thick coating on both the fuel spheres as well as the high aspect ratio channels of the fuel elements. The primary focus of this study was the coating material to be deposited via ALD. Potential candidates were selected based on low neutron cross sections and high melting points. Density functional theory (DFT), a computational quantum mechanical modelling method, was then employed to calculate energy barriers for a hydrogen atom to move through both the surface unit cells and the bulk unit cells of these candidate materials. Those with the highest barriers were deposited on surrogate materials and compared to materials previously investigated for this application. Samples were characterized to ensure deposition was successful and determine film thickness. Stability of materials in a high hydrogen environment was also studied with DFT. This predicted stability was then tested experimentally in a high temperature furnace system. Through this process, it was shown that nitrides are predicted to outperform pure metals when used as EBCs in NTP engines and many will remain stable at very high temperatures. It was also demonstrated that both tungsten nitride (WN) and molybdenum nitride (MoN) are not stable enough species for this application. Candidate materials from this work will be deposited on surrogate substrates to be experimentally compared to each other, as well as barrier materials previously investigated for this application, in future work.

585906 Theoretical Design and Analysis of High-Spin Ground-State a-D-a-π Conjugated Polymers

Theoretical design and analysis of high-spin ground-state A-D-A-π conjugated polymers

John D. Teas, Md Abdus Sabuj and Neeraj Rai

Dave C. Swalm School of Chemical Engineering and Center for Advanced Vehicular Systems, Mississippi State University, MS, 39762, USA.

High-spin ground-state donor-acceptor (DA) conjugated polymers (CPs) used as organic semiconductors (OSCs) are prevalent in research spanning the fields of physics, chemistry, and materials science. The OSCs with open-shell character have applications in many modern technologies, including organic electronics, nonlinear optics, and energy conversion and storage. Alternating arrangement of DA CPs exhibit various properties, such as interactions with light, the ability to transport charges, and tunability through chemical modification, which also allow for the active usage of these OSCs in optoelectronic technology (London *et al*, 2019). However, due to the presence of unpaired electrons, these high-spin organic molecules are unstable and highly localized. Despite these challenges, computer simulation allows for the design and analysis of these high-spin organic materials. Here, we have designed three conjugated A-D-A-π polymer semiconductor structures that show signs of having a triplet ground state ($S = 1$) in the long polymer chain limit. Density functional theory (DFT) calculations at the unrestricted (U)B3LYP/6-31G** level of theory were performed on the oligomers ($n = 1$ to 5) to identify the evolution of their electronic properties. The diradical character index ($0 < y < 1$), which describes the contribution of the open-shell resonance form to the ground-state, was found to be $y = 0.537$ for $n = 1$ and $y = 0.999$ for $n = 5$. This implies the formation of a pure diradical ($y = 1$) for longer polymer chain lengths. The singlet-triplet energy gap (ΔE_{ST}) decreases with an increase in the length of the polymer chain ($\Delta E_{ST} = 0.114$ eV for $n = 1$ and $\Delta E_{ST} = 2.7\text{x}10^{-6}$ eV for $n = 5$), which shows the extent of electron exchange interactions from admixing of the HOMO and LUMO in the ground state. This study shows the possibility to synthesize new A-D-A-π polymers, which are stable at ambient conditions, with intriguing molecular and electronic properties.

587446 Generalized Antoine and Wagner Equations Consistent with the Predictions of the Peng-Robinson Equation of State Using Similarity Variables

Chemical engineers use the Antoine and Wagner vapor pressure equations because of their relative simplicity; however, their accuracy is limited to restricted temperature ranges and they require substance-specific constants. Generalized methods were developed to find Antoine and Wagner equation constants consistent with the Peng-Robinson (PR) equation of state. Two similarity variables were introduced to simplify calculations. Adjusted temperature, η, combined reduced temperature and acentric factor. Adjusted pressure, Ψ, combined reduced pressure and reduced temperature. A PR vapor pressure algorithm generated a data set of these similarity variables, and a nonlinear least squares method was used to find generalized Antoine constants. When all three generalized Antoine constants were independently fit, the results matched PR vapor pressure predictions within 0.05 percent for twelve subranges of adjusted temperatures from the critical point down to 0.23. Fitting only the C constant by least squares with the A and B constants calculated by exact fits of the adjusted pressure at the endpoints of each subrange, only five adjusted temperature subranges gave the desired accuracy from the critical point down to 0.23. For the Wagner equation, only three adjusted temperature subranges gave the desired accuracy from the critical point down to 0.23.

588238 A Comparison of Methods for Predicting the Melting Point of Perarylphosphonium and Perarylsulfonium Ionic Liquids Using Molecular Dynamics

Ionic liquids (ILs) are organic salts that have melting points below 100°C. Their unusual properties, such as non-flammability, negligible vapor pressure, and high tunability have made them of interest for industrial applications such as heat storage, fuel cells, and batteries. In most applications melting point is desired so that the IL remains pumpable at moderate temperature. Oftentimes the IL structure can be modified, which affects the melting point, however knowing what modifications must be made to lower the melting point remains a challenge. Molecular dynamics (MD) is a method to simulate the physical interactions and movements among a molecule and its atoms. Several different methods exist in MD to calculate the melting point of a compound; however, it is not entirely clear how well each of these methods perform for ionic liquids. Knowing which method performs the best would allow us to connect the changes in chemical structure of ILs with the melting point. In this work we compare the performance of three popular methods for computing the melting point of a set of perarylphosphonium and sulfonium ionic liquids: the void density method, the hysteresis method, and the solid-liquid interface-based (SLIB) method. Each of these methods requires prior knowledge of the crystal structure, which was obtained experimentally from X-Ray scattering. Furthermore, the experimental melting points were also obtained using differential scanning calorimetry, allowing us to compare these melting points with melting point predictions from each method. Once a model provides an adequate prediction of the melting point, the molecular interactions provided by that model are further studied to understand the molecular-level causes of melting point depression in these compounds. Understanding the molecular interactions of a complex ionic compound will assist in generating new ILs and altering newly formed complex ionic compounds.

585238 Performing Quantum Chemical Computations at Different Levels of Theory to Predict Am(III) over Eu(III) Selectivity for DTPA-Based Ligands

The presence of minor actinides (especially Americium) in spent nuclear fuel has posed a veritable challenge for the long-term storage of nuclear waste due to their high radiotoxicity and thermal emission. Minor actinides also have the potential to be reused in nuclear fuel cycle, making it further desirable for their extraction. However, despite the development of polyaminocarboxylate ligands that can achieve the problematic separation of Am(III) from trivalent lanthanides via selective complexation of the former, insufficient theoretical understanding of the molecular chemistry behind such complexation has hindered the design of more efficient and industrially attractive ligands. These theoretical obstacles arise primarily from challenges in computationally describing (1) heavy metal electronic structures and (2) the increase in covalent interactions between polyaminocarboxylate ligands and Am(III) compared to Ln(III).

This work aims to determine a reliable computational protocol in predicting the Am(III) over Eu(III) selectivity of a polyaminocarboxylate ligand by evaluating the performance of *ab initio* quantum chemistry computations at various levels of theory, including density functional theory, Post-Hartree-Fock, and multiconfigurational methods. We also employed basis sets of varying degrees of flexibility for light and heavy atoms each. In our search for a protocol that provides reasonable accuracy while expending as few resources as possible, our results have potentially shown not only the importance not only of utilizing basis sets of sufficient size, but also of examining how a particular computational method accounts for electron correlation and relativistic effects. Further evaluation of these computational methods to other similar systems as well as comparing their results with experiment are needed in order to systematically corroborate their predictive ability.

586340 Computational Analysis of C-H Hydroxylation in Bio-Inspired Monometallic Complexes

With sustainable energy innovations on the rise but logistical and technological barriers preventing their widespread implementation, it is of paramount importance to increase the sustainability of conventional energy generation for the remainder of its lifecycle. Natural gas, despite being the most environmentally benign of the fossil fuels, is underutilized due to difficulties in storage and transportation. Catalytic C-H hydroxylation is achieved in nature by enzymes known as lytic polysaccharide monooxygenases (LPMOs) . The goals of this research project are twofold:

1. To quantify the difference in energy barriers faced by a singlet and triplet state catalytic complex
2. To determine whether a one step or two step reaction mechanism is energetically favorable for this C-H hydroxylation process

Computational methods are applied to bio-inspired catalytic systems, utilizing quantum mechanical algorithms to optimize initial and final molecular geometries and determine stable transition state structures. The relevant energy and entropy differences are then calculated and compared between the spin states and pathways, shedding further light on the nuanced and controversial mechanisms by which these biological complexes catalyze C-H hydroxylation, and potentially enabling more effective catalyst design. Tentative preliminary results are consistent with literature values.

588052 Examining Potential Intermediates in the Electrochemical Synthesis of Ozone from Water

Ozone can be produced by electrochemical water oxidation, yet there are many questions about how to engineer its reaction mechanism that involves six electron transfer steps. Understanding this reaction can offer insight into properties of intermediates along the reaction pathway, which can influence the activation energies and rate determining step of this reaction. Once this information is known, it may be used to determine the properties of a catalyst, which can ultimately improve the rate and the energy cost for this reaction. To properly understand this reaction, it is imperative to have an accurate understanding of the energetics and thermodynamic properties (e.g. acidity constants and standard redox potentials) of hypothetical reaction intermediates along the reaction pathway. We have used Kohn-Sham density functional theory (DFT) along with implicit and mixed solvation models to determine these acidity constants and reduction potentials for a collection of oxygen containing molecules. A mixed implicit/explicit solvation model and the accompanying thermodynamic cycle has also been found to reduce errors in calculations compared to the available experimental data. After benchmarking this approach with available experimental data, we predict thermodynamic properties for different solvated reaction intermediates and determine which are relevant for the electrochemical ozone production reaction.

586847 Generalized Antoine and Wagner Equations Consistent with the Predictions of the Soave-Redlich-Kwong Equation of State Using Similarity Variables

Chemical engineers use the Wagner and Antoine equations to correlate vapor pressure as a function of temperature for various substances; however, they do not describe the entire vapor pressure curve from the triple point to the critical point. Methods of finding generalized Antoine and Wagner equation constants consistent with the predictions of the Soave-Redlich-Kwong (SRK) equation of state were studied. Calculations were simplified using two similarity variables. Common adjusted temperature, η^*, combined reduced temperature and acentric factor, and common adjusted pressure, ψ^*, combined reduced pressure and reduced temperature. Results of an iterative SRK vapor pressure algorithm were transformed into a data set of the new similarity variables. Antoine and Wagner constants were empirically determined by nonlinear least squares to achieve less than 0.05 percent relative deviation from SRK predictions. Three common adjusted temperature subranges, from the critical point down to 0.23, were needed for the Wagner equation. Only the C constant was fitted by least squares for the Antoine equation, with the A and B constants determined by exact calculation at the endpoints of common adjusted temperature ranges. Five common adjusted temperature subranges, from the critical point down to 0.18, were needed for the Antoine equation.

584856 Rapid Assessment of Chemical Toxicity Using Machine Learning and Combinatorial Quantitative-Structural Analysis Relationship Models

Due to the nature of their work, United States Airmen and other members of the armed services can be exposed to harmful chemicals that may degrade their physical and/or cognitive performance. Additionally, on-the-ground personnel may require rapid assessment of a chemical's toxicity; laboratory testing is often too slow to provide an adequate response. Therefore, there is a real need for rapid, robust, and accurate *in-silico* models that can computationally predict key toxicological activities. Quantitative structural-activity relationship models, or QSAR, are becoming popular as models of choice with the rise of machine learning in the field. The developed Python QSAR package, named as QSARBayesOpt, employs combinatorial QSAR, an emerging type of model in the cheminformatics community. Using consensus voting, the predicted values from each independent model are combined to increase the predictive power of the package as a whole. Through this automatic tuning framework, the package allows the end-user to easily change datasets using Bayesian optimization. Bayesian optimization is also an intelligent tuning method, learning from previous iterations to make smarter searches in the parameter space; therefore, significant amount of computational time and cost can be saved using this approach. Furthermore, we have tested this package against several datasets, including inhibition activity of acetylcholinesterase by organophosphates, where we saw 78% regression accuracy and 97% classification accuracy.

584180 Role of Confinement, Molecular Connectivity and Flexibility on Entropic Driven Surface Segregation of Polymer-Colloid Mixtures

Relative surface affinity between polymers and colloids are leveraged in many applications like filtration, adhesion, bio-sensing etc. The surface affinity is governed by both enthalpic (relative interactions between the species and surface) and entropic (excluded volume) effects. Neglecting enthalpic effects i.e. for purely athermal systems, entropy is the only driving force that controls the surface affinity of the species in a binary or multi-component mixture. Many intensive (relative size of colloids, chain length, equilibrium bond angle, chain flexibility) and extensive (confinement, temperature) factors can dramatically change the entropy of the system and thus enhance or decrease the surface affinities of the constituent species. In this article, we use coarse grained metropolis Monte Carlo simulations to delineate the role of these factors on entropic surface segregation in binary mixture of polymers and colloids. At low number densities, excluded volume effects are negligible and we do not observe any entropic driven surface segregation. Therefore our system of interest is binary polymer-colloids mixtures at moderate to high number densities where excluded volume effects are predominant. Our results indicate that for flexible polymer chains, the surface is always enriched with colloids compared to polymers and this effect is enhanced for longer polymer chains. The configurational entropy of the flexible polymers is significantly reduced near the surface and therefore they prefer to stay in bulk (away from the surface). However this behavior can be completely reversed by introducing large degree of confinement and making the chains relatively rigid (less flexible). Our results show that polymer segregation of long stiff chains in slit pore geometry is driven by nematization near the surface while looping of polymers is observed under large degree of confinement. We observe that for longer polymer chains with an equilibrium bond angle (q = p), both confinement and chain stiffness enhance the surface segregation of polymers relative to colloids.However, the segregation behavior within a confinement is dependent on the polymer chain length. The surface segregation of polymers is dramatically decreased for chains with equilibrium bond angles; pi/3, pi/2, 2pi/3 and 5pi/6 independent of chain length and flexibility due to excluded volume effects as this leads to inefficient packing near the surface.

585715 Simulation and Optimization of Epoxypropane Preparation Process Via Hppo Based on Aspen Plus

A continuous HPPO lab-scale pilot plant was designed for the production of propene oxide via HPPO (Hydrogen Peroxide Propene Oxide) process. It is necessary to optimize the technical parameters for improving epoxypropane yield. In this paper, Aspen Plus software was introduced to simulate the main units of Hydrogen Peroxide Propene Oxide, such as Epoxidation Reaction unit, Propylene Circulation unit, Epoxypropane Purification unit and Propylene glycol Purification unit.

In the reaction, propylene is epoxidized with H_2O_2 over a TS-1 catalyst using methanol/water as a solvent. The reaction occurs at a temperature and pressure of 52°C and 3 MPa, respectively. HAZOP and P&ID were conducted for the reactor and distillation column.

Aspen Sensitivity was employed to proceed the sensitivity analysis of dividing wall column (DWC), double effect distillation column, extract distillation column and azeotropic distillation column. The optimized key technical parameters were obtained by Aspen simulation.

The main objective of the simulation is to complete the design and optimization of a plant that produces 337,000 metric tonnes per year (8000hr) of propylene oxide (PO) and 21,000 metric tonnes per year (8000hr) of propylene glycol (PG).

Through simulation, the mass fraction of propylene oxide in the product is greater than 99.3%, and the mass fraction of by-product propylene glycol is greater than 99.99%. The model can be considered of high utility in designing and optimizing HPPO process, to achieve high reactants conversion and propene oxide yields.

586848 Volume Change of Vaporization from the SRK Equation of State Using a Similarity Variable

Equations of state such as the Soave-Redlich-Kwong (SRK) equation are used to predict thermodynamic properties of chemical substances. Most commonly, PVT properties are directly calculated, but equilibrium properties such as volume change of vaporization may be calculated using more complex algorithms including fugacity calculations. In this project, direct calculation methods for estimating volume change of vaporization consistent with the SRK equation were studied. Reduced volume change of vaporization, ΔV_R, was expressed as a power series of natural adjusted temperature η, a similarity variable combining reduced temperature and acentric factor. Starting from a previously determined reduced density power series, twenty series coefficients were derived. Series results were compared to exact SRK data for volume change of vaporization, and the relative deviation was less than 0.22 percent for natural adjusted temperatures between 0.0516 and the critical point. Work continues to find additional series coefficients and increase the range of natural adjusted temperatures while keeping the relative deviation within 0.5 percent, and to combine the reduced volume change of vaporization series with the Clapeyron equation to calculate heat of vaporization.

584833 Evaluation of Solvent Recovery Options for Economic Feasibility through a Superstructure-Based Framework

As the chemical market continues to expand, the concerns with regards to environmental sustainability have increased due to excessive disposal of chemical solvents. To date, there is no proper mitigation plan to handle such volume of chemical waste. Solvents are used in many industries for purposes such as reaction medium and as extraction, cleaning, and purification agents. These solvents can account for up to 90% of the process by mass and are often discarded after a single use. Incineration, the most widely used process for solvent disposal, is not a "green" method due to the potential detrimental effects to the environment. A systematic framework for solvent recovery has been developed which considers multiple alternative recovery and purification technologies to recover solvents simultaneously through a superstructure-based approach. These technologies are represented as mathematical models that are composed of material and energy balances, utility requirements, equipment design, and costs to assist in determining the most economical method for solvent recovery of any chemical waste stream. Two case studies involving simple binary solvent system and a more complex multi-component system were analyzed and formulated as mixed-integer non-linear programming problems. Selecting solvent recovery options required considerably lower cost in comparison to typical incineration. The solutions obtained from the case studies presented in this work demonstrated the capability of the solvent recovery framework to obtain environmentally friendly and economically viable solvent recovery pathways as an alternative to incineration.

585902 Binding of Lignin Dimers on MWW-2D Zeolite: Solvothermal Effect Study

Lignin as the potential renewable source of biofuels, chemicals, and other value-added products has gained significant attention. However, due to its complex structural network, developing efficient depolymerization techniques are challenging. Lignin depolymerization is governed by reaction conditions (pressure, temperature, etc.), catalyst used, and the presence of solvent. In this work, the binding modes of different ß-O-4 linkages have been analyzed using periodic density functional theory (DFT) with the optB88-vdW functional. The effect of how changing the ratio of H to OH on the catalytic surface, different temperatures, and different solvents affects the binding energy has been analyzed. Overall, model MWW framework zeolites are characterized by *ab-initio* first principle simulations, with the goal of gaining molecular level insights into the different binding modes of ß-O-4 linkages on the catalytic surface.

585378 Analysis and Simulation of FCC Unit Using Aspen HYSYS

In the preliminary step of petroleum refinery, crude oil is processed and fractionated into lighter hydrocarbons which have lower ignition point i.e., petroleum naphtha, gasoline or diesel. Inherently, product stream of fractional distillation column other than the lighter hydrocarbons also contain heavier hydrocarbons which is often referred to as residue (gasoil). Due to the higher ignition point, it is deemed to be less economically valuable product. To serve this purpose, Fluid catalytic cracking (FCC) is employed to add more value to the heavier products. FCC is one of the most important processes in the petroleum refining industry for the conversion of heavy gasoil that provides about 35% to 45% of the blending stocks in refinery of gasoline. Various alkenes like ethylene and propylene are also produced by FCC for polymer applications. This work describes the model of FCC in ASPEN HYSYS that can simulate the behavior of FCC close to real work conditions. The FCC models mainly consists of riser, stripper, reactor and regenerator. In this work, a basic refinery process is designed and the vacuum gas oil from the vacuum distillation column is used as feed in the FCC unit. A sensitivity analysis is carried on parameters of interest i.e., *pressure difference between regenerator and reactor, catalyst heat capacity, feed temperature*, to determine the correlation that affect the FCC performance *(mole-fraction of Butane is used to measure the performance)*. The typical data in refinery is adopted, in which few changes were made in order to achieve proper simulation. The results of simulation predict the parameters which are more sensitive towards the performance of FCC, for example, performance of FCC is more sensitive to change in heat capacity of catalyst as compared to feed temperature. Hence, with this model, set of parameters are determined which have pre-dominant effect to FCC's performance.

587519 Uncertainty Analysis with Geothermal Techno-Economic Tool Geophires v2.0: Model Development, Validation and Case Studies

GEOPHIRES v2.0 is an open-source techno-economic simulator used for modeling geothermal reservoirs. One of the challenges in geothermal reservoir modeling is handling uncertain parameters that are difficult to measure or characterize. Over the course of this project, three modules have been developed for GEOPHIRES v2.0 to address uncertainty in geothermal parameters: multiparameter tornado diagram analysis, single parameter sensitivity analysis and Monte Carlo analysis. These modules provide the user with tools to assess the impact of parameter uncertainty on the technical and economic performance of a geothermal system. High performance computing capabilities have also been implemented to effectively perform the large number of GEOPHIRES v2.0 simulations required for Monte Carlo analysis. A geothermal plant in Paris was used as a case study, first to validate the GEOPHIRES v2.0 tool by comparing the output with the results from the System Advisor Model (SAM) and actual plant performance, and secondly to test the three uncertainty analysis modules. The uncertainty analysis indicated that drilling costs and geothermal gradient have a major impact on overall plant levelized cost of electricity (LCOE) or heat (LCOH), while reservoir rock properties only have a minor impact.

587812 Analysis of Model Application Using CFD in Transition Mixing

Mixing is a common practice used in a vast majority of industrial processing and manufacturing applications. Mixing is typically studied in either the laminar or turbulent flow regimes because the equations governing these regimes are well understood. Due to the complexity of transition mixing, minimal modeling is done in this regime. Certain industries naturally find themselves operating in and around the transition regime, but are unable to fully understand the process due to the inability of accurate modeling. The mixing of many personal care products fall into this category because they typically deal with complex non-Newtonian fluids. Computational Fluid Dynamics (CFD) is a tool that enables the modeling of various physical systems to analyze velocity profiles, streamlines, and numerous other fluid and flow parameters. Transient simulations of the mixing of Carbopol (a high molecular weight, crosslinked polyacrylic acid polymer) were run using CFD in order to compare the blend time results with real data. Analysis was done by looking at concentration profiles and volume fractions at various locations in the tank throughout the mixing process. The viscous models compared were the laminar and realizable k-epsilon. The mixture model was used to compare the two different phases identified by different dye color. The Carbopol viscosity was modeled using a non-Newtonian power law. The two different viscous models yield different blending results, neither of which match up with the real data, indicating a discrepancy between the models and mixing within the transition flow regime.

585716 Mathematical Modeling of Nephrin Loss in Diabetic Kidney Disease

Diabetic kidney disease is the leading cause of kidney failure. The job of the kidneys is to filter the blood. However, when the kidneys are damaged, the filtration system is less efficient. In this project we focused on podocyte cells, which compose the outer filtration barrier. In between podocyte cells, a protein called nephrin connects the podocytes like Velcro. The blood is filtered through this area to excrete waste and excess water. Diabetes can cause the nephrin to lose its Velcro abilities causing it to disconnect from the podocyte cells and be excreted in the urine. The loss of nephrin causes the mountain and valley landscape of the podocytes and the gaps between them to become distorted. Diabetic complications can cause the nephrin to lose the Velcro ability and be excreted. We will present an empirical model for predicting category of proteinuria (protein in the urine) based on loss on nephrin expression using human biopsy data for Type 1 and Type 2 Diabetes patients. Type 1 and Type 2 have different blood sugar landscapes, this potentially causes differences in nephrin loss. We will also present images from a graphical user interface that we developed, with graphs presenting nephrin expression in Type 1 and Type 2 Diabetes. The model relates the percentage of nephrin lost to the mRNA expression of nephrin and to the urinary protein concentration. This noninvasive modeling research provides quantitative guidance for the threshold where nephrin loss causes long-term damage to the kidney clinically detectable by the presence of protein in the urine.

585702 Mathematical Modeling of Mesangial Cells in Diabetic Kidney Disease

Diabetes is the leading cause of kidney disease. About one out of four adults with diabetes has kidney disease. The kidney is supposed to filter wastes and extra water out of the blood stream to produce urine. When the kidney is damaged, it cannot filter blood like it should, which leads to waste build up and excess water in the body and eventually to death. One of the key cell types damaged in the kidney is mesangial cells. The mesangial cells and the surrounding extracellular matrix hold together the bundles of capillaries where blood is filtered in the kidneys. Experiments have shown multiple chemical pathways that change mesangial cells in diabetic kidney disease (DKD). It is challenging to describe how these chemical pathways interact with each other in mesangial cells over time. I have used mathematical modeling of differential equations to represent the chemical pathways in mesangial cells that are known to be affected by DKD. I have combined the separate pathways into a systems biology model that can consider the interactions between pathways and between multiple mesangial cells in the kidney. I will present model results for conditions that lead to expansion of the mesangial tissue (cells and extracellular proteins) during diabetic hyperglycemia.

584823 Quantifying United Nations Sustainable Development Goals Using Greenscope Indicators

The 2015 United Nations (UN) Sustainable Development Goals were created with ambition to construct a better world by the year 2030. These goals suggest how we should live, and how companies should conduct their businesses. However, they are quite broad with limited ways of quantification. This work proposes a framework that could be used to quantify such goals so that systems' evolutions can be evaluated at different times in their operating cycles. Using the chemical process sustainability assessment tool GREENSCOPE, sets of indicators that represent specific UN Sustainable Development Goals are identified and used to quantify a given target within a goal. The identified indicators may be grouped in categories: resource efficiency, recycling, waste management, and resilience. After choosing a set of GREENSCOPE indicators, a case study of an Acetic Acid manufacturing simulation in CHEMCAD is analyzed. Sections of the manufacturing process are studied and discussed to determine their performance in terms of resilience and sustainability.

585711 Computer Modeling of Aerosol Diffusion through Lung Mucosa

Diseases of the lung are some of the most common and deadly in the world. Existing treatments, if any exist, tend to be extremely invasive and time-consuming. Further, due to the poor bioavailability of drugs traditionally administered orally or through injection, these treatments are not very effective. Technology is emerging that allows an aerosolized drug dosage to be delivered directly to the diseased area; however, the mucus layer separating the airways from the tissue remains a barrier to this method.

The mucociliary clearance mechanism in the lungs clears foreign particles (including cigarette ash, dust, and bacteria) and prevents infection. A layer of mucus on the surface of the inner lung is constantly pushed upward towards the throat by a bed of cilia, and most particles that impact on the mucus are cleared from the lungs quickly and without incident. Although mucociliary clearance aims to prevent a disease from being contracted, it remains a barrier to drug delivery, as drugs must cross the same thick mucus layer that is being swept away from the tissue.

We used COMSOL Multiphysics software to develop a model of the mucosal interface between the airways and the lung tissues. Data from the literature is used to determine details like dimensions, velocity, and viscosity. The mucus layer moves upwards towards the throat in laminar flow, imitating the mucociliary effect, and the underlying periciliary layer has no net movement due to the regular beating of the cilia that move the mucus. The "drug" enters from the airway side and moves through the fluid by convection. Given these inputs, the model outputs the concentration distribution of the administered drug molecule or particle subject to diffusion through the moving mucosa. We aim to maximize the amount of drug that reaches the target lung tissue. The model is customizable, easily modified to simulate other drugs or any other particle so long as some properties are known. Variables have been specifically parametrized to find more complex relationships, like effective diffusivity, from an input of more readily available information, like particle radius. Lung conditions can also be altered to meet the needs of the application (for example, the mucus layer is much thinner in the alveolar region, and the mucus of cystic fibrosis patients is much denser than average). Thus, the model can provide insight into the efficacy of new lung treatments and aid in the design of drug transport properties.

584435 Applying Computer Science to the Development of Chemical Reactions: Applications in Heck-Type Cross-Coupling Reactions

Organic chemistry and computer science have historically been immiscible. My research is focused on harnessing the power of computer science to streamline the development of chemical reactions, in particular using high throughput experimentation (HTE) to study Heck-type coupling reactions. These reactions require a metal center supported by a ligand, and reactivity and selectivity are dictated by the multifaceted interactions of the pairing. These reactions often produce mixtures of nearly identical products, commonly including enantiomers whose separation and classification prove uniquely challenging. Through the design of a novel program and user interface, we are able to parse large data sets of yield and selectivity - particularly those generated by HTE - and present the information in a meaningful way. Graphical representations, partially analyzed with MATLAB and Python and coupled with intuition and prior experience, aid chemists in rapid interpretation of these data sets. Challenges emerged in the design of the program: representations need to be generated quickly, be meaningful at a glance, and be created by users unaccustomed to data science. These considerations are being addressed in the design process.

588014 Developing Matlab Code for Analysis of Heterogeneous Engineered Cardiac Tissue during Contraction

Title: Developing MATLAB Code for Analysis of Heterogeneous Engineered Cardiac Tissue during Contraction

Primary Author (and presenter): Bush, Michaela, Z

Additional Authors: Ellis, Morgan. Finklea, Ferdous. Lipke, Elizabeth

Department: Department of Chemical Engineering

College/School: Auburn University

Description:

Engineered cardiac tissue (ECT) is needed for treating various cardiovascular diseases, studying development and disease mechanisms, and for high-throughput drug screening, so we are producing ECT directly from human induced pluripotent stem cells. The ECT must have the appropriate phenotype, maturation, and functionality for downstream applications. One way we assess these properties of ECT is by analyzing the speed and duration of calcium propagation, necessary for cardiac function. Calcium propagation is measured using optical mapping; we record high-speed videos of ECT treated with a calcium-sensitive fluorescent dye that brightens in the presence of calcium, creating flashes that follow calcium propagation. The purpose of this project is to create a custom MATLAB® program that can analyze these videos and extract the calcium transient duration (CTD), conduction velocity, and the heterogeneity index of the tissue. Due to the heterogeneous nature of 3D-ECT, not all pixels contain useable data. New features have been incorporated to remove unusable pixels based on their lower fluorescence or range. The program can filter data temporally to smooth the propagation waves and normalize it to remove variations in base fluorescence. Pixels that lack consistent propagation are then removed based on their dissimilarity to pixels with consistent propagation. From the remaining usable pixels, the number of calcium propagation events and the time these events occur is quantified and visualized using isochrone maps. The conduction velocity is calculated by choosing several pairs of points and calculating the time calcium travels between the points divided by the distance between the points. A new feature incorporating the MATLAB® gradient function allows direction of the velocity for groupings of pixels to be plotted to aid in visualization of the wave propagation and the optimal pairs that best represent the overall conduction velocity. A method of calculating the CTD, which measures the time between depolarization and repolarization, at any repolarization percentage was implemented using the luminosity at the peak and trough of each wave. The heterogeneity index measures the variance of the CTD which is relevant as cells in 3D-ECT are not distributed evenly. Using this data, we can determine the functionality of ECT and the quality of our current methods to create ECT.

585251 Application of Machine Learning Methods to Improve Leukemia Diagnostics

In 2019, there will be an estimated 1,762,450 new cancer cases diagnosed and 606,880 cancer deaths in the United States [1] . With both numbers increasing every year it is very important to be able to diagnose each and every patient in the most cost efficient and timely manner. Every patient that is suspicious of having any type of cancer will undergo multiple tests to find a correct diagnosis. It has become extremely important to assess the signs of cancer in patients with the most accurate test possible for efficient diagnosis. Biomarkers, external symptoms, and other important indicators allow medical professionals to evaluate the cancer status upon examination. One of the most prominent markers that appear within the blast cells of patients with Acute Myeloid Leukemia (AML) is myeloperoxidase (MPO) [2] . This project has focused on the use of machine learning methods to evaluate the chances of a patient having AML or Acute Lymphoblastic Leukemia (ALL) based on the clinical variables obtained via morphological, and cytochemistry tests. The two machine learning methods used in this study were decision trees and neural networks. These methods require a data set to train the neural network or decision tree and then a testing data set to evaluate how accurately these machine learning methods identify each patient as having ALL or AML. The immunophenotyping (IPT) results serve as the confirmatory test for checking the solution accuracy. These machine learning approaches can possibly help to determine which symptoms/markers are most important to search for when diagnosing a patient with either AML or ALL. A marker is deemed to be significant when it is used as a node within the tree and as for the neural network there are weights assigned to each possible marker with the more important markers resulting in a higher weight. It will also help to inform doctors which tests will be most efficient in determining a diagnosis for patients. It was found that the decision tree approach was able to diagnose patients with AML 97.14% of the time when considering myeloperoxidase (MPO). The neural network that was created using the same dataset was able to classify patients correctly 100% of the time while assigning the highest weight to the MPO marker. To check for the significance of the MPO marker both the decision tree and neural network were recreated neglecting the MPO marker. Both machine learning methods had a drop in accuracy when classifying a patient with AML with the accuracy of decision trees falling to 91.43% and the neural networks decreasing to 90.91%.

585904 Electronic Properties of Dicationic Ionic Liquids for Corrosion Inhibition

Ionic liquids have been found to impart corrosion inhibition performance in industrial applications. In order to assess the corrosion potential of several imidazolium-based dicationic ionic liquids, quantum mechanical calculations were performed to predict their electronic properties. The structure of these dicationic ionic liquids was R_1-Im-R_2-Im-R_3 where R_1, R_2, R_3 are alkyl chains and Im are imidazolium rings. Three alkyl chain lengths (butyl, hexyl, and octyl) were studied for R_1 and R_3, whereas ethyl, butyl, and hexyl groups were considered for the linker domain R_2 for a total of 15 cations. These cations were paired with two chloride Cl^-, two tetrafluoroborate $[BF_4]^-$ and an equimolar mixture of Cl^--$[BF_4]^-$ resulting into a total of 51 cation-anion pairs. Population analysis was carried to obtain electrostatic potential maps and HOMO-LUMO energy gaps. The HOMO-LUMO gap and optimized structures were analyzed to discern patterns in their gas phase behavior based on length of alkyl chains and the configuration of anions. The HOMO-LUMO energy gap was found to be the most dependent on the anion, with pure Cl^- exhibiting the lowest gap, while the pure $[BF_4]^-$ showing the highest when comparing structures with equivalent cations. The most important factor found to affect the structure of the studied ionic liquids was the length of the alkyl chain linking the two imidazolium moieties. From the structural respective, Cl^- location in the optimized geometry was decided by the acidic hydrogen atoms on the imidazolium rings due to its tendency to hydrogen bond. $[BF_4]^-$ also participated in the hydrogen bonding, but increased area of negative charge in the anion allowed it to bridge the imidazolium rings more effectively. A novel behavior was found when the $[BF_4]^-$ anion was present along with the butyl group as the linker in which the cation curled up on itself, creating a large region of positive charge within the center of the molecule.

585999 Analytical Prediction of Non-Newtonian Concentration Profiles Using a Correction Factor

Mathematical models are a vital part of chemical engineering. Modelling systems allows for predictions to be made in specific cases without a dependence on empirical data alone. In fluid mechanics, models of systems containing Newtonian fluids can be easily obtained by established methods. However, Non-Newtonian fluids are much more difficult to model. The introduction of a correction factor could allow for Newtonian models to be used for Non-Newtonian fluids. Due to the nature of dimensionless or reduced states, the non-dimensionalized models for two different fluids under the same conditions are equivalent. By non-dimensionalizing a Newtonian model and then applying a correction factor, the characteristics of the model could be adjusted to match the hypothetical non-Newtonian case. With this method, the documented solution for a system containing a Newtonian fluid could be used to generate a model of the same system with a non-Newtonian fluid. The use of a correction factor to obtain a model of a non-Newtonian system significantly reduces the amount of time and computational power that would typically be required for such a task. This contribution proposes a method to obtain velocity and concentration profiles of a system containing a non-Newtonian fluid through the use of a correction factor and the analytical solution to the same system with a Newtonian fluid. The proposed analytical solution will be compared to the numerical solution to prove the accuracy of this method.

587813 Nonlinear Dynamic System Identification of Monoclonal Antibody Production

The demand for monoclonal antibodies (mAbs) has increased due to their high affinity and specificity to thousands of biological targets. Immunotherapy using mAbs are widespread in the treatment of various diseases including cancer and autoimmune disorders. The entry of biosimilars (variations of previously approved mAbs with similar efficacy and safety to its predecessor) into the market creates a need for production process intensification, and computational models are valuable tools which can be used for optimizations purposes. In this work, we identify data driven models that simulate the system dynamics of mAb (IgG4 cB72.3) generation in GS-NS0 cells in a semi-batch operation. Hammerstein Wiener and Nonlinear AutoRegressive models with eXogeneous (NARX), a subclass of Recurrent Neural Networks, are fitted to simulated data and compared for performance. The trained data driven models demonstrate good accuracy and reduce computational cost in comparison to the rigorous dynamic model. This potentially makes the data-driven approach a versatile tool for computationally demanding optimization studies. Results favor the Hammerstein Wiener model for multiple simulation datasets, and the NARX model for a single simulation dataset. Further exploration is under progress to determine the best surrogate model, which can be used for process optimization.

*The first and second authors contributed equally to this work.

585208 Mathematical Modeling of the Influence of Toxin Exposure on Rheumatoid Arthritis

Rheumatoid arthritis is a debilitating autoimmune disease that attacks the joints in the body. These attacks cause chronic inflammation that degrades the surrounding bone overtime. This disease effects around one percent of the population globally, and women are three times more likely to be diagnosed with this disease than men.

Rheumatoid arthritis research is broad with many scientific studies underway to identify causes, cures, and preventions. Previous studies have determined the mechanism of the disease once it becomes established in the body, but only limited published results focus on identifying triggers and mapping the physiological responses at the onset of arthritis. It is understood that environmental factors account for seventy percent of the disease generation, while genetics accounts for the other thirty percent. Recent studies have shown that the immune responses to environmental factors that lead to arthritis may begin in the gut, gums, or lungs years before symptoms appear. However, the links between the timing and extent of various environmental exposures on the onset and progression of arthritis are not clearly understood. Mathematical models can provide insights into explaining the complex linkages between molecular level actions of chemicals and their effects on the cellular, tissue, and whole body levels of physiology in a disease.

We are building a model for the chemical and biological processes that are theorized to be involved in the onset of arthritis. To model how environmental toxins, such as herbicides and bacteria, stimulate chronic inflammation, we have combined two previously published models into a more complex and informative model. This model tracks how the concentration of tumor necrosis factor (TNF), a chemical that triggers inflammation, changes in the joints, the bloodstream, and the surrounding tissue. The model predicts TNF regulation by the inflammatory stress on the immune system. We will explore possible explanations for how specific toxins can alter TNF regulation to lead to arthritis. This work will impact society by improving fundamental understanding of the effects of toxins on arthritic autoimmune inflammation, laying the scientific foundation for advances in disease treatment and prevention.

585366 The Interaction of Azole Bio-Isosteres with Amino Acid Side Chains for Drug Design

There is significant interest in the design of new drugs with novel binding sites for interactions with specific peptides in proteins. The hydrogen bond is a key noncovalent interaction governing what happens between a drug molecule and its local environment. For drug molecules, this local environment can be a biological target, a biological off-target, aqueous solution, a lipid membrane, or even a crystalline solid. Consequently, hydrogen bonding impacts a wide range of molecular properties critical to drug design including potency, selectivity, membrane permeability, and solubility. This project is investigating the energetics of hydrogen bonding at the fundamental level to gain new insights into the basicity of a molecular site and its role in controlling hydrogen bond interactions as well as the possibility of proton transfer from one site to another. We are using the amide group as a model for the protein backbone as it can work both as a hydrogen donor and acceptor. At this point, there is no single model that explains the large range of binding energies exhibited by such hydrogen bonds. We are modeling the interactions of the basic amino acids, lysine, arginine and histidine with the non-classical amide bio-isosteres 3,5 dimethyl-1,2,4-triazole and 2,5-dimethyl-2H-tetrazole using computational electronic structure methods at the ab initio correlated molecular orbital theory and density functional theory levels. The structures are initially optimized using density functional theory to probe different conformations and binding sites. This information is used as input into more computationally demanding correlated molecular orbital theory calculations at the G3(MP2) level to predict the energetic properties of these azoles and their hydrogen bonding interactions with different amino acid side chains. Besides the gas phase calculations, the basicities and hydrogen bond energies in different solutions are predicted using self-consistent reaction field approaches to provide better insights into what is happening in a biochemical environment.

585265 Carbon Monoxide in Electronic Cigarette Effluent

The electronic cigarette market has an estimated market value of over $10 billion dollars (2018). While some e-cigarette users are former smokers trying to quit, many youth report e-cigarettes being their only source of nicotine. E-cigarettes are still relatively new devices and their design continues to evolve rapidly. There are many different types of electronic cigarettes on the market, ranging from miniature disposable electronic cigarettes to Advanced Personal Vaporizers (APVs). Some modern e-cigarettes offer customizable temperature/power levels as well. Research on the health effects of these devices is still emerging. However, e-cigarette effluent has been shown to contain formaldehyde, other aldehydes and ketones, other carbonyl compounds, reactive oxygen species and chloropropanols, volatile PAHs and aluminum.

It has now been shown, using diode laser absorption spectroscopy, that e-cigarette effluent also contains carbon monoxide (CO). The concentration of CO produced varies as a function of the power of the e-cigarette and the flavor of the e-liquid heated. The CO concentrations ranged from 0 ppm to more than 180 ppm at a high power. CO is a well-known toxicant which when inhaled at high concentration over long periods can lead to death. Short duration exposures are not as well characterized but have shown deleterious health effects.

587178 Increasing Chemical Engineering K-14 Involvement By Tackling Water Conservation Using Human-Centered Design

AIChE 2019 Chemical Engineering Division

Increasing Chemical Engineering K-14 Involvement by Tackling Water Conservation using Human-Centered Design

Anays Hernandez[1] and Sindia Rivera-Jiménez, Ph.D. [2]

[1]*Department of Chemical Engineering, University of Florida and* [2]*Institute for Excellence in Engineering Education, University of Florida*

ABSTRACT

How much water does the average person use at home per day? The U.S. Geological Survey reports that old showers may use up to 5 gallons of water per minute and water-saving shower heads produce about 2 gallons per minute. The agency recommends that taking a *shorter* shower using a low-flow showerhead saves lots of water. Even if an individual invests in a low-flow showerhead, they still need to figure for how long they are showering. A water consumption sensor that relatively anyone can install in their shower head can help to solve this problem. In addition, global market research reported that the environmental sensor market was valued at USD 1.22 billion in 2017 and is expected to reach USD 2.19 billion by 2023.

The use of sensors in the K-14 classroom has shown to be of significant educational value because they can help address societies problems while integrating many disciplines such as chemistry, physics, mathematics, computer science, engineering, civics, among others. Exposure to the use of sensors to solve real problems can encourage students to follow career paths in Science Technology, Engineering, and Mathematics (STEM), particularly for underrepresented communities. The primary objective of this pilot project is to create a water consumption sensor that is easy to use, and cost-effective for the everyday individual. The integrated sensor system uses simple electrical equipment including photo-resistors, light sources, waterproof cabling, Legos, 3D printed parts, and Arduino. Ideally, this work will become an instructional module that can be used to increase the interest of K-14 students into the chemical engineering field by tackling environmental problems using a human-centered design approach where the needs and wants of the user are paramount for the designer.

585395 Understanding the Relationship between Engineering Identity and Academic Motivation

The primary goal in this research is to explore the connection between motivation and identity within undergraduate engineering students. Uncovering the relationship between these two constructs has the potential to inform techniques that encourage academic motivation. In turn, this could encourage a greater correlation between identity and engineering identity. Previous work shows that having a greater intersection between one's career identity and other identities (i.e., social, demographic, etc.) increases motivation. We hypothesize that this could translate into an academic setting, in which academic motivation is more prevalent among those who have a secure sense of engineering identity. In the end, we believe this will correlate to overall higher success.

The method used in this study involves analyzing an individual's previous responses to failure as a measure of a student's motivational behavior and comparing this measure to the strength of their engineering identity. The sample population consists of upperclassmen aeronautic and astronautic engineering students. Engineering identity is measured via an instrument created by Dr. Allison Godwin of Purdue University, West Lafayette. The instrument centers on uncovering how strongly students feel they identify as engineers, within categories of recognition, interest, and competence. This study also uses the critical incident technique. Students are asked to reflect upon a time where they successfully and unsuccessfully recovered from an academic or professional failure (the critical incident). Failure is a sensitive subject that provokes strong reactions within people, some greater than others. Thus, in analyzing these reactions, we can gain insight into different patterns of coping mechanisms and determine what patterns hinder or help motivational growth. We are interested specifically in the disciplinary distinction between astronautic and aeronautic engineering students. The subject population in this study is drawn from an upper-level systems dynamics course required for both majors. For students on the astronautic track, it serves as an integral part of their curriculum, while students on the aeronautic track take the course for a greater perspective. In analyzing the difference between majors within the same course, we demonstrate how the intersection between course content and major identity play in subsequent motivation to succeed in the class.

This study aims to give educators greater insight into student behavior so they can adjust their teaching techniques to optimize student motivation or encourage greater identity correlation. In future work, we will continue to work through what exactly defines identity, manipulating certain variables to greater understand relationships.

587962 Lower Extremity Flailing Motions in Response to a Hypothetical Rear-End Vehicular Impact

Occasionally a victim of a rear-end car crash will experience a lower extremity injury (for example, a patella fracture). One mechanism that has been suggested for such injuries is the result of "flailing" motion (involuntary motions) of the lower extremities that may occur during the impulse phase of the rear end collision. To date, there has been no peer-reviewed literature that indicates that such flailing motions occur that can be causally related to any lower extremity injuries. The purpose of this study is to explore various combinations of driver body geometry (particularly lower extremity dimensions and angles), vehicular interior structure locations (particularly underneath the dash area, steering wheel, and other interior surfaces directly ahead of the driver, and seat positioning. Measurements will be made under two conditions: 1) normal driving position; and 2) after subjects rotate their legs upwards as a simulation of "flailing". First contact point and (if it occurs) 2nd contact points will be documented. Note that a rear end collision is not actually occurring, i.e., only a single stationary unpowered vehicle is employed.

585677 Seek and You Shall Find: Where Do Faculty Obtain and Engage with Active Learning Resources

Different types of active learning are commonly used within classrooms throughout primary and sometimes secondary schools, but upon reaching college classrooms active learning is still gaining traction. The use of active learning techniques can provide an energetic learning environment and has been shown to have a positive effect on student learning. Kinesthetic learning is a branch of active learning that actively and physically engages students with the material being taught through physical movement and/or exercises. Kinesthetic learning specifically moves students around the classroom instead of the traditional sitting for a 50 to 80 minute lecture. Some examples would be rotating stations around the classroom, acting out a process, or students drawing out a process or problem on the board.

Kinesthetic and other active learning techniques face many barriers in being adopted in classrooms within the collegiate community. Faculty members can be resistant to changes to the traditional lecture style of teaching, which is possibly connected to a lack of former pedagogical training and awareness of the resources available to support active learning strategies in the classroom. There is also a difficulty in knowing where and how to effectively share materials with other educators when creating active-learning materials. This leads us to the research question: how do faculty members find resources to develop teaching methods for their classrooms?

In terms of understanding how to integrate kinesthetic and active learning teaching techniques into college classrooms, information on how faculty members obtain pedagogical skills and methods should be considered. The study conducted seeks to identify where faculty members within engineering disciplines look for resources regarding teaching methods. With this information we can be determine what the best ways are to distribute active learning resources engineering faculty. Data collection through survey and interviews looked at the resources for active learning and how faculty are currently using or are willing to use them. The results lead to information on how to approach developing and disseminating resources to help and improve classroom engagement.

587501 Investigation of Particle Noise Produced By Tip Sonication

The tip sonication of nanoparticle solutions can create particulates in test tubes resulting in significant data misinterpretation and faulty results for researchers analyzing nanoparticles. The process of tip sonication is implemented by many researchers who work with nanoparticles to disperse particles. This method is fundamentally applied when nanoparticles are synthesized by nanoemulsion techniques. Low retention, plastic microcentrifuge tubes are typically implemented to store particles, however, previous research has discerned contamination following tip sonication in these tubes. The contamination generated by particulate noise from the tube potentially can result in major inaccuracies. We seek to optimize the process of sonication and reduce data misinterpretation through determining conditions which minimize the contamination particulate formation. In these experiments, ultrapure water is tip sonicated using a Qsonica sonicator and the resulting concentration of contamination particles is measured by nanoparticle tracking analysis. The conditions we have tested include brand of tube used, amplitude and process time. These all impact the concentration of particulate noise produced. Furthermore, several cleaning procedures on the tubes were tested to determine if debris in the tube was the source of the contamination. In addition, we have conducted experimentation to confirm that the particulates we are examining are not bubbles by measuring each sample twice following varying time gaps of ten, twenty and thirty minutes.

585725 3D Printed Kidney Filters for Hands-on Learning

In this project, I designed an anatomically correct 3D printed model of a part of the kidney and zoomed in at a more detailed level for a 3D printed component that mimics the filtration of blood by the kidney. I describe the motivations, the models, and their use in hands-on demonstrations for learning science and engineering concepts.

The glomerulus, the functional unit of filtration in the kidney, removes waste from the blood while retaining important compounds needed for a healthy body. Glomeruli are a bundle of capillaries surrounded by a capsule and lined with specialized filtration cells. Through physical openings and chemical diffusion, the blood traveling through the capillary is filtered of its waste, which is then excreted as urine. Because the glomerulus is so important in blood filtration, it is a major component in kidney health and important for research studies related to filtration in the body. The glomeruli are found in the thousands in the human kidney but are invisible to the naked eye. With computer aided microscopy, images have been developed and circulated through current literature. However, this data has not yet been used with 3D modeling software to represent the anatomically correct microscopic glomerulus. I used SolidWorks was to model the glomerulus at different spatial scales for demonstration tools. The 3D model of a glomerulus capsule and capillary matrix follows current literature describing the size and volume of an average human glomerulus, which is approximately 200 micrometers in diameter, and considers both artistic renderings and microscopy images currently available. Standard 3D printers and splicing software were used to print each physical model. In this poster, I will justify the development of the 3D glomerulus model highlighting its purpose and use in demonstrations to increase scientific literacy in a variety of venues. I will show the accuracy of my model with respect to current findings on glomeruli size and features. I will also show how a simplified kidney filter was designed and 3D printed for engineering education workshops. Using descriptions and images of glomerular pores, I devised a simple cylindrical filter that when used in tandem with beads representing the various molecules and cells in the blood provides a simple hands-on demonstration of the process of glomerular filtration and how it changes between healthy and diseased conditions.

588158 Effect of Gamma Irradiation on Antioxidant Activities of Molle (*Schinus molle*) and Guaviduca (*Piper carpunya*) Essential Oils.

Molle (*Schinus molle*) and *guaviduca* (*Piper carpunya*) are Ecuadorian endemic species traditionally used as medicinal plants since they have several biologically active components and they are even attributed culinary properties. In this research, the effect of a treatment with gamma radiation on the antioxidant capacity of *molle* and *guaviduca* essential oils was evaluated. The essential oils were extracted from dry leaves through hydro-distillation. The irradiation was carried out with a Cobalt-60 source in the raw materials or in the essential oils. The evaluated doses were 10, 15 and 20 kGy. The antioxidant capacity of the oils was determined with the ABTS radical cation and with the DPPH radical methods. In the *Schinus molle* essential oil irradiated with a dose of 20 kGy before the extraction, the antioxidant capacity increased from 7,71% to 11,07%, while in the *Piper carpunya* essential oil, the antioxidant activity decreased in all cases due to the irradiation. These results suggest that the ionizing radiations are able to modify the composition of essential oils and modify their properties.

585737 Why Do I Have to KNOW This? Engineering in a Globalized Society

As society is becoming increasingly globalized, engineers must be able to integrate non-technical factors into their problem-solving and design. This research study investigates whether engineering students, engineering faculty, and professional engineers have similar thinking when it comes to solving problems within a global context. This is aligned with the ABET outcome that students must have "an ability to apply engineering design to produce solutions that meet specified needs with consideration of public health, safety, and welfare, as well as global, cultural, social, environmental, and economic factors,". This student outcome is referred to as a professional skill. Professional skills were developed as a response to old accreditation standards being too stringent and that they stifled innovation. For engineers, professional skills includes the implicit knowledge of how one's work impacts a globalized society. At most universities, bringing in global contexts into problem-solving is neglected until students' senior-year capstone course. This study compares three populations, undergraduate seniors, engineering faculty, and practicing engineers, understanding of the significance of engineering solutions within a global context. Comparing the results from the three populations will provide insight into whether there are discernable differences between the group's beliefs. Comparing the populations will also provide insight into whether ABET requirements are being met. The results from this study could be used to change how courses or programs are structured to help best prepare students professionally.

586003 Metrics of Student Success and Problem Difficulty Using Auto-Graded Online Homework

Online homework and interactive textbooks are becoming more prevalent in university classrooms. Research suggests that students appreciate the immediate feedback as well as being given multiple attempts to earn points for a problem. However, determining the quality of auto-graded problems is still an open question. To develop a metric, over 700 questions were evaluated from a second-semester freshman course in Material and Energy Balances. Using three years of data from over 250 students and more than 100,000 problem attempts were analyzed. Problems consist of multiple parts with each part increasing in difficulty. A variety of problem types were used in including multiple choice, numerical response, and graph reading. Metrics used to analyze the questions included number of attempts before a correct answer was given, total attempts, and success rate. Using these quantities, a point system was created to determine the quality and difficulty of problems based on student performance. Additionally, online homework questions were analyzed to correlate student success in the course with completion of the challenge activity questions. Students who completed more of the challenge activities had better success in the course as measured by final grade. Overall, web-based analytics of student success and attempts-before-correct can quantify the quality of hundreds of auto-graded problems.

586006 Analytical Solution of the Heat Transfer Problem for Power Law Non-Newtonian Fluid

Accurate predictions of velocity profiles for Non-Newtonian fluids has received considerable attention for about half a century. The specific rheology of these type fluids is well known has an impact on the concentration of species and the corresponding temperature profile in the bulk of the system when mass and/or heat transport occur. Significant research effort has been devoted to the analysis of kinematic viscosity and shear rate to characterize and model the deviation from the Newtonian postulate. However, Non-Newtonian behavior not only directly affects shear stress Newtonian viscosity but also transport of heat in its parameters, thermal diffusion and convective film coefficient.

This contribution focuses on the analytical solution of the Energy and Navier-Stokes equations for the case of a Non-Newtonian behavior affecting the main transport physicochemical properties. The power law model, Ostwald-de Waele equation, has been adopted for the solution of the momentum equation. A similar power law expression has been use in the modeling and simulation of the energy equation. The solution strategy uses a similarity approach and the principle of recursion to obtain an explicit mathematical function. Furthermore, this study concentrates on the comparison of the numerical solution and the combined similarity-recursion solution to validate the accuracy of this last method and its predictions. A few different geometries have been chosen to further validate the approach and highlight its simplicity. Concluding remarks are presented to highlight the potential applications of this mathematical tool in the design of heat exchangers.

585954 Effect of Tank Bottom Shapes on N_{js} and Power Dissipation in Stirred Vessels Under Different Baffling Configurations

Mechanically stirred mixing vessels are widely used in the chemical, pharmaceutical, biotechnology, food, and other related industries to homogenize the liquid content of the vessel and any phase dispersed in it, such as finely divided solid particles. The minimum agitation speed, N_{js}, to just suspend finely divided solid particles in the liquid and the power, P, dissipated by the impeller in these systems are two of the most critical parameters for scaling up mixing processes from laboratory scale to industrial scale.

In agitated vessels, the degree of solid suspension can be classified into three levels: on-bottom motion, complete off-bottom suspension, and uniform suspension. Enough agitation is important to completely suspend the solids off the vessel bottom. Below this needed agitation, the total surface area of the solid suspended within the liquid is not completely or efficiently utilized. Therefore, it is important to be able to determine the impeller agitation speed, denoted by N_{js}, at which the just suspended state is achieved.

Both N_{js} and P (and its non-dimensional value P_{o}) depend on several factors including type of impeller, its diameter, vessel geometry, and the dynamic conditions at which the mixing is operated (agitation speed). Although most correlations available in the literature for N_{js} and P_o were obtained at the laboratory scale using flat-bottom vessels, vessels with hemispherical or torispherical bottoms are much more commonly used in industrial applications. Until now, the effect of the shape of the vessel bottom on N_{js} and P_o has received little attention in the literature. Therefore, the main objective of this work was to experimentally determine N_{js} and the power dissipation at these measured N_{js} values for different tank bottom shapes using an axial-flow impeller (Lightnin A310).

By using a method recently developed in the mixing lab at NJIT to measure N_{js}, it was found that the hemispherical-bottomed tank dissipated about 15% more power than flat-bottomed tank when tested at the same agitation speed. Additionally, it was found that N_{js} was higher in the hemispherical-bottomed tank than in the flat-bottomed tank. Therefore, the power dissipated in hemispherical bottomed to just suspend the particles was determined to be nearly 50% higher than that of the flat-bottomed system. The results obtained here can be expected to be of significant applicability to industrial systems, such as glass-lined vessels, typically used in the pharmaceutical industry.

584042 Microbial Electrolysis Cell Bio-Degradation of Alkane Compounds

Alkanes are simple carbon chain molecules that are used in the fuel industry to be combusted to utilize their stored energy. Alkanes are pumped at gas stations across the world for consumers to use, but a majority of gas stations have some amount of alkane contamination that occurs in the surrounding soil that pollutes the environment. In this study, we test a mixed culture of bacteria in a Microbial Electrolysis Cell (MEC) to bioremediate two alkanes, octane and decane, to produce a current to ensure there is activity. The MECs were 120 mL reactor vessels ran in duplicates that contained a graphite anode connected to a Titanium and a stainless steel mesh cathode connected to a 316 stainless steel wire. A voltage of .7 volts was applied to the MECs by a power source and the output voltage from the exoelectrogenic organisms was recorded by Kiethly recording software. We found that a current was produced when the microorganisms found in the soil obtained from a closed gas station near Raleigh that was assumed to have contaminated soil were fed alkanes. The 2 types of alkane reactors produced a current while the control reactor vessels containing acetate, a commonly oxidizable compound, produced no current alluding to the presence of exoelectrogenic microorganisms that could only oxidize alkanes being present in the contaminated soil. These results show promise for bioremediation of alkane polluted soil but it also gives possibility to the potential for co-metabolization to bioremediate other pollutant compounds with similar chemical structures.

586871 Aqueous Film-Forming Foams Containing per- and Polyfluoroalkyl Substances and Their Precursors: Characterization and Treatment Options

Per- and polyfluoroalkyl substances (PFAS) are chemical compounds that have been used in numerous industries, commercial applications, and extensively in aqueous film-forming foams (AFFF) for fire-fighting activities. They are environmentally persistent compounds and reported to cause adverse effects on human health. According to the USEPA, currently available industrial formulations of AFFF should be free from long-chain perfluoroalkyl acids (PFAAs) which are the most commonly found PFAS in the environment and thought to be the most problematic. This project was centered on characterizing three industrially formulated and currently available AFFF samples to determine whether PFAAs can be destroyed by oxidizing the PFAS precursors using various advanced oxidation processes (AOPs) and utilizing plasma treatment for the oxidized samples. Preliminary measurements of 12 PFAA and 10 precursors with an LC-MS/MS have shown high concentrations of short-chained PFAA, traces of long-chain PFAA, and a significant amount of PFAS precursors. The 6:2 fluorotelomer sulfonate (6:2 FTS) was present at high concentrations in the three samples. Preliminary treatment experiments were performed by adjusting the dosage of the oxidation treatments in diluted AFFF samples. Once the optimization of the oxidation treatments is completed, efforts towards oxidizing PFAA for complete treatment of AFFF with preliminary plasma treatments will be realized.

Support for Sopuruchi Uwakweh was provided by NSF Award No. 1659623 (Advancing Sustainable Systems and Environmental Technologies to Serve Humanity (ASSETs to Serve Humanity).

587216 Modeling UV-Visible Absorbance Spectra of Multiple-Component Carbonyl-Containing Aqueous Aerosol Mimics

Proper characterization of atmospheric brown carbon (BrC), its formation, and its absorbance effects are critical to understating the aggregate effect that these materials have on overall aerosol climate forcing. Since a vast of chromophoric compounds can be produced from a single dicarbonyl/ammonium pairs, it is impractical to identify every specific chromophoric species that comprises an aerosol's overall absorbance spectrum as a reaction system ages. Therefore, we present an alternative method of representing UV-visible absorbance spectra as a composite of multiple, dynamic spectral line shapes. Previous work demonstrated that a model with four fitted Gaussian curves with magnitude following first-order kinetics achieves accuracy within 66% in the 200-350 nm range to track the absorbance behavior of single carbonyl/ammonium reaction systems (methylglyoxal and glyoxal) with different reagents concentrations and aging times. This research aims to verify the validity of this model on a complex carbonyl mixture (methylglyoxal and glyoxal)/ammonium reaction system. Absorbance data for varying concentrations of methylglyoxal and glyoxal in ammonium sulfate solutions were measured and modeled. Results were compared to the model gained by directly adding up the model of individual reaction systems. Residual analysis indicated that instead of fix at 220 nm, the peak of one of curve was within the range of 215 to 220 nm. Finally, a new peak was formed at around 225 to 230 nm. Therefore, a new Gaussian curve and the interaction parameters were added to track the result of the interaction between methylglyoxal and glyoxal.

584233 Ultrasound-Assisted Post-Pyrolysis Magnetization of Biochar without Adsorption Reduction: Effective Removal of Heavy Metals

Low cost and efficient adsorption of heavy metals from wastewater and thorough removal of adsorbent after water treatment has become the two essential needs. Hence, the main objective of a series of our studies is to develop an advanced acoustic-based method for activation of carbonaceous structures which can be effectively used for adsorption and easily removed from the aqueous solution. Biochar, the solid byproduct of pyrolysis with microporous carbonaceous structure, has been increasingly recognized as an efficient adsorbent for a vast number of pollutants. Magnetization, though eases the separation and reuse of biochar, significantly reduces its adsorption capacity to a comparatively much higher extent.. In this study, a hybrid post-pyrolysis modification was developed which sustained and even significantly increased the adsorption capacity of biochar. The process included i) structural modification of biochar under ultrasound waves, ii) magnetization with magnetite (Fe_3O_4) nanoparticles and iii) functionalization with 3-aminopropyl triethoxysilane (APTES). Low frequency ultrasound irradiation exfoliates and breaks apart the irregular graphite layers of biochar, and creates new/opens the blocked micropores, thus enhancing the biochar's porosity. On the other hand, APTES stabilizes the nanoparticles on the biochar surface, while it participates in water treatment. Preliminary results showed that acoustic-based magnetic biochar exhibited a much greater ability to remove Ni and Pb, with 124% and 31% higher adsorption compared to raw biochar. This high performance can be attributed to the synergistic effect of ultrasound activation on increasing the porosity and surface area of biochar and its further interaction with amine functionalization using APTES.

587749 Wastewater Treatment of Urea-Formaldehyde Resins Reaction By Heterogeneous Fenton with Goethite and Coupled By Distillation

In this study, degradability of contaminants present in effluents from urea-formaldehyde reaction resin was investigated. The degradation process consisted in distillation combined with Fenton Heterogeneous. In this research, temperature, distillation time, pH, reaction time, and catalyst dosages were varied in order to obtain the best set of conditions. This set was stablished to be 100°C, 2h for distillation, pH of 3,8; goethite and/or goethite impregnated in silice sand concentration equal to 3 g/L. Fenton Heterogenous process was performed at local conditions of pressure and temperature, and stirring speed was 250 rpm. Degradation was measured with chemical oxygen demand (COD). For distillation without Fenton Heterogeneous, the COD was found to be 71,01%. For the process with Fenton Heterogeneous after distillation, this value was 57,61%. The COD values were evaluated in conjunction with the reduction of total organic carbon (TOC). For distillation combined with Heterogenous Fenton, the reduction in TOC was about 49,77% more. Something important to mention, separation time of water and goethite impregnated in silica sand is much shorter than separation time of water and just goethite. This investigation demonstrated that distillation combined with Fenton Heterogenous process has a potential for efficient mineralization of waste from of urea-formaldehyde resins production.

585795 Conceptual Design of Glycol Recovery Method for the Treatment and Disposal of De-Icing Fluid Run-Off and Stormwater Management at Vancouver International Airport

During the winter months, Vancouver International Airport (YVR) performs de-icing operations during which aircrafts are sprayed with de-icing fluids containing ethylene glycol. The wastewater generated from this process contains ethylene glycol, a component of de-icing fluid, and consequently has a high biochemical oxygen demand (BOD). This wastewater is harmful if it is released into the environment untreated, as high BOD levels reduce the dissolved oxygen (DO) levels and adversely affects the ecosystem. YVR currently implements a periodic offsite disposal system, in which the spent de-icing fluid is collected and sent to Annacis Island Wastewater Treatment plant for further treatment. In addition, the airport's stormwater runoff periodically exceeds the Canadian Glycol guideline, which mandates that the glycol runoff must not exceed 100 mg/L.

The purpose of this project is to present a conceptual design option for the treatment, transport, and disposal of de-icing fluid run-off and management of stormwater runoff. This project analyzes alternative options to YVR's current waste management method of offsite disposal, with a focus on methods of recovering glycol from spent de-icing fluid. The solution meets environmental regulations, current and future capacity until the year 2050, and manages stormwater run-off outside of de-icing periods. The effects of climate change, airport expansion plans, and flow and loads are taken into to account. A thorough economic study is performed to ensure that the solution is a feasible alternative to YVR's current operations. The system that is determined to be the most optimal path forward consists of recovering the spent de-icing fluid onsite by utilizing an electric heater and mechanical vapor recompression system, as well as selling the recovered product to a third-party glycol supplier.

588153 Carbon Dioxide Removal and Oxygen Replenishment While Producing Proteins within a Mars Human Space Camp

NASA plans to have humans living on Mars for long periods of time within the next 20 years. There are many challenges that must be addressed before this vision is realized. Humans within confined cabins, like a Mars planetary surface camp, produce carbon dioxide as they respire and take in the oxygen from within the provided cabin air. The University of Louisiana working in concert with NASA and the Louisiana LASPACE program are researching methods to remove carbon dioxide as it builds up within the cabin air and potentially converting it to oxygen for reuse. Another potential source of carbon dioxide within the cabin is generated during the aerobic biotreatment of wastewaters produced during the support of humans on Mars. Additionally, the team hopes to produce protein from microalgal systems performing the carbon dioxide to oxygen photosynthetic conversion. Two primary methods are under consideration. One involves precipitation of carbon dioxide using sodium hydroxide solutions. The solid-precipitated carbonates produced are fed to *chlorella* microalgae to yield proteins (non-gas phase transfer of carbon to the algae). The second method is direct sparging of the spent cabin air into the *chlorella* where the gas-phase carbon dioxide is utilized for autotrophic growth. All cultures are grown in photobioreactors that are irradiated using fiber optics simulated sunlight (initially earth irradiation is tested then simulated Mars sun-irradiated will be tested using specialized lamps). The overall process flow diagram, theory of conversion, simulated yields, and initial results will be presented in the poster.

588266 Phosphate Adsorption Using Nitrogen Doped Biochar from Egg Shells and Egg Carton Feedstock

The project discusses the potential use of biochars produced from waste for phosphate(P) removal. Different series of biochar were produced from different ratios of eggshells and egg cartons to find the optimal ratio of eggshells to egg carton for the adsorption of phosphates. The biochars were produced through pyrolysis of feedstock under a nitrogen gas atmosphere at 750 C. In order to characterize the biochar, the samples were analyzed using TGA, BET surface analysis, CHN, FTIR and pH. Elemental and proximate analysis showed that eggshells have a high concentration of ashes and nitrogen. The pyrolysis decreased the carbon content in the 1:1 shell to carton biochar but increased the nitrogen content. The best variation of biochar produced was an eggshell biochar under nitrogen at 750 C and had a peak adsorption capacity of around 955 mg PO_4/g of biochar. This was promoted by the basicity of the solution which was measured to be 12. The eggshells promoted chemical adsorption through precipitation. Langmuir isotherms were fitted to all data sets but only ratios with equal or more carton fit the isotherm. The BET analysis showed increased surface area in samples produced from higher levels of egg carton. The mixture of 1:3 shell to carton had a surface area of 241.24 m^2 / g and the eggshell alone had a surface area of 12.82m^2 / g. Eggshell samples had smaller surface area but high adsorption capacity due to presence of Calcium (Ca) that allowed for the formation Ca-P crystals. This project shows that eggshells could be applied as a source of calcium to produce biochar with high phosphate adsorption capacity.

585417 Study of How the Concentration of Ozone and a-Pinene Precursor Gasses Affect the Hygroscopicity of the Resultant Secondary Organic Aerosol Particles

Cloud-related phenomena are the largest source of uncertainty in climate models. Thus, many researchers are investigating how various particles interact with water (their hygroscopicity) to form clouds in order to improve these climate models. While the interactions of some chemical species, specifically inorganic salts such as sodium chloride or ammonium sulfate, are well understood, other chemical species are less well understood. One class of particles, called secondary organic aerosols (SOA) which forms from the reactions of gasses in the atmosphere, fall into this class of poorly understood interactions. The reaction of alpha-pinene with ozone to form aerosol particles is a suitable model as both gases are abundant in the atmosphere. Alpha-pinene is a biogenic species that is emitted in by vegetation, while ozone, an inorganic molecule, has both natural and anthropogenic origins. Various methods have already been used by different research groups to investigate the hygroscopicity of the SOA model. However, due to a lack of standardized experimental set-up, comparing results is difficult. To relieve ambiguity, our research group established and used a standardized procedure to explore the effect of different initial concentrations of alpha-pinene and ozone on hygroscopicity.

585990 Study of the Degradation of Chlorinated Pesticides Atrazine and 2,4-D By Continuous System Ozonation

Atrazine and 2,4-Dichlorophenoxyacetic Acid (2,4-D) are chlorinated pesticides used for weed control in corn and sugarcane crops; Its presence in water can cause an environmental problem and damage to human health. In Ecuador, the presence of these compounds or their treatment in wastewater has not been investigated, but it is known that the treatments are not specific for their degradation, because these molecules are relatively stable. The project included the study of treatment of synthetic effluents containing atrazine and 2,4-D, through advanced oxidation processes. A continuous ozonification reactor was designed and built with a longitudinal diffuser consisting of two concentric cylinders. The outer cylinder of PVC, 8 mm in diameter and 200 cm in length will be divided into four sections of 50 cm by means of valves for the collection of samples. The ozone diffusion system consists on a silicone cylinder with a diameter equal to 4 mm, with holes of 0.5 mm in diameter to bubble the ozone along the reactor. The effluent entered the reactor with a peristaltic pump and ozone was supplied at rate of 0.84 mg/s through a "PHILAQUA" ozone generator. Atrazine solutions 1.53x10-4 M (33.0 mg / L) and 2.4 D 1.53x10-4 M (33.8 mg / L) were prepared, which were treated in "Batch" and continuous system at different pH values. In continuous system, degradation of atrazine equal to 53.8% was reached at pH 11, and 77.3% of degraded value was mineralized. 250 mL of Atrazine solution were treated in the "Batch" system, in 285.7 s it was affected by the same degradation as in the continuous system and 240.86 mg of Ozone was used. In the continuous system a degradation of 2,4-D equal to 54.4% is detected at pH 11, and 45% of degraded value was mineralized. 250 mL of 2,4-D solution were treated in the "Batch" system and in 424.33 s it was affected by the same degradation as in the continuous system and 357.7 mg of ozone was used. Therefore, continuous system was 2.28 times more efficient than "Batch" system for the treatment of Atrazine and 3.38 times more efficient for the treatment of 2,4-D.

587811 Greenhouse Studies for Phytoremediation of Soil at Legacy Uranium Mining Site

Historical mining practices have contributed to long term environmental impacts to soil and water and may pose a risk to public health if not managed appropriately. The focus of this study is to phytoremediate soil and water with elevated levels of radionuclides, including radium and uranium, at a legacy uranium mine in northern New Mexico. Uranium present is undergoing radioactive decay, producing radium as one of decay products at levels that require remediation before the land can be returned to other uses. Part of a proposed closed-loop system to remediate the water and soil affected involves using plants to remove uranium and radium from the soil. Here, greenhouse studies are being conducted to determine which plants are most effective for uptake of the heavy metals. In the first greenhouse study (summer 2019), spring canola (Brassica napus var. L140), safflower (Carthamus tinctorius var PI544006), guar (Cyamopsis tetragonoloba var. Kinman), carinata (Brassica carinata0, confectionary sunflower (Helianthus annus var. Red River 2215), and several native grass species were germinated and placed in pots containing contaminated soil collected from the legacy mine site. Of these plants, the confectionary sunflower, guar, spring canola, and carinata grew sufficient biomass for analysis. One challenge noted was the heavy texture of the soil, which limited root penetration of the transplanted seedlings into the contaminated soil and which affected plant-available water. Soil properties, and nutrient and metal levels, were measured using inductively-coupled plasma optical emission spectroscopy (ICP-OES), inductively-coupled plasma mass spectrometry (ICP-MS), gamma spectroscopy, and liquid scintillation counting before transplanting and after harvest. For the second greenhouse study (winter 2019-2020), winter canola (Brassica napus) and hemp (Cannabis sativa) will be added to the studied species, a new contaminated soil sample with lighter texture will be used in the pots, and plants will be directly seeded. Special attention will be paid to the soil moisture content. Results from the greenhouse studies will be used to inform design of an on-site study at the legacy mine site.

584551 Performance Evaluation of Two Cost-Effective Particulate Matter Sensors Under Ambient and Controlled Conditions

NASA's Multi-Angle Imager for Aerosols (MAIA) investigation is currently in development for the purpose of studying linkages between different types of airborne particulate matter (PM) and human health. The investigation requires the MAIA satellite instrument to be complemented by networks of surface PM monitors, and will focus on a selected set of major metropolitan areas around the world. In target areas with insufficient surface monitor coverage, the MAIA project will supplement existing reference-grade monitors with cost-effective sensors. To support this objective, we evaluated the performance of PurpleAir-II (PA) and Alphasense OPC-N3 (AS) optical particle counters (OPCs) against size-binned particle counts measured by GRIMM EDM 164, a transportable version of the EPA-approved GRIMM EDM 180, as well as a Differential Mobility Analyzer/Condensation Particle Counter (DMA/CPC).

Experiments were performed in indoor and outdoor (i.e., ambient) environments with the PA, AS, and GRIMM, and under controlled laboratory conditions with DMA/CPC in addition to the 3 OPCs. The DMA allows for differentiation of particles based on size, and the CPC grows the particles to larger size for counting. For laboratory evaluations, narrow size distributions of NaCl particles were generated and an airtight aluminum chamber was used to isolate these particles from the external environment.

In general, correlations between the low-cost sensors and the GRIMM were a function of particle size and concentration and were generally linear with varying slopes and offsets. In the outdoor environment, PA and AS consistently had $R^2 > 0.9$ for PM 1 and PM 2.5 (particles below 1 µm and 2.5 µm in diameter respectively). PA demonstrated the highest R^2 for most of the PM and counts measurements. In the laboratory, high R^2 was observed in correlations of total counts between GRIMM and DMA/CPC, supporting the use of GRIMM as a reference for size-binned comparisons. For NaCl particles generated with a narrow size distribution (0.5 ± 0.1µm), the OPCs put most of the counts in the smallest size bin (0.3-0.5 µm), but some counts were misclassified to into larger bins. To our knowledge, this study is unique in comparing the response of multiple OPCs to aerosols with controlled narrow size distributions.

586027 Sub-Surface Flow Engineered Wetland Design for Treatment and Disposal of Spent Aircraft De-Icing Fluid at Vancouver International Airport

Vancouver International Airport (YVR) airside operations de-ice aircraft during winter and utilize an offsite disposal system to manage the spent de-icing fluid and storm-water runoff. The aircraft de-icing fluid (ADF) contains ethylene glycol which is collected on the de-icing pad and conveyed to a ten million litre lagoon and is then transported to the Annacis Island Wastewater Treatment Plant for further treatment. Without being adequately treated, the de-icing fluid and storm-water run-off result in wastewater being generated with a high biochemical oxygen demand (BOD) and is toxic if released into the environment.

The purpose of this project is to design and analyze biological treatment as an alternative to offsite disposal of the ADF and present the treatment, conveyance and disposal of fluid and present an economic and environmental justification to support the implementation of the solution. The design process involves examining several biological treatment options to meet design constraints of not interfering with airside operations, maximizing the use of existing assets when possible, not attracting wildlife and can be built within the confines of the airside boundaries. A hydraulic feasibility study, life cycle analysis and economic study are conducted to determine the most ideal and viable option. The biological treatment chosen and designed is an artificially aerated engineered wetland system which can withstand current and future capacity until the year 2050, meets all municipal provincial and federal regulatory requirements and can store storm-water runoff outside the de-icing periods.

587538 Investigation of Presence and Role of Polyphosphate Accumulating Organisms (PAOs) in Lake and Sediment

The international community faces an imminent global phosphorus crisis. There exists a growing need for phosphate as fertilizer for agricultural food production, but large amounts of phosphate are wasted and end up polluting water resources. Polyphosphate accumulating organisms (PAOs) are known to uptake phosphorus under aerobic conditions and can store polyphosphate. When switched to anaerobic conditions, PAOs will then release phosphate and obtain energy through the cleavage of polyphosphate. The Gu Research Group in the Department of Civil and Environmental Engineering at Cornell University partnered with the Cornell Biological Field Station to examine the presence and abundance of polyphosphate and PAOs in Oneida Lake. A series of core sediment samples were collected weekly over a one-month period and transported to Cornell University for preparation and subsequent analysis using Raman spectroscopy, flow cytometry, and nuclear magnetic resonance (NMR) spectroscopy. The bacteria were isolated from sediment samples using density and mechanical separation methods. Both flow cytometry and subsequent analysis using Raman showed that polyphosphate-containing cells are present in the sediment and they appeared to correlate with DO levels at the surface of the sediment. Further experiments, such as colorimetric analysis, ion chromatography, and NMR, need to be conducted to confirm the absolute concentration of polyphosphate in the sediment samples. The findings from this study can benefit the search for PAOs to mitigate the global phosphorus crisis.

588178 Quantification of Uranium and Radium in Soil and Water Phytoremediation

Leachates from spent ore piles and mine tailings, and small particles moved by the wind, have enriched uranium and its decay products (radium) in the soil and water near legacy uranium mines in northern New Mexico. To be effective, combined remediation of both soil and water is needed. In this study, algae are being tested for their uptake of radionuclides and heavy metals from groundwater and several species of terrestrial plants are being evaluated for their ability to extract radionuclides from the contaminated soil. Two categories of terrestrial plants have been selected for this study, cellulosicoil-producing. Evaluation of the phytoremediation efforts requires quantification of uranium, radium, and other metals in the water, soil, and produced biomass. Given the low concentrations of the metals and the occasionally small sample sizes, quantification involves multiple strategies. Plants are partitioned into roots, stems, leaves, flowers, and (oil)seeds. Each part is analyzed to help understand the movement and aggregation of radionuclides in plants. The overarching goal is to be able to produce fuels from the plant material that remains uncontaminated while reducing waste volume and preventing further spread of contamination. The analytical techniques being considered are inductively-coupled plasma optical emission spectroscopy (ICP-OES) for heavy metals and higher levels of uranium; high-resolution, dual channel ICP mass spectroscopy (ICP-MS) for low metal concentrations and radium; gamma spectroscopy, and liquid scintillation counting. Most of these analyses require multi-step sample preparation including drying, ashing, acid digestion, filtering, ion exchange chromatography fractionation, and spiking with standards. Here, we present preliminary results about phytoremediation efficiency and the potential for value-added use of the produced biomass.

588295 Chemical Model of Raritan River Sediment for Lead Sensor System

In collaboration with the Environmental Science department at Rutgers University and the Center for Advanced Infrastructure and Technology (CAIT) this project seeks to develop a fast-acting sensor system and MINTEQ model program capable of producing a total Pb concentration from "fresh" sediment on the order of minutes. Pb concentrations in various chemical fractions of Raritan River sediment and environmental data was used to optimize a sediment geochemical model in MINTEQ. The MINTEQ geochemical model is used to determine the environmental conditions at various river sites that influence the fractions of lead bound to: dissolved organic matter (DOM), sulfide, carbonates, and hydrated ferrous oxide (HFO), an important fraction in the Raritan River. The MINTEQ model calibrated at a pH of 7.4 generates Pb^{2+} fractions of 11.805% bound to DOM, 65.172% bound to sulfide, 0.677% bound to carbonate, and 19.816% bound to HFO similar to those seen in field samples. In a pH sweep of the model from 6.4 to 8.4, the Pb fraction bound to sulfides decreases while the fraction bound to HFO and DOM increased. A Cl sweep from 0.03 to 0.5M, indicates that additional Pb-Cl species form in higher concentrations of Cl with the fractions of each species increasing with Cl. The effect of HFO concentration was tested in a sweep from 0.00036 to 0.00056 g/L increasing the Pb fraction bound to HFO and decreasing the fractions bound to sulfide. These initial results indicate a MINTEQ model can improve our Pb speciation model from laboratory measurements in atypical chemical conditions when it errors significantly from total sediment Pb concentrations and provide data of Pb partitioning from sites not accessible by boat. Our sensor system and geochemical model opens the opportunity for accurate on-site Pb measurements and the potential to measure sediment Pb concentrations of an entire watershed within days. This point-of-use technology has the potential to quickly identify Pb contaminated areas for remediation and augment treatment processes for Pb in polluted sediment.

586335 Electrospun Polymer-Based Visible Colorimetric Sensor for Selective and Sensitive on-Site Determination of Polycyclic Aromatics Hydrocarbons in Aquatic Ecosystems

Water is an integral part of our society. Polycyclic aromatic hydrocarbons (PAHs) resulting from oil spills and industry pollution are of particular concern to the Gulf Coast area, as these areas represent a critical nexus of food-energy-water for the region and the greater United States. Through this work an electrospun polymer-based PAH-in-water sensor will be devised. This will provide insight into the underlying phenomena of the advanced preconcentration and sensing methods used. Here, silica nanofibers with a formaldehyde detection chemistry were used to detect PAHs down to the ppb-scale using a small sample (~10 mL) wihtout the need for analytical equipment. This allows water samples to be tested on-site and in real time. The research has addressed the advanced material and interfacial challenges of the polymer-based sensors and the operational challenges of sensors for aquatic ecosystem monitoring. This work is applicable to a wide array of water sensing functionalities and integrations.

582957 Competitive Adsorption and Photocatalytic Degradation of Alizarin Red S and Bromocresol Green Anionic Dyes Using TiO_2 nanoparticles

Water contamination is a widespread problem that jeopardizing life on earth. The fact that water is capable of dissolving many different substances makes it uniquely vulnerable to pollution. Many human activities are causing direct and indirect water pollution such as landfill, industrial runoff, leaks and spills, mining, etc. Therefore, major steps forward should be taken to treat those water bodies. Cutting edge nanotechnologies are currently invested in order to mitigate water contamination problems and provide environmentally sound solutions. Adsorption and photodegradation using photoactive nanomaterials are considered one of the states of technologies that are used in industrial water purification.

This study evaluates the effective photodegradation properties of TiO_2 brookite nanoparticles for the removal of binary aqueous mixtures of the anionic dyes: Alizarin Red S (ARS) and Bromocresol green (BCG). Batch adsorption experiments were conducted to investigate the effect of dyes concentration, contact time, and temperature effect. The adsorption isotherms were fitted for the single adsorption using the models of Langmuir, Freundlich and Sips, while the binary adsorption isotherms were fitted using the Extended-Sips model. The photodegradation experiments for the adsorbed dyes were achieved using UV light irradiation (6 W, λ_{max} = 312 and 365 nm). Overall, post-adsorption photodegradation preliminary results showed high selectivity towards ARS molecules in the presence of BCG molecules. In addition, the adsorption process was spontaneous in nature with endothermic behavior for both anionic dyes. The TiO_2 nanoparticles were successfully regenerated using UV irradiation. Moreover, molecular dynamic computational modeling was performed to understand the molecules optimum coordination, TiO_2 geometry, adsorption selectivity, and binary solution adsorption energies. The outcomes of this study will be showcased and discussed in detail.

584791 Point of Entry Water Purification System in Quiché, Guatemala

The design of a community-wide point of entry water purification system is presented. The system will provide water for up to 600 people in a rural Mayan community in the Department of Quiche, Guatemala. In the past year, a piping system was installed from a spring located 13 kilometers away. This system relies on gravity to provide water to the homes in the community. The large-scale purifier will be installed at the point of entry to a storage tank. It utilizes solar panels to power the unit and an ultra-violet photoreactor to purify the water and inactivate waterborne pathogens.

The design considerations for the purification system included: volumetric flowrate through the system, prefiltration for sediment removal, photovoltaic and battery system for power, and the ability to stop water flow in case of photoreactor failure. The system is designed to exceed a 14 gallon per minute flow rate such that a need for a pump to overcome pressure drop is obviated. When designing the prefiltration, which is necessary for maximum sterilization through the photoreactor, pressure drop measurements were taken to designate which option would best fit the flowrate and pressure drop considerations. The photovoltaic and battery system was designed so that the purification system would be able to run without power from the solar panels for a day and a half. The worst-case scenario for the global horizontal irradiance for the Quiche, Guatemala location was used to calculate how many solar panels and what size battery would be required to do this. A solenoid valve was added to the design to stop the flow of water in case of photoreactor failure. A bypass valve was added prior to the unit, per the communities request, so water could continue to flow in the event of full unit failure.

This project was designed and implemented in just under one year. Throughout the year-long journey, one goal was sought after; providing clean water to a developing and well-deserving community in Guatemala as clean water is a human right.

584543 Rapid Photo-Catalytic Oxidation of Dissolved Manganese with TiO_2

Manganese(III/IV) (hydr)oxides (hereafter Mn oxides) occur ubiquitously in terrestrial and marine systems, and play crucial roles in natural electron and element cycles. However, the understanding of their occurrence is limited due to the slow oxidation of Mn^{2+}(aq) to Mn(IV) which precedes the formation of Mn oxides. Through the photocatalytic process with TiO_2 (the 9th most abundant compound in the Earth's crust), we found very fast oxidation of Mn^{2+}(aq) to Mn(IV). The reaction rate is even much faster than microbial processes, which has been believed to be the primary mechanism for the oxidation. We studied the oxidation process under a variety of conditions, such as different TiO_2 structures (anatase vs rutile), different solvents (saltwater vs freshwater), and different atmospheric conditions (open air vs nitrogen). These analyses show that the fast photocatalytic oxidation occurs primarily due to direct electron transfer between Mn^{2+}(aq) and photo-excited TiO_2. This understanding will enable us to explain the abundance of natural Mn oxides on rock surfaces. In addition, because the redox process of Cr and Tl, which are newly emerging paleoredox proxie, is significantly governed by Mn oxides, understanding of Mn oxide formation will help to calibrate the redox proxy systems for oxygen evolution on early Earth.

587331 Extracting and Detecting Uranyl Species Using Aliquat-PAN Mats and Gold Nanostars

The ability to detect and specify trace concentrations of uranium in groundwater is crucial to addressing public health concerns associated with uranium in groundwater and restoring adversely impacted lands. Surface-enhanced Raman scattering (SERS) detection of trace concentrations of uranium on polymer mats has been previously reported using amidoximated-polyacrylonitrile (AO-PAN) mats and carboxylated gold nanostars, but a reproducible method of determining uranyl species in aqueous solution is still needed. In this project, we seek to utilize Aliquat 336 polyacrylonitrile (Alq-PAN) mats to isolate the original uranyl species from solution, followed by separation analysis at trace concentrations based on unique vibrational frequencies in SERS spectra with gold nanostars. To observe how efficiently Alq-PAN mats isolate and uptake negatively-charged uranyl species, we varied the uranyl-to-oxalate concentration ratios (5:50, 5:100, 5:300) of the aqueous solution and incubated Alq-PAN mats within the aqueous solution. We hypothesize that increasing the number of oxalate ligands, or increasing the uranyl-to-oxalate concentration ratio, will increase the number of negatively-charged uranyl-oxalate species in solution, thus promoting uranyl uptake and SERS detection. Understanding the interaction between Aliquat and uranyl-oxalate complexes will allow us to extend the speciation analysis to other aqueous samples with different negatively-charged uranyl species, such as uranyl-carbonate or uranyl-phosphate.

587389 Open Borehole Cross-Contamination: A Unique Opportunity to Evaluate Natural Attenuation of Mixed Organic Contaminants in a Sedimentary Rock Aquifer System

At a site in Wisconsin, organic chemicals were released into the subsurface resulting in the formation of a persistent, dense non-aqueous phase liquid (DNAPL) source between 46 and 55 m below ground surface (bgs) in a fractured sandstone. Since 2003, researchers have applied various hydrogeological measurement and modeling techniques to characterize the flow system, contaminant mass, and phase distributions. In July 2017, three holes were continuously cored to 100 m bgs, downgradient of the source zone (MP-23D, MP-24D, and MP-25D). The cores were geologically logged and sampled (about 5 per 1.5 m) for contaminant analyses by removing a puck from the core, crushing it in a hydraulic press, and placing about 15 g of the rock into a pre-prepared vial of high purity methanol for contaminant extraction. The rock core results quantify the total (i.e., dissolved, sorbed, NAPL) contaminant mass in the rock matrix pore space. Then the coreholes were logged using geophysical and hydrophysical methods in the open and FLUTe™ sealed boreholes. Prior to multilevel system (MLS) installation, the holes were sealed with FLUTe™ liners unless open hole measurements were being made. Each borehole was instrumented with high resolution (20 monitoring intervals over 40 m) Westbay® MLS in November 2017 and used to sample groundwater in June 2018 and 2019. These MLS samples represent the water and contaminant mass in the fractures. The rock core was expected to have high, quantifiable results through 55 m bgs. Therefore, permanent casings were set at 60 m bgs before proceeding with coring. The rock core data showed consistent, high concentrations through 62 m bgs and sporadic quantifiable concentrations separated by non-detects were identified to 83 m bgs. In contrast, high, quantifiable concentrations were observed from 60–98 m bgs in groundwater samples from all three MLS. The discrepancy between the rock core and MLS groundwater data suggests the boreholes became cross-connected during open hole conditions. Repeat groundwater sampling in the MLSs show concentration declines in the deepest portion of all three holes and drastic compositional changes for all MP-25D ports. The cross contamination of these boreholes provides an opportunity to study natural attenuation of these chemicals in a sedimentary rock aquifer system with baseline conditions obtained from rock core VOC data. Future work will consist of comparing the depth discrete groundwater concentration trends to hydrophysical data quantifying active flow regions in each hole and analyzing mechanisms and rates of the natural attenuation.

585412 Destabilizing Nanosized Biochar

Biochar has been widely developed as a low-cost and environmental-friendly carbon material for its application in energy, environment, and agriculture. Its applications are largely due to its physiochemical characteristics such as its outstanding porous structure, large surface area, and abundant functional groups which all contribute to its adsorption abilities of contaminants. Recently, a novel method, ball milling has been investigated to enhance the adsorption abilities of biochar in removing organic and inorganic pollutants from contaminated sources. As a new type of engineered biochar, ball-milled biochar (BMB) presents outstanding porous structure, larger surface area, and more abundant functional groups than pristine biochar. Due to its nanosized structure, however, BMB also displays increasing stability and suspension in aqueous environments, which hinders its separation in regard to regeneration and waste disposal. The goal of this research is to investigate to what extent conventional coagulation and flocculation methods could effectively destabilize BMB from aqueous solutions.

587518 Environmental Fate and Transformation of Dichloroacetamide Herbicide Safeners

Dichloroacetamide safeners are a class of agrochemicals co-applied in large scale with chloroacetamide herbicides to protect crops from herbicide toxicity. Although all four dichloroacetamide species have been detected in Midwestern surface waters, safeners are currently regulated as "inert" constituents under the U.S Federal Insecticide, Fungicide, and Rodenticide Act, and as such, have received meager attention in the peer-reviewed literature. Recent studies suggest that some dichloroacetamide safeners are reactive in aquatic systems and can yield products with increased bioactivity which may pose risks to human and environmental health. To evaluate the fate and transformation of dichloroacetamide safeners, we measured timescales (i.e. half-lives, sorption rates) for predicting safener persistence, determined the effects of water chemistry (i.e. pH, DOC) on the rate and extent of hydrolysis and sorption processes, and quantified sorption parameters for active herbicide co-formulants and mixtures. We found that most dichloroacetamide safeners undergo both acid-catalyzed and base-catalyzed hydrolysis, and we identified two hydrolysis products for benoxacor, the most widely-used dichloroacetamide safener. Sorption studies revealed limited uptake of safeners and herbicide co-formulants, with more sorption to soil with a high organic carbon content. In general, dichloroacetamide safeners demonstrated reduced uptake onto soil compared to their chloroacetamide herbicide co-formulants.

587644 *Analyzing the Effects of Surface Energy on Algae with Specialized Attachment Mechanisms in Static and Flow Environments*

The goal of this research is to observe the effects of surface energy on the attachment of algae with specialized attachment mechanisms to substrata in stationary environments and small, controlled flow environments. The substrate materials tested are glass and polylactic acid (PLA). Reactor channels were 3D-printed from PLA filament and a standard glass microscope slide and pressed PLA were used as the substrates for attachment. The algae species in focus for this experiment are two types of filamentous green algae, Oedogonium and Stigeoclonium. Oedogonium is known to have root-like structures, called holdfasts, and mucilage basal pads as a specialized attachment mechanism. *Stigeoclonium* is a freshwater algae that has a specialized “star-like” attachment mechanism that is known as basal rhizoids. In nature, multiple algae types grow together on rocks and surfaces, but the factors affecting algal attachment are not fully understood. Different types of algae are preferred for different commercial applications, such as pollutant removal, nutrient supplements, and biofuel production. The economic viability of algae cultivation could be improved if the details governing algal attachment were better understood.

585581 Economic Analysis of Waste-to-Energy Technologies for Urban and Rural Areas

The average amount of waste generated in the United States is 4.4 lbs/day/person, and the majority is landfilled. The energy contained in these wastes presents an opportunity for renewable energy production. Renewable energy production methods that are more environmentally friendly are needed. Therefore, it is crucial to implement waste to energy technologies to reduce landfilling and minimize environmental impact.

Municipal Solid Waste can undergo processes like Anaerobic Digestion (AD), Incineration or Gasification to produce biogas or syngas. Harmful gases produced in landfill can also be used as an energy source. These gases can be upgraded to be used in energy recovery methods like combined heat and power units (CHP), Fischer-Tropsch Synthesis (FTS) for the production of liquid fuels, or the production of Compressed Natural Gas (CNG). All of these technologies are new and minimal data has been reported for their operation. A limitation encountered when trying to implement/build one of these plants is that there is no set tool that compiles the techno-economic and environmental data of existing plants and can provide fast cost estimation for the development of these projects. The goal of this research is to create a software tool based in Excel to assist decision makers such as municipalities and waste management companies. Using this tool, the user is able to select the technology desired: AD, Gasification or Landfilling, as well as the energy recovery method: CHP, liquid fuel or CNG production. The user will also provide the feedstock type (Municipal Solid Waste, Food Waste, Animal Manure, Farm Waste), feed properties and plant capacity. The software tool is set to provide the user with the capital and operational cost requirements and energy profits from the process. Additionally, Carbon balances, biogas production rate and its composition are calculated. The calculator also includes the cost of biogas purification based on the purity requirements for each energy recovery technique.

The use of the tool is illustrated using a base case AD plant that generates biogas used to produce liquid fuels via FTS. The plant processes 60,000 dry tons/year of MSW with a 30% moisture content. Using the calculator, the capital cost of the entire system is $17.0 MM and the total operational expenses are $4.46 MM /year. The revenue from tipping fees and liquid fuel sales are $8.6 MM/year. The software tool provides the user with the ability to compare and contrast alternative waste-to-energy options and incentivize building these plants.

587949 Scale-up of Moringa Coated Sand Filters

Clean drinking water is a global issue, affecting more than 844 million people around the world. The proposed project is to scale-up and redesign a Moringa seed coated sand filter to remove bacteria from wastewater with locally accessible and affordable materials. Moringa seeds contain a cationic protein that can create electrostatic forces between negatively charged microbes, removing the pathogens present in the water. For the purpose of this study, the lab-scale glass columns used for the filters were replaced by PVC pipes, using a binary mixture of collector size and increasing the column diameter and flow rate according to the Clean Bed Filtration Model to obtain the EPA standard approved removal. To prove the functionality of the filters, fluorescent *E. coli* bacteria was first utilized to quantify the experimental removal. It was then noted that when treating water spiked with *E. coli* and TOC, the organic matter (TOC) competes with the bacteria for protein sites, impeding the proper removal. To address this problem, an activated carbon pre-filter was implemented. The results showed that, when this pre-filter is applied, the TOC sticks to the activated carbon particles, allowing the f-sand filter to function properly and remove bacteria above the EPA standard approved removal.

588142 Exploration of the Efficiency of Surfactant Extraction from High Internal Phase Emulsions

Clean water and clean air filtration are essential for environmental sustainability, in which there is a need to provide efficient ways to process both fluids to help communities have healthy lifestyles. One component that can improve the filtration is the use of foams. Foams are porous substances that have a combination of a continuous phase and an internal (or dispersed) phase. Pores of the foams consist of the dispersed phase of the foam formed in conjunction with surfactants. Surfactants are substances that lower surface tension between two immiscible phases, which results in a more stable interface between both phases. In more general applications, surfactants are used in conventional products such as soaps and detergents, in which the surfactants help with filtering any unwanted chemicals and compounds. In the emulsification process, the surfactant gets mixed in with the high internal phase emulsions and gets washed off using soxhlet extraction. However, there has not been any studies on the quantity of surfactant that is washed off. This is an issue due to the fact that surfactants will contain unwanted contaminants in which will cause toxicity in water quality. In this project, we investigated the process of polymerization of a water/monomer system, more specifically the step of washing the surfactant off. This step of the process is essential because, in most applications, the surfactant is washed off to decrease in the possibility of contaminates still present and can cause impure products. The parameters that will be tested are (1) how long does it take to wash out the surfactant, (2) what solvent works best to wash out the surfactant as well as to analyze what surfactant we are using (for this experiment we will be using Polysorbate 80, Polysorbate 20, Span 20, and Span 80), and (3) how can we quantify how well the surfactant is washed off. UV-visible spectrometry was in this experiment as it can analyze the absorbance of a solution with light. With this and Beer-Lambert's law, which states that the absorbance of the compound is directly proportional to the concentration of the solution, we can produce conclusions on the washing off of surfactants and how we can manipulate the step to efficiently provide filtration mechanisms. The resolution before and after the wash of the surfactant can easily be measured and help understand the parameters of this part of the process related to emulsion-based polymerization.

588359 Depolymerization of Polyethylene Terephthalate Plastics By Ozonolysis

Polyethylene terephthalate (PET) is the most abundant of the seven categories of recycled plastics due to its affordability and desirable chemical properties for use in diverse applications such as the ubiquitous containers and synthetic fibers. The improper disposal of plastics contributes to environmental pollution adversely affecting nearly all types of ecosystems. Beyond washing and reusing consumer plastics, chemical or tertiary recycling breaks plastics down into constituent monomers which may be transformed to make new consumer products. However, such techniques are still in the research and development stages. Herein, we investigate a new method that attempts to break down PET using ozone (a powerful oxidant that is also used in water purification and deconstruction lignin, a naturally occurring biopolymer) to decompose the polymers into smaller molecules for reuse. To perform ozonolysis, a solvent that dissolves PET and is itself inert to ozone is essential. Among various solvents [acetic acid, tetrahydrofuran (THF) and trifluoroacetic acid (TFA)] in preliminary studies, TFA demonstrates high PET solubility as well as compatibility with ozone. Bubbling ozone (9.4 mole% in O_2) through such a PET solution produces several smaller molecules including terephtalic acid (TPA) as inferred by NMR spectroscopy. These initial results are promising as TPA is a starting material for producing PET plastic. They provide guidance for future research towards making chemical recycling practically viable, thereby improving environmental sustainability.

583387 Narcotics Consumption Trends at a Southwestern U.S. University Campus in 2018 Tracked By Wastewater-Based Epidemiology

Current methods measuring the consumption of prescription and illicit drugs are often hampered by innate limitations, which can impact the relevance and robustness of the associated data. Here we applied wastewater-based epidemiology (WBE) as a metric to measure trends in the consumption of twelve narcotics within a collegiate setting from January 2018 to May 2018 at a Southwestern U.S. university. Local legislation, put into place because of the opioid crisis, has decreased the number of opioid prescriptions in 2018 by 36% compared to 2016. We hypothesize that per-capita estimates of heroin and fentanyl (illicit alternatives) derived from campus-generated wastewater will be significantly higher (a = 0.05) from the previously sampled academic semester. An initial study, completed for the campus from August 2017 to December 2017 (Fall semester) identified consumption of select narcotics, in units of mg/day/1000 persons in the following order: cocaine (551 ± 49), heroin (474 ± 32), amphetamine (256 ±12), methylphenidate (236 ± 28), MDMA (88 ± 35), oxycodone (80 ± 6), methadone (72 ± 8), alprazolam (60 ± 2), codeine (50 ± 4), and morphine (18 ± 3). The present follow-up study was designed to identify potential changes in the consumption patterns of prescription and illicit drugs as the academic year progresses as well as the occurrence of narcotics not previously investigated at this location. Seven consecutive 24-hour composite raw wastewater samples were collected each month (n = 70) from two sites that capture nearly 100% of campus-generated wastewater. Samples were extracted using a Dionex 280 AutoTrace solid phase extraction (SPE) system with Oasis Hydrophilic-Lipophilic Balance (HLB) SPE cartridges. Internal standards for target analytes were used to quantify losses during sample processing. Extracted samples were analyzed by liquid chromatography tandem mass spectrometry for the following parent drugs and metabolites: alprazolam, amphetamine, buprenorphine, cocaine, codeine, fentanyl, heroin, MDMA, methadone, methylphenidate, morphine, and oxycodone. Chemical analyses are in progress for the January 2018-May 2018 samples. Forthcoming data will be shared immediately with public health officials to help address the health crisis and to showcase the use of WBE as a tool to directly measure the impact of an implemented public health intervention (i.e., opioid prescription reduction).

583566 On-Site Extraction and Electrochemical Analysis of Heavy Metal Pollution in Soil, Plant, Air, and Water Samples

Heavy metal pollution is a worldwide problem. Within the US these metals are leached into the environment from landfills and from the thousands of mines across numerous mountain ranges (e.g. such as in the infamous Gold King Mine Spill). It is crucial to be able to quantify and identify metals in water, soil, plant, and air to reduce exposures and inform remediation efforts. Currently, the metal testing process often takes days to weeks and requires an equipped laboratory with trained technicians. But imagine if there was a simple integrated device that could not only give you results in minutes but is able to be used on-site. Through a collaboration between Colorado State University and Access Sensor Technologies, we have created a system to do just that. Soil and air filter samples can be immediately placed in a solution capable of extracting metals. Once the extraction is done and adjusted to optimal conditions for the experiment, stripping voltammetry is performed to identify and quantify several different metals including lead and cadmium. In the past, electrochemical techniques have been limited because of the many interferences within the solution, leading to nonsensical results or masking the presence of the metals entirely. However, by modifying the extraction solution and holding at the correct pH we have found many of these interferences disappear, making stripping voltammetry practical for use not only in the lab but also out in the field, where we can determine concentrations down to single part per billion (ppb) levels.

585759 Oxidation of Reactive Blue 4 through the Use of a Heterogeneous Fenton Catalyst

Organic aqueous dyes are commonly found in the wastewater of textile companies. Unfortunately, the treatment of these chemicals tends to be complicated, due to their natural resistance to degradation, filtration, and centrifugation. For industrial use, it is important to develop a fast, reusable, and green method to treat and degrade these organic dyes. Therefore, we chose to make use of a Fenton catalytic pathway in the form of Fe^{3+} ions. Loading these iron ions on a mesoporous structure, such as SBA-15, was explored to find a way to make the Fenton catalyst more reusable in industry. In other words, to turn a homogeneous catalyst into a heterogeneous catalyst. The heterogeneous catalyst will theoretically be able to retain more of the iron ions than the homogeneous ions after filtration, while still providing active sites for the Fenton pathway. It was found that the iron loaded SBA-15 worked better at oxidizing the reactive blue 4 dye, around 70% oxidation, than just the reagents without the catalyst; however, the experiments ran into complications when an unexpected product was formed within the reaction chamber. With this unexpected product in play, more research will be needed to increase the efficacy of the catalyst as it failed to match up to the homogeneous catalyst.

588354 E.*coli* Removal in Moringa-Coated Sand Using Binary Mixtures

Providing access to clean water is one of the 14 grand challenges on engineering. 1 billion people do not have access to clean drinking water more than 4,000 children under the age of five die daily due to diarrhea diseases linked to drinking unclean water. This study offers *moringa oleifera* seed coated sand solution as an option for sustainable water treatment. The moringa seed is readily available in countries that need safe drinking water the most. One tree produces 15,000 seeds in one year and the seeds contains 1% cationic protein allowing it to act as a coating for sand. The charge on the seed attracts to the sand, reversing the charge of the sand surface. Coating very fine sand (106 micron) with moringa seeds removes 100% of *E. coli* from our lab-scale columns and satisfies the EPA standard. However, the removal decreases significantly with larger sand size and the limited access to this size sand defeats the purpose of our solution. One solution is to mix this small sand with practical sieved sand to produce binary mixtures. Our hypothesis was that this mixture will provide the same hydraulic conductivity and bacterial removal. Our experiments confirmed this hypothesis. The hydraulic conductivity experiments showed that binary mixtures of 40% sieved sand and higher should behave the same as the successful 100% 106-micron columns. The *E. coli* removal experiments proved this hypothesis correct, showing binary mixtures of 40% and higher have similar removal as 100% 106-micron sand. The next steps in this project include scaling up the column containing the binary mixtures, and using activated carbon to remove organic matter in real water.

587664 Study of a Treatment System Based on a Biofilter Using "*Eisenia Foetide*" Worms, for the Treatment of Effluents from a Slaughter Company

This study is based on the evaluation of a wastewater treatment system of a slaughtering industry based on a vermifilter built with low-cost recyclable material using *Eisenia foetide* worms as degrading organisms with a concentration of 200 units/m^2, the analyzed effluent has high concentrations of contaminants such as BOD with 1401 mg/L, a COD of 3642 mg/L, total suspended solids 1437 mg/L and oils and fats of 1313 mg/L. For this support, the constants of the Schulze equation that relates the residence time to the height of the filter were determined, the average dimensional constants obtained were n=1,0940 and C = 0.0305. Six 0,27 m^2 filters were built and kinetics were analyzed with residence times of 4, 8, 14, 17, 21 and 24 h with a hydraulic load of 3 L/day. For this effluent the vermifilter presented a first-order degradation kinetics with a constant k=0.1078 h-1. Wood chips were used as a filter bed due to this material is a waste in the local timber industry and has a low cost. In addition, recycled material was used for the construction of vermifilters which allows the system to be low cost of implementation and high duration. These systems have several advantages such as the formation of high quality solid fertilizer as a by-product and low energy consumption. The evaluation of this system will allow the design of a wastewater treatment plant for the total flow of the company.

587140 Removal of Antibiotic-Resistant Bacteria in Synthetic Stormwater Via Biochar Columns

Water is one of the world's most valued commodities, and antibiotic-resistant bacteria are one of the emerging pollutants of concern. Stormwater runoff picks up pollutants such as chemicals, bacteria, and sediment as it runs over impervious surfaces and into waterbodies. The aim of this project is to test the use of biochar columns as an effective and inexpensive form of stormwater treatment. Biochar is biomass that has undergone pyrolysis. It serves as an adsorbent for contaminants in water including antibiotic-resistant bacteria. It is commonly used in land applications as a soil amendment, but there are concerns as to how long biochar is effective and what occurs once the biochar is saturated. Three infiltration columns will be tested; one containing biochar, one with soil and biochar mixed in a ratio typically seen in a soil amendment, and one with soil. Due to the unpredictability of stormwater, semi-natural stormwater will be synthesized by inoculating stormwater collected from the OSU-Benton County Green Stormwater Infrastructure Research Facility with several *E. coli* strains. Using culture-based methods, *E. coli* in the synthetic stormwater will be quantified before and after treatment through columns in triplicates. The ability of the columns to both remove and retain an adequate concentration of *E. coli* will be analyzed to determine the best step for future use of biochar.

586385 Ionic Strength Effects on Flocculation Performance in Taylor-Couette Flow

The removal of flocs, which are aggregates consisting of solid particulate, from source water is an essential part of the drinking water treatment process. To improve flocculation, salt can be added to increase the solution ionic strength. This creates a charge screening effect, which allows for aggregation of particulate. However, increasing the salt concentration would increase the load on reverse osmosis filters down the line, which is cost prohibitive. The addition of long-chained, charged polymer flocculants makes the flocculation possible without need for excess salt. Flocculation in a treatment tank is subject to a variety of hydrodynamic flow states, and isolating how they affect flocculation is difficult. A Taylor-Couette (TC) cell, which consists of two concentric cylinders where the gap between the cylinders can be filled with fluid, can generate precisely defined flow states at specific rotational speeds. The Dutcher group has built a TC cell with radial injection capabilities from the inner cylinder, therefore flocculation can be captured via image analysis from the start of the process. This TC cell will be used to study how different concentrations of NaCl impact flow dynamics and floc size, growth rate, and configuration.

588353 Microbial EPS Impact on Residual Water Saturation in Emulated Soil Micromodels

Microbial communities are an integral part of soil environmental conditions. Within soil, bacteria contribute to moisture retention by secreting extracellular polysaccharides (EPS). EPS helps regulate moisture retention by reducing water evaporation, in turn promoting plant growth. EPS mitigates water loss at pore-scale by (i) acting as a hydrogel, which will swell during wet conditions and retain moisture in dry conditions, (ii) promoting soil aggregation, resulting in porous microstructures that increase the interparticle capillary forces, and (iii) altering soil surface properties such as hydrophobicity. Studies on moisture retention at bulk scale have few chances to uncover these intricate interactions among microbes, soil and solution systematically. Here, we employed an emulated soil micromodel to examine the mechanisms of moisture retention on EPS collected from *Sinorhizobium meliloti* cultures and compared them to the drying effects of deionized water, dilute glucose solution, and artificial groundwater (AGW). The results showed that EPS reduce the evaporation rate and promoted residual saturation at equilibrium compared with AGW and DI water within identical physical geometry. Dilute EPS (0.006%) in AGW solution had an average residual saturation of 2.8 ± 2.94% after 16 hours drying. Comparatively, deionized water reached 0% residual saturation after 6 hours drying, artificial groundwater reached 1.6 ± 3.19% saturation after 6.5 hours, and dilute glucose in AGW solution (0.006%) reached 3.6 ± 3.61% saturation after 8 hours. In this study, we confirmed the effect of EPS on moisture retention at pore-scale and provide a platform to understand microbial-mediated moisture retention. Ultimately, we hope this work to aid in implementing sustainable agriculture leading to increased food production and less water usage.

583017 A Study of the Physical Properties of Plastic Derived Fuel Oil Produced from Waste Plastic

Plastic waste around the globe has become a massively detrimental issue that can lead to health complications and destruction of the environment. Studies and prior experimentation suggest that thermally decomposing plastic will cause the carbon bonds to break apart and form a fuel oil similar in composition to diesel. Using a low-cost and simple processor called the Trash-to-Tank (3T) processor developed by the University of Kentucky Appropriate Technology and Sustainability (UKATS) Research Team, waste plastic can be converted into fuel, known as Plastic Derived Fuel Oil (PDFO) and sold as a diesel alternative.

The 3T processor has been specifically designed to be appropriate for underdeveloped regions where access to sophisticated literature are often lacking. Currently, the UKATS Research Team is conducting a pilot study in Uganda to assess the performance of the 3T process. In order to ensure that the fuel meet all appropriate regulatory requirements for use as a motor vehicle fuel in Uganda, emissions testing has been commissioned from the Ugandan Bureau of Standards. The laboratory results for the PDFO will be presented and compared with traditional number 2 fuel oil. The 3T process has been implemented in Uganda and has the potential to provide a low cost way for rural communities in developing regions to take ownership of the waste plastic problem in their communities. This process is part of the establishment of locally managed decentralized circular economies in underdeveloped regions and further can give women and other people living in poorer regions the ability to become entrepreneurs.

585931 Characterizing Microplastic Degradation in Simulated Marine Environments

As more plastic materials accumulate in the ocean, there is a growing concern for their impacts on marine ecology and human beings. This study focuses on establishing characterization methods to investigate the potentially hazardous degradation products of plastics found floating on the ocean surface. Unprocessed plastic pellets (polyethylene, polypropylene, and polystyrene) and consumer drinking straw pieces (non-biodegradable straws and compostable straws) were placed in simulated seawater and examined under different weathering conditions. The effects of ultraviolet light and physical mixing on the rate of degradation using scanning electron imaging was performed to visualize surface degradation. The mass balance of the pellets before and after the degradation process was also analyzed to confirm the scale of degradation. A GCMS protocol is proposed to identify organic degradation species by extracting them from salt water to hexane to confirm the presence and identity of organic degradation byproducts.

585162 Multi-Modal Resins for the Adsorption of Nitroaromatics

Legacy munition constituents (MCs) including nitroaromatic compounds (TNT, RDX, HMX), along with insensitive high explosives (IHE; NQ, DNAN, NTO), and their degradation products are hazardous molecules that could negatively impact environmental waterways if not removed from wastewater. Although there are remediation techniques in place, new generations of MCs and IHEs are not readily treated. Advanced oxidation processes, such as the Fenton reaction, have been utilized for treatment but are not cost-effective, especially in dilute concentrations, so new technologies are needed. In this work, we propose synthesizing porous, multimodal resins that utilize coulombic interactions, hydrophobic interactions, and hydrogen bonding to improve selectivity for nitroaromatics and overall adsorbent lifetime.

Resins synthesized from the monomer glycidyl methacrylate are particularly attractive because they contain reactive epoxy groups which provide a tailorable surface chemistry. Resins are synthesized through suspension polymerization. Properties such as bead size, pore volume, and surface area were manipulated by changing variables such as polymerization temperature, agitation speed, crosslinker ratio, and presence of a porogen. We hypothesize that (1) macroporous resins will provide greater access to binding sites than nanoporous resins and (2) multimodal ligands will exhibit higher selectivity for nitroaromatics than traditional ion-exchange groups. To test the first hypothesis, porous resins are synthesized by including a cyclohexanol and tetradecanol mix in the suspension polymerization reaction. To test the second hypothesis, ligands with multiple modes of binding (i.e. varying degrees of hydrophobicity, H-bond donors, and polarity) were incorporated via epoxide ring opening reactions. Resin functionalization was monitored with Fourier transform infrared spectroscopy (FTIR) by collecting spectra before and after reaction. Resin capacity and binding affinity were studied through equilibrium batch adsorption tests with organic adsorbates phenol, pyridine, picolylamine and benzene. While these are not explosives, they represent adsorbates with varying propensity for H-bonding. Adsorbate concentration in solution is measured by ultraviolet-visible spectroscopy. This work can be extended beyond explosives, because understanding the role of molecular-level interactions in binding affinity will allow for the design of adsorbents that can capture other emerging organic contaminants.

583793 Synthesis and Characterization of Poly(hydroxyethyl methacrylate) (PHEMA) Hydrogels for Drug Delivery Applications

Nanoparticle based drugs offer immense possibilities as a next-generation medical technology, but currently lack a nontoxic, reliable method of targeted delivery. Due to its biocompatible nature, flexible cross-linked structure, and numerous accessible hydrogen bonding sites, PHEMA hydrogels may be viable candidates as encapsulating materials for targeted nanodrug delivery. The primary objectives of this study are the systemic synthesis of PHEMA hydrogels via radical polymerization of and the multimodal materials characterization of PHEMA hydrogels. Material and structural properties of the PHEMA hydrogels were evaluated using scanning electron microscopy, fourier-transform infrared spectroscopy, and ultraviolet visible spectroscopy. The influence of pH on hydrogel swelling behavior over a period of 48 hours was investigated in this study. To verify its efficacy as a targeted drug-delivery agent, PHEMA hydrogel's overall retention and release of iron oxide nanoparticles was measured through dynamic light scattering and ultraviolet visible spectroscopy. The results of this study will be highly significant in the design of higher-efficacy vehicles for targeted drug delivery.

585389 Sustained Delivery of Phenotypic Regulators of T-Cells

T-cells are the critical helper, effector and regulatory cells of the adaptive immunity and are essential for life. However, defects in their number and function can contribute to tissue and organ specific inflammatory autoimmune diseases (e.g., rheumatoid arthritis, multiple sclerosis, etc.). A shared feature of T-cell mediated autoimmune disorders is an upregulation of pro-inflammatory T-cell subsets, often accompanied by a deficiency in the number of regulatory T-cells (T_{regs}). T_{regs} lose their capacity to modulate the function of other immune cells and adopt a pro-inflammatory phenotype, secreting pro-inflammatory cytokines. This process is mediated, in part, through the IL-6/IL-6R axis in a pro-inflammatory microenvironment. We hypothesize that the local metabolic regulation of T-cells in affected tissues can restore their numbers and function, thereby mitigating the symptoms and underlying cause of inflammation. To test the hypothesis, we use small molecule immunomodulators (SMIs) of T-cells that have been previously shown to modulate T-cell phenotype in vitro. However, their small size and hydrophobicity precludes efficient delivery and therefore require high doses which contributes to undesired systemic side effects in vivo. To enhance the efficacy of SMIs, we designed a biomaterial-based system to sustain the local delivery of SMIs and suppress autoreactive T-cells and improve T_{reg} function. We developed a polymer hydrogel to retain and release the bioactive SMIs by directly conjugating the SMIs to a biopolymer via an ester linkage in a crosslinked hydrogel. We used spectrophotometry to quantify SMI loading onto the scaffold and cell counting and flow cytometry to determine any expansion and phenotypic changes. We characterized the conjugate using spectrophotometry (λ = 360 nm) and quantified a ~5% degree of substitution of SMI. Furthermore, the SMI-polymer conjugate reduced IL-17$^+$ T-cells by ~15% and expanded IL-10$^+$ T-cells by ~2 fold compared to the controls without any SMI stimulation. The reduction in IL-17$^+$ T-cells was directly proportional to the reduction in IL-6 receptor expression on T-cells. In summary, the SMI-polymer conjugate modulated the activity of autoreactive T-cells and improved T_{reg} function, supporting its utility as a phenotypic regulator of T-cells.

583984 Systematic Exploration of Peg-Grafted Polyelectrolytes for Intracellular Drug Delivery

The cellular and endosomal membranes represent a significant delivery barrier in cytosolically active therapeutics. To address this challenge, several investigators have developed cationic and pH-responsive self-assembling polymeric nanoparticles as a drug delivery vehicle to enhance cellular uptake and endosomal escape of drug cargoes into the cytosol. To date, these approaches have largely leveraged homopolymers or block copolymer architectures – as such, design rules for other polymer architectures remain poorly characterized. For this project, we fill that gap with exploration of a PEG-grafted copolymer system. Libraries of poly[poly(ethylene glycol) methyl ether methacrylate-co-butyl methacrylate-co-(2-diethylamino) ethyl methacrylate] (PEGMA-co-EB) graft copolymers with varying hydrophobic molecular weight, hydrophilic volume fraction and graft lengths was synthesized and characterized. They were then screened for hydrodynamic sizes, endosomolytic activity and drug delivery competency *in vitro*. Promising formulations were applied in a murine B16.F10 melanoma utilizing 2'4'-cGAMP, a STING agonist, as the therapeutic agent and cargo. From our *in vitro* experiments, we discovered that formulations with low hydrophilic volume fraction (W_{PEG} ~ 20%), moderate hydrophobic molecular weight (MW ~ 30 kDa), and short graft lengths (MW_{PEG} ~ 300 Da) self-assembles into larger-sized particles and exhibit superior bioactivity. These formulations, when applied to our *in vivo* tumor model, were able to better slow tumor growth compared to our free drug control. Collectively, our results highlighted design rules for application of a cationic graft copolymer structure for drug delivery, and adding understanding of another tool in the drug delivery arsenal.

586175 Characterizing a Novel Brain-Targeted Polymeric Nanoparticle Platform

Parkinson's Disease (PD) is a chronic neurodegenerative disorder that leads to significant brain damage. Currently approved therapeutics treat symptoms, but do not slow down the progression of disease. There are several drugs that can slow progression, but delivery is challenging due to the blood-brain barrier (BBB). Polyanhydride nanoparticles (NPs) can improve therapeutic efficacy for brain delivery by crossing the BBB and releasing the drug at the right time. The goal of this project is to develop and characterize targeted NP-based drug delivery methods that will slow down the progression of neurodegenerative diseases, such as PD.

There are multiple hurdles to delivering therapeutics to the brain. They need to cross the BBB, target the diseased area of the brain, and often deliver payloads to specific organelles or pathological features associated with cells in this area. Using a novel targeted functionalization polyanhydride NP platform, each of these hurdles can be addressed effectively and in a cascading manner. In order to optimize therapeutic efficacy, however, NP-drug interaction properties need to be well characterized. By better understanding drug-NP interactions, the NP synthesis conditions can be precisely tuned so that the therapeutics can be released at the right time, with the correct dose needed to slow disease progression.

In this study, the interaction of NPs functionalized with a cationic ligand, triphenylphosphonium, known to have BBB-targeting capabilities, with several anti-neurodegenerative drugs was characterized and compared to the same formulations without functionalization. A correlation with drug hydrophobicity was found regardless of functionalization, where more sustained release was observed for hydrophobic drugs over multiple days, while hydrophilic drugs experienced a large "burst" release within the first day with trace amounts released thereafter. However, the difference in encapsulation efficiency (EE) between hydrophilic and hydrophobic drugs was not significant. Using the surfactant span-80 during NP synthesis, however, raised the EE of hydrophobic drugs in non-functionalized NPs but had no apparent effect on functionalized NPs, nor on the overall release kinetics for either formulation, regardless of the encapsulated drug. Lastly, the use of surfactant lowered the EE of hydrophobic drugs in functionalized NPs but had no apparent effect on non-functionalized NPs, nor on the overall release kinetics for either formulation. Ongoing and future studies are focusing more deeply on drug-NP interaction properties, surface charge, and cellular internalization efficiency using neurons, microglia, astrocytes and endothelial cells.

585668 Design & Development of 3D Printed Polymeric Structure for Pharmaceutical(s) Delivery

3D printing as an advanced manufacturing process provides a unique opportunity to design and develop complex structure as opposed to using traditional manufacturing techniques. 3D printing encompasses the benefits: agile and robust manufacturing, reduce the cost, reduction in turnaround times, and reliability. 3D printing of pharmaceuticals is especially advantageous for personalized medicine, making multi-drug pills where current manufacturing processes unable to do that. Traditional pharmaceutical processes and tools for the manufacturing of pills have changed relatively little for over a century. Traditional pharmaceutical manufacturing focus on large-scale batches and little flexibility in the manufacturing process due to the multi-step process. Current direct compression manufacturing process unable to make multi-drug pills due to challenges of multiple active material processing, limitation on compression, and pill size control. As an advanced manufacturing technique, 3D printing has the potential to develop personalized medicine, multi-drug pills. It is an exciting branch of research and it requires research and development to utilize the full potential.

The research goal for this work is the design and development of a 3D printed polymeric structure for pharmaceutical(s) delivery. There are several 3D printing processes available and we will focus on pressure-assisted microsyringe (PAM) printing. PAM is well suited for the pharmaceutical industry due to its capability of handling materials at low temperature. Specific research objectives will be (i) develop a methodology for manufacturing 3D printed structure suitable for pharmaceutical(s) delivery, (ii) study the process parameters (nozzle size, printing pressure) that will affect the printed structure. The research outcome is expected to contribute to 3D printing fundamentals and develop personalized medicine.

587520 Nanoparticle-Mediated Oral Delivery of GLP-1 for the Treatment of Diabetes

Diabetes is a tremendous health epidemic; it is the seventh leading cause of death in the United States, affecting more than 30 million Americans [1]. The vast majority of cases are Type 2 Diabetes (T2D) which is characterized by insulin resistance and high blood sugar levels that can cause severe damage to the eyes, nerves, and organs, as well as increase the risk of cardiovascular disease [1]. A promising treatment for T2D utilizes incretin glucagon-like peptide 1 (GLP-1), which serves to enhance and promote insulin secretion while decreasing glucagon secretion by the pancreas [2]. Currently available treatments deliver GLP-1 via subcutaneous injections, which present complications including patient noncompliance, injection pain, and needle phobia [3]. Therefore, development of an oral delivery method would improve patient compliance and offer a safer, easier way to deliver drugs, improving outcomes and quality of life.

We are developing a novel technique for oral delivery of GLP-1 using PLGA-PEG nanoparticles as a delivery device to remove the barriers caused by daily injections. The major challenges in administering oral therapeutics are protecting the drug against degradation in the harsh environment and fluctuating pH levels throughout the gastrointestinal tract, while facilitating drug delivery across the intestinal epithelial barrier layer.

In this research, we synthesized and optimized the formulation of PLGA-PEG nanoparticles, then encapsulated GLP-1 in PLGA-PEG using a water-in-oil-in-water double emulsion. We characterized the size distribution of the GLP-1 loaded nanoparticles using Dynamic Light Scattering and tested the encapsulation efficiency and drug loading capacity using High Performance Liquid Chromatography. A Caco-2 epithelial cell monolayer was used to perform nanoparticle transport experiments. Transepithelial/transendothelial electrical resistance (TEER) was performed to characterize the cell seeding density and optimize the monolayer's readiness for the transport experiments. Nanoparticles were conjugated with fluorescence and permeability through the cell monolayer was tested. Based on these results, we will move forward with conjugating IgG antibodies onto the nanoparticles to facilitate transport across the intestinal epithelium.

References:

1. *Type 2 Diabetes*. [Web] May 30, 2019; Available from: https://www.cdc.gov/diabetes/basics/type2.html.
2. Trujillo, J.M., W. Nuffer, and S.L. Ellis, *GLP-1 receptor agonists: a review of head-to-head clinical studies.* Therapeutic Advances in Endocrinology and Metabolism, 2015. **6**(1): p. 19-28.
3. Wong, C.Y., J. Martinez, and C.R. Dass, *Oral delivery of insulin for treatment of diabetes: status quo, challenges and opportunities.* Journal of Pharmacy and Pharmacology, 2016. **68**(9): p. 1093-1108.

584933 Modeling the Dissolution of Prednisolone Loaded Polyvinyl Alcohol Films

Prednisolone loaded polyvinyl alcohol (PVA) films were prepared and characterized to find density and glass transition temperature. Drug release and mass loss studies of these films were conducted in order to establish a predictive and tunable model that could be used in order to formulate pills specific to meet unique release requirements. The studies were conducted at different solution volumes (10 mL and 50 mL) at pH 6.8, as well as with and without the surfactant Tween 80 (thought to imitate conditions in the intestines) in order to create a comprehensive drug release system. The addition of Tween 80 allowed for the complete dissolution of the drug in supernatant within 2 hours, while other changes in the surroundings, such as change in volume, and changes to the film, such as film thickness and drug loading, did not largely influence the release kinetics. The experimental results (concentration profile with time) were fitted using the Noyes-Whitney Approximation, a surface erosion drug dissolution model based on Fick's first law, which produces a good correlation ($R^2 > 0.9$) with simultaneous prediction of physical parameters, solubility and lumped rate constant. The developed model and experimental procedures for the release of prednisolone from the PVA films will be significant for future film based drug delivery development and design.

587088 Implantable Air-Brushed Nonfouling Microfiber Mats for Drug Delivery

A significant challenge in the field of biomaterials is the nonspecific adsorption of proteins to implants. Upon implantation, this nonspecific protein adsorption triggers the natural foreign body response leading to encapsulation and failure of the device. Zwitterionic materials are excellent at resisting protein adsorption. For this reason, we are investigating the zwitterionic polymer poly(sulfobetaine methacrylate) (polySBMA). Our goal is to produce nonfouling polymer microfibers by air-brush spraying. We hypothesize that nonfouling air-brushed polySBMA microfiber mats will provide a breakthrough technology for implantable drug delivery vehicles. To date we have explored the influence of spraying pressure, nozzle diameter, distance to collector, polymer molecular weight, and solvent. We have also optimized the use of a photo-polymerization reaction to reduce the water solubility of the resulting microfibers, even though they are sprayed from a water based solution. To date we have demonstrated the formation of water insoluble microfibers with diameters ranging from 2.8 to 43.1 mm. On-going work includes verification of the nonfouling properties and incorporation of pseudo-drug molecules.

587658 Supramolecular Filaments As Inhalable Drug Carriers for Lung Disease Therapies

Drug carrier systems that are capable of increased deposition, prolonged retention, and drug release in the lungs are extremely powerful in the treatment of diseases such as lung cancer, COPD, asthma, and tuberculosis. Supramolecular filaments (SFs) are a class of nanostructures formed by units such as peptide amphiphiles (PAs); peptides conjugated to alkyl chains and assembled through non-covalent forces. SFs can be "loaded" with therapeutic agents and controllably release these drugs upon filament disassembly. We take advantage of this technology to produce biocompatible PA-based drug carrier systems. Here, we provide key insights into PA design parameters necessary to deliver therapeutic agents effectively, a proof-of-concept in jet nebulization's ability to effectively deliver drug-loaded supramolecular filaments, as well as a proposed mechanism of SF breakdown during nebulization.

586577 Targeted Delivery of Parathyroid Hormones for Accelerated Fracture Repair

The goal of this project is to complete the lab-to-clinic translation of a drug that efficiently targets and promotes the rapid healing of fractured bones. The annual frequency of bone fractures in the U.S. is approximately 6.3 Million.1 Costs associated with conventional treatments and subsequent lost productivity carry a major economic burden--a problem compounded by an aging population. This increasing age demographic will continue to raise the frequency of osteoporosis and the associated rate of complicated and life-threatening fractures, with hip fractures expected to increase 160% to 500,000/year by 2040. The total estimated expenses due to hip fractures in the elderly are $20 billion/year in the U.S. alone, with half of patients unable to regain full mobility and a quarter dying within a year from associated complications. We are developing a promising solution to this problem; namely, a potent but nontoxic fracture-targeted bone anabolic agent that is injected systemically but accumulates selectively on a bone fracture surface. The targeted therapy avoids the requirement for invasive surgery and eliminates the danger of ectopic bone growth while improving the rate and quality of bone fracture repair.

583648 Fiber Density Promotes Tip Cell Formation to Initiate Angiogenesis

Angiogenesis is the process of extending microvasculature from pre-existing blood vessels and plays a critical role in development, tissue homeostasis, and disease progression. This process is initiated when normally quiescent endothelial cells lining the walls of blood vessels transition into activated tip cells that can degrade and invade the surrounding extracellular matrix (ECM). Invading tip cells then lead trailing stalk cells to collectively migrate, eventually resulting in the formation of patent neovessels. While many pro-angiogenic soluble factors have been identified, how physical properties of the ECM influence tip cell formation remains less explored. Here, we engineered a microfluidic 3D tip cell formation assay composed of a lumenized endothelialized arteriole-scale channel fully embedded in a user-defined ECM. To examine the influence of matrix fibers on mediating tip cell formation, we integrated fiber-reinforced fibrin hydrogels with this assay. Briefly, linear high-molecular-weight dextran (86 kDa) was functionalized with vinyl sulfone (a UV-initiated photocrosslinkable group) and dissolved at 0.6 g ml^{-1} in Milli-Q water and dimethylformamide with 0.015% Irgacure 2959 photoinitiator. This polymer solution was electrospun into fiber mats and selectively photocrosslinked and processed into fiber segments with tunable diameter, length, and stiffness that were functionalized for cell adhesion with the peptide CGRGDS. Embedding fiber segments within fibrin hydrogels enabled tunability of fiber density independent of fibrin weight % (allowing us to maintain a constant bulk hydrogel density, stiffness, and pore size). We found that increasing the density of fibers enhanced tip cell formation and invasion into fibrin hydrogels. Supportive of an endothelial-to-mesenchymal transition, tip cells in our assay exhibited traits associated with mesenchymal cells including migratory capacity, the ability to degrade the surrounding matrix, development of actomyosin-protrusions, and loss of vascular endothelial cadherin. Taken together, this work investigates how fiber density, an important physical attribute of the ECM, regulates tip cell formation and suggests that a fibrous topography promotes the transition of quiescent endothelial cells towards a mesenchymal phenotype. A greater understanding of the mechanobiology underlying tip cell formation and function is critical to our understanding of angiogenesis and in the future could aid in the development of pro-angiogenic therapies and implantable biomaterials.

584158 Role of Tissue Geometry in Regulating Liver Bud Morphogenesis

Fetal liver development is a multicellular morphogenetic process that transforms an initial hepatic bud into liver tissue. A pivotal point during development occurs when the hepatic bud forms and migrates into the surrounding mesenchyme to begin rapid spatiotemporal liver growth. This particular transient stage of liver development is due to many local biomechanical signaling mechanisms that regulate hepatic architecture and migration. However, complete knowledge of this phenomenon is still insufficient. Through the application of bioengineering principles, current research looks to examine how micro-environmental matrix stiffness stimulates hepatic bud formation and migration into the mesenchyme. Our approach utilizes an in vitro model of the fetal liver bud via 3D spheroidal and toroidal culture of a model hepatic cell line (HepG2) to determine how liver bud geometry influences subsequent cell proliferation. HepG2 cells are seeded into round bottom ULA (ultra-low attachment) plates and microtissue 3D ™ devices to form compact spheroids and toroid tissues (72 hours) respectively. These tissues after harvesting are embedded into varying concentrations of Matrigel (growth factor positive) as a model of the local hepatic mesenchyme to determine an association between matrix stiffness and cellular migration. Preliminary data suggests a strong correlation between tissue geometry and migration of cells in varying concentrations of matrix. This project proposes new insights into the biomechanical mechanisms that promote liver development.

585730 Salt Loading Effects in Cathepsin B KO Mice on Renal Enac and Blood Pressure

Cathepsin B is among a myriad of proteases that cleaves and activates the renal epithelial sodium channel (ENaC). ENaC expressed in the distal tubule and collecting duct cells of the kidney play a key role in sodium balance and blood pressure regulation. Salt loading results in the development of hypertension in the salt sensitive kidney. We investigated the mechanism by which salt loading regulates the activity of Cathepsin B and ENaC in the kidney. We hypothesized that Cathepsin B knock out mice develop hypertension after salt loading, but the effect is attenuated when compared to wild type mice. We examined ENaC subunit protein expression in kidney cortex lysates and native tissues by Western blotting and immunohistochemistry, respectively. We performed activity assays for cathepsin B and other proteases that cleave and activate ENaC including furin and kallikrein. We measured urinary sodium and potassium levels using a SmartLyte analyzer to correlate changes in blood pressure in these mice. Taken together, our data suggest cathepsin B plays an essential role in regulating ENaC activity and blood pressure, but other proteases contribute to this regulation.

584510 Engineering *Escherichia coli* for Methanol-Dependent Growth and Production

An abundance of methane from natural gas reserves has raised interest in engineering synthetic methylotrophs that can utilize methane or methanol, the oxidized form of methane, for growth and chemical production. Synthetic methylotrophs are non-methylotrophic platform organisms that have been engineered to utilize reduced C1 compounds. Since native methylotrophs lack many of the well-developed genetic engineering tools that platform organisms possess, the development of robust synthetic methylotrophs is of considerable interest. Our lab previously developed a synthetic *Escherichia coli* methylotroph that consumes 1 mol of methanol with 1 mol of glucose concurrently, making this strain "methanol-dependent." This allows the sugar feedstock cost to be reduced by supplying a portion of the carbon substrate as methanol, which is less expensive. Using adaptive laboratory evolution and process optimization, we report improvements to growth and acetone production. One such evolved strain exhibited robust methanol-dependent growth in glucose minimal medium, whereas negligible growth was observed using the ancestral strain. To improve acetone production (from acetyl-CoA via enzymes from *Clostridium*), yeast extract was supplemented to this minimal medium to provide amino acids for biomass synthesis, allowing methanol and glucose to be used primarily for acetone production. As expected, our data demonstrate that acetone production increases with increasing yeast extract concentration (from 2 to 6 g/L). Future work will involve scale-up to a recently developed cell-recycle bioreactor system for high-titer acetone production at bench-scale. This work expands the knowledge of the synthetic methylotrophy field by highlighting the importance of using reduced C1 compounds, alone or together with sugar-based feedstocks, for the production of liquid fuels and chemicals.

584546 Optimizing Production of EV71 Virus-like-Particle from CHO Cells

EV71 virus-like-particle (VLP) mimics the enterovirus causing hand-foot-and-mouth disease (HFMD), an illness that is prevalent in east and southeast Asia and currently has no specific prevention method or treatment. The EV71 VLP has been shown to elicit the immune response necessary for a promising vaccine. To test the VLP's feasibility as a vaccine candidate for HFMD, large quantities of it must be isolated. Chinese hamster ovary (CHO) cells transfected with the EV71 VLP vector used are a unique system in that the cells are capable of continuously producing EV71 VLP without inducing cell lysis, a common problem in cells transfected with other variations of the VLP vector. To optimize EV71 VLP production further, we conditioned expressing CHO cells with dimethyl sulfoxide (DMSO) or sodium butyrate (NaB) and observed differences in EV71 VLP amounts qualitatively via western blot imaging. Additionally, we performed microtiter BCA assays for total protein quantification to see if this method could be used to quantify the differences in VLP production.

The optimal percentage of DMSO for small volume plate cultures (3ml of media) was found to be 2.0%, and the optimal concentration of NaB for 3ml and 10ml plate cultures was found to be 1.0mM. We concluded the BCA assay is an inappropriate method for quantifying VLP production due to its inherent relativity and lack of specificity for the VLP. In the future, obtaining EV71 VLP may be further improved by conditioning simultaneously with DMSO and NaB, by developing a cost-effective method for specifically quantifying EV71 VLP, by scaling up production with CHO cells to larger volumes or suspended culture, and by optimizing purification of the VLP via ion exchange chromatography.

584550 Utilizing Endotracheal Tubes to Modulate Particle Deposition Profiles in a 3D-Printed Lung Model

Design of therapeutics for aerosol delivery relies heavily on the use of deposition profiles of healthy, male patients to predict drug performance and efficiency. Delivery of aerosolized therapeutic using a device such as a nebulizer may be needed while a patient is intubated; however, it is unknown whether the presence of an endotracheal (ET) tube alters the deposition profile of a drug in the lungs. Our objective is to investigate the differences in airflow through the lungs in the presence of an ET tube and determine whether these effects can be tuned through altering parameters such as tube diameter, cuff inflation, and release outlet location to inform the formulation and administration of aerosolized therapeutics. By leveraging our 3D-printing capabilities, we can quantify particle deposition for each of the five lung lobes in an anatomically accurate model derived from patient computed tomography (CT) scans printed using the Carbon M1 3D-Printer. ET tubes with inner diameters (ID) of 7, 8 or 9 mm were compared to a clinical flow profile of a healthy male without an ET tube. Additionally, ET tube depth and cuff shape were altered to skew deposition toward one side of the lung. We found that, while altering depth had little effect on the deposition profile, altering ET tube ID did have a significant impact with the 9 mm ET tube skewing 74% ($p = 0.001$) of the fluorescent polystyrene 1 μm diameter particles delivered towards the right lung. A similar effect was achieved by altering the inflation of the cuff to direct 70% ($p = 0.03$) of the delivered particles to the right lung. These effects were also investigated at various flow rates to provide information on the optimal point of release in the breath cycle. Overall, the ID and orientation of an ET tube can have a significant effect on the efficiency of the delivery of aerosolized therapeutics to a specific region of the lung and should be considered when designing pharmaceutics and dosing for intubated patients.

584574 Rewiring Amino Acid Biosynthesis Via Regulator Modulation and Operon Overexpression Improves Methanol Utilization in a Synthetic *Escherichia coli* Methylotroph

The abundance of natural gas has made reduced one carbon compounds like methane and methanol attractive feedstocks for commodity chemical and fuel production, particularly in bioprocesses where they can replace more expensive sugar substrates. While many organisms natively consume one carbon substrates, these microbes (known as methylotrophs) lack many of the well-developed genetic engineering tools that platform organisms possess, which limits their production capacity. Engineering methylotrophy in industrially popular organisms is a promising alternative – in particular, progress has been made to engineer *Escherichia coli* to consume methanol. However, to date, no *E. coli* strain has been developed that can grow on methanol as a sole carbon substrate. We have identified amino acid biosynthesis as a crucial bottleneck to sole methylotrophy using ^{13}C labelling – while most metabolites in glycolysis and the TCA cycle show substantial methanol incorporation, labelling in amino acids is significantly lower, and many amino acids lack methanol carbon entirely. Starting with an *E. coli ΔfrmA* variant expressing heterologous ribulose monophosphate (RuMP) pathway enzymes, we achieved greater methanol incorporation into amino acids using a variety of metabolic engineering strategies. Improvements were made by modulating regulators of amino acid biosynthesis, overexpressing enzymes for the biosynthesis of specific amino acids, and utilizing enzyme variants that do not possess traditional allosteric inhibition. The best variants from each method were then combined, yielding additive benefits to methanol incorporation. The resulting strains represent unique starting points for achieving sole methylotrophy via adaptive evolution or further metabolic engineering.

584944 Investigating Factors Affecting Oxygen Reduction Reaction on Boron Doped Diamond Electrode and Its Possible Effects on Neural Stimulation

The oxygen reduction reaction (ORR) in neural stimulation electrochemically generates reactive oxygen species, which if diffused away into the implanted tissue, might cause tissue damage. Therefore the ORR should be as minimized in neural stimulation (JT.Morton, 1994; S.F Morgan, 2010; EM Hudak, 2017). Boron-doped diamond electrodes have attracted considerable interest as a bio-interfacing electrode material, because of their biocompatibility, wide water window, and electrochemical stability; they also have been reported to have slow kinetics for the ORR. However, few investigations have focused on how its sp^2 carbon content, or non-diamond content, can contribute to oxygen reduction. Moreover, ultrananocrystaline diamond (sp^2 abundant) was recently reported in neural interface applications, justifying the need for further understanding the effect of sp^2 carbon.

In this work, we investigate how sp^2 coverage affects the ORR on polycrystalline diamond film electrodes, and also how other factors such as anodic oxidation increase ORR activity. Electrochemical redox activity of $Fe^{2+/3+}$ and Raman spectroscopy were used to characterize sp^2 content, and X-ray Photo electron spectroscopy to probe surface oxygen functional groups, before and after anodic oxidation. The electrode's cathodic charge storage capacity (cCSC), as a measure of the relative contribution of oxygen electrochemistry in charge injection, was compared with cyclic voltammetry (200mV/s scan rate) in N2 and O2 saturated phosphate buffered saline (PBS). Preliminary results showed that diamond electrodes containing surface sp^2 carbon showed more activity towards oxygen reduction than those containing less sp^2 carbon. In addition, we found that if sp^2 containing electrode was anodically oxidized by potential scanning in dilute H_2SO_4, the cathodic current (presumably oxygen reduction current) increased. The increase in ORR activity after anodic oxidation could be explained by surface chemistry change or increased surface roughness (thus effective surface area). Since anodic polarization can oxidize an as-grown, primarily hydrogen-terminated diamond surface, and reaching anodic potentials during neural stimulation is inevitable, the effect of anodic oxidation should also be considered when evaluating the effect of sp^2 carbon. In the future, possible products of ORR will be investigated, and the ORR process investigated using faster scan rates to match the time scale of neural stimulation methods.

585047 Critical Role of Lipin-2 in the Development of Chronic Multifocal Osteomyelitis

Chronic Recurrent Multifocal Osteomyelitis (CRMO) is an autoinflammatory bone disorder that causes patients to exhibit chronically active or recurrent sterile bone lesions at multiple sites. The chronic multifocal osteomyelitis (cmo) mouse carries a mutation in the *Pstpip2* gene and develops IL-1b mediated sterile bone disorder reminiscent of human CRMO. On the other hand, mutation in the *LPIN2* gene that encodes for , a vital molecule in lipid metabolism, is known to cause an autoinflammatory disease called Majeed syndrome which is characterized by CRMO, dyserythropoietic anemia, and may have neutrophilic skin inflammation. Previous reports have shown that cholesterol levels in the cells regulate IL1b production by the Lpin2 deficient immune cells. To investigate the cholesterol synthesis in the bone marrow cells from cmo and Lpin2 deficient mice, expression of genes involved in cholesterol biosynthesis was performed using quantitative PCR method. We found that the expression of several cholesterol biosynthesis pathway genes is attenuated in cmo mice-derived cells, suggesting that difference in the levels of cholesterol in the cells might be involved in the susceptibility/resistance from the cmo disease.

584906 Development of *Mycobacterium Smegmatis* As a Host Strain for Polyketide Synthase Production

Polyketide synthases (PKSs) are enzymes that synthesize a biologically important family of complex chemicals called polyketides, which include antibiotics, anticancer, and immunosuppressant drugs. Due to their highly complex architectures, polyketides are difficult to synthesize via traditional organic chemistry means. Through bioengineering, scientists are able to genetically modify PKSs to selectively produce polyketides of interest such as adipic acid, methyl ketones, and chiral building blocks. However, commonly used heterologous hosts for PKS production, such as *Escherichia coli* and *Streptomyces,* often produce misfolded, truncated, and non-functional PKSs or they can be challenging to genetically manipulate and grow slowly. In response to this need, we seek to develop *Mycobacterium smegmatis* as a new PKS host due to its more easily manipulated genome, availability of genetic tools, relatively fast growth rate, and ability to naturally produce precursors for biosynthesis. To assay for phosphopantetheinylation, an essential post-translational modification of PKSs made by phosphopantetheinyl transferase (PPTase), we introduced the gene *bpsA*, which encodes a blue-pigment indigoidine synthetase, into wild type *M. smegmatis*. We found that wild type *M. smegmatis* produced indigoidine, which indicates the presence of an active PPTase. Furthermore, we genomically integrated a second copy of a PPTase under the control of a constitutive promoter to improve the ability of the host to phosphopanteheinylate heterologous PKSs and integrated T7 RNA polymerase to make the host compatible with IPTG inducible T7 promoter expression systems. Additionally, we developed an IPTG inducible T7 promoter expression plasmid, containing a his-tagged RFP under an inducible T7 promoter for protein purification. Flow cytometry was used to study the inducibility of the system by measuring fluorescence levels of induced and non-induced *M. smegmatis* containing the T7 expression plasmid, further providing evidence that these genetic tools will be a useful platform for purifying PKSs or for fermentative production of polyketides.

587911 Metabolic and Regulatory Profiling of *Caldicellulosiruptor Bescii* reveals Promising Avenues for Lignocellulose Decomposition Optimization

Lignocellulose is the most abundant terrestrial carbon source, providing an attractive alternative to petroleum-based industrial chemicals. Harnessing the native cellulolytic ability of microorganisms is a promising avenue for industrial chemical production from renewable plant biomass; however, the need for pretreatment of lignocellulose limits its use in industrial processes. *Caldicellulosiruptor bescii,* a Gram-positive, anaerobic hyperthermophile growing optimally at 78°C, can deconstruct unpretreated lignocellulose and ferment both C5 and C6 sugars in a single step, making it a strong candidate for biologically mediated consolidated bioprocessing (CBP). While some chemicals such as ethanol have been demonstrated in *C. bescii* at moderate titers[1], a deeper understanding of the physiology and regulatory networks of *C. bescii* is critical for the development of a scalable industrial process. Through metabolic reconstruction, biosynthetic pathways can be validated, prioritized, and optimized to redirect lignocellulosic sugars to target products. Here, we employed continuous culturing, transcriptome analysis, and *in vitro* characterization of regulatory proteins to investigate fermentation profiles, sugar uptake, and metabolic regulation. Fermentation profiles from continuous cultures of *C. bescii* on 5 g/L glucose or xylose at a dilution rate of 0.2 hr^{-1} show nearly the same theoretical yield of acetate (56.1% xylose, 55.7% glucose); however, glucose fermentation produces 4-fold more lactate. Because lactate production in *C. bescii* fermentation is due to buildup of H_2, we investigated the influence of agitation rate on lactate production. Increased agitation rate from 150 rpm to 400 rpm led to a 10-fold reduction in lactate, a 1.7-fold increase in acetate, and exhaustion of carbon in the medium. By tuning the reactor operation, product yield and sugar consumption can be further controlled. Preliminary transcriptomics comparing glucose and xylose fermentation reveal upregulation of genes responsible for xylose transport and metabolism representing several distinct regulons, necessitating a robust metabolic reconstruction for sugar uptake and metabolism to account for crosstalk between pathways. Through *in vitro* characterization of *C. bescii* xylose metabolism regulators, further understanding of regulator binding and affinity will be determined, which is essential for the development of efficient, biologically mediated biofuel production.

References:

1. Williams-Rhaesa et al. Engineering redox-balanced ethanol production in the cellulolytic and extremely thermophilic bacterium, *Caldicellulosiruptor bescii, Metabolic Engineering Communications,* Volume 7, 2018

587920 Engineering a Split Ubiquibody for Inducible Control of Protein Interface

Similar to humans throwing away trash, the cells in our body need to discard proteins that aren't being used to prevent buildup. This process of protein degradation is accomplished through the ubiquitin-proteasome pathway (UPP), which tags the proteins to be degraded with a marker called ubiquitin (Portnoff, Stephens, Varner, DeLisa 2014). This pathway has been harnessed by several techniques to reduce levels of selectively targeted proteins by providing a method of eliminating proteins in the cell that are stable, but might be detrimental to cell function and life. Such techniques successfully enable degradation of targeted proteins, but are not modular, which requires an optimization for each new target protein. The goal of this project was to engineer a split ubiquibody, where the two ubiquibody domains (catalytic and binding) are expressed separately and dimerize in the presence of a chemical rapamycin. The results yielded a successful engineering of a split ubiquibody fused with either the FRB and FKBP protein. Before testing the dimerization of the two split complexes, the individual expression of each construct was tested to ensure that the individual complexes themselves could express well in mammalian cells. When testing these split ubiquibodies for expression using a Western Blot using both the FLAG and HIS tag, the complexes containing the binding domain expressed well consistently; however, catalytic domain had low expression indicating that the structure itself was not yet optimized for dimerization of the overall complex. To address this problem in the future, the complex with the catalytic domain is being redesigned with a linker and different orientations to improve stability and optimize expression.

583973 Elucidating Metabolism of *E. coli* double Knockout Strains

Determining the metabolic effects of gene knockouts in workhorse microorganisms such as *Escherichia coli* is an important topic for metabolic engineering since it allows for the elucidation of kinetic limitations within those organisms that are essential to understand when designing them in industrial contexts. Most studies in the past have focused on strains with single gene knockouts; this project extends this analysis to double knockouts, which are significant since they allow for the restriction of fluxes to one of the three glycolytic pathways present in *E. coli* and the complete elimination of reactions catalyzed by multiple isozymes. In this study, adaptive laboratory evolution (ALE) was performed on Δ*pgi*Δ*gnd*, Δ*pgi*Δ*edd*, and Δ*rpiA*Δ*rpiB* double knockouts for around 1 month to maximize their growth rates on minimal medium with glucose. At the end of this time, they were grown with ^{13}C-containing glucose tracers, their amino acids and carbohydrates were collected and analyzed with GC-MS, and the isotopic labelling implied by the GC-MS data was used to reconstruct metabolic fluxes using the Metran metabolic flux analysis software. A comparison was made between the glucose uptake-normalized and absolute fluxes of the Δ*pgi*Δ*edd* strain. Future work will include continuing this ALE for a longer time period and with more sets of gene knockouts, as well as sequencing the genomes of the evolved cells to determine which mutations allow them to grow faster.

587938 Design Principles of Microparticle Backpacks for Control of Cellular Phenotype

An emerging platform for immunotherapy is the use of adoptively transferred cells of the innate immune system, such as macrophages, which can affect diseased tissues in an antigen-agnostic fashion. Macrophages display the most phenotypic plasticity out of any circulatory cell, which allows them to perform several related and often opposing biological functions, including promoting and suppressing inflammation, wound healing, and participating in host defense. However, the very plasticity that makes macrophages a versatile candidate for immunotherapy also hinders their ability to perform an intended function once transferred to the patient, as they can shift their phenotype in response to conflicting signals in the surrounding microenvironment. To address this issue, we have developed a disc-shaped microparticle "backpack" that is capable of attaching to macrophages and regulating their phenotype through the release of immunomodulatory cytokines. Backpacks are fabricated via microcontact printing of multiple layers of biodegradable polymers, which include a cell-binding layer and a cytokine-encapsulating layer (e.g., interferon-gamma, IFN-g). The thickness of the cytokine-encapsulating layer was tuned to balance stability and activity of IFN-g with microcontact-printing efficiency. The stiffness and size of the backpacks were confirmed with atomic force microscopy. Once fabricated, we show that the backpacks are able bind to the surface of primary murine-derived macrophages for at least 5 days. We also show that they release IFN-g over several hours to polarize the cells to which they are bound to a desired phenotype. Overall, the ability to control cellular phenotypes may allow for new cell-based therapies for a variety of immune-regulated conditions, including atherosclerosis, wound healing, and cancer.

586035 Elucidation of the Ochratoxin A Biosynthetic Pathway via Heterologous Expression in Aspergillus nidulans

Ochratoxin A (OTA) is a secondary metabolite produced by the *Penicillium* and *Aspergillus* fungi genera. OTA is a food toxin that has been shown in previous studies to be carcinogenic, immunotoxic, and genotoxic, and has been found to be produced under shipping conditions in cereals and fruits. Consensus biosynthetic pathways have been found in *Aspergillus carbonarius*, but biosynthetic pathways in other species have not been verified. The current biosynthetic pathway utilizes polyketide synthase (PKS), non-ribosomal peptide synthetase (NRPS), Cytochrome P450 monooxygenases and halogenase enzymes to synthesize OTA. We first verify the biosynthetic pathway of OTA via heterologous expression in *Aspergillus nidulans*. Our experiments yield the OTA intermediates 7-carboxymellein and 7-methylmellein via liquid chromatography-mass spectrometry (LC-MS), indicating that the proposed biosynthetic pathway failed to yield ochratoxin B at the peptide sythetase step. The absence of phenylalanine ethyl ester indicates either the failure of phenylalanine to esterify properly or the absence of non-ribosomal peptide synthetase for lack of transcription factors to activate the gene encoding NRPS. Transcription factors were found in *Aspergillus carbonarius* and OTA was heterologously expressed again in *Aspergillus nidulans.* Future directions include investigation of the possibility of hybrid PKS-NRPS genes in OTA production. Furthermore, characterization of the domains of PKS and NRPS can elucidate the mechanisms by which the 7-methylmellein and ochratoxin B is formed, respectively. Upon elucidation of these mechanisms, possibilities of gene regulation via gene therapy or similar approaches can be used to mediate OTA production in food production and transport.

587945 The Role of dach1 in *De Novo* Formation of Collateral Arteries in the Uninjured Heart

Coronary artery disease (CAD), is the leading cause of death in the US, accounting for 1 in 3 deaths annually. CAD occurs when coronary arteries (CAs) that supply the heart with oxygen become diseased. Collateral arteries, vessels that can interconnect CAs independent branches, can re-supply oxygen to the deprived muscle tissue, thereby maintaining heart function. Understanding the molecular factors involved in the development of the CAs and collaterals can lead to potential treatments for CAD. The transcription factor, DACH1, is necessary for CA development. In order to evaluate the role of DACH1 in collateral artery formation, we used an inducible transgenic murine line where DACH1 has been over-expressed (OE). Mice were induced at day 5 and processed for immunostaining at day 11 and week 12. iDISCO, a whole-mount immunolabeling clearing technique, and ImageJ were used to visualize, trace, and quantify the number of collaterals. We observed a slight decline in the number of collaterals in the OE vs WT at day 11 and expect this trend to repeat in the adults. Interestingly, a preliminary investigation on post-injury collateral formation yielded higher numbers in WT mice than OE. A qualitative analysis of the hearts suggest that the OE mice have more branches at the tips of the arteries. This indicates that the OE model may have a higher supply of oxygen to the heart and is a potential reason for the lower collateral count. In conclusion, this work brings us closer to understanding DACH1's role in collateral artery formation.

587625 Development of an in Vitro Model for Extravillous Trophoblast Invasion over a Gradient

It is currently difficult to establish the effect of external stimuli on extravillous trophoblast (EVT) invasion. Improper invasion of EVTs can cause problems with pregnancy, including the most common pregnancy complication, preeclampsia. Microfluidic devices are an option of modeling this early invasion; however, most trophoblast labs are not equipped to produce these devices through lithography. We are developing a device that will model the invasion of EVTs over gradients of maternal stimuli. We established novel methods to create the PDMS structure of the device, insert a 1 cm matrix to model the maternal decidua, establish a gradient that can be consistently confirmed, and track the movement of EVTs in 3 dimensions over time. This device will allow us to establish the effect of environmental factors, drugs, and other molecular regulators of placental development on EVT invasion.

586384 Determining the Effect of Zygotic Cactus on the Dorsal/Cactus Feedback Loop

Feedback loops in biology help regulate cell signaling, cell responses, and cell differentiation. The basis for differentiation lies within cell-to-cell communication through signaling. The signal of interest in this experiment is Dorsal (a NF-κB homolog). It was hypothesized that Cactus, an inhibitor of Dorsal, provides a negative feedback loop to ensure robustness of signaling during differentiation. Cactus binds to Dorsal to create a complex and prevent Dorsal from entering the nuclei to regulate target genes. The goal was to determine the role of zygotic Cactus in the fruit fly *Drosophila melanogaster*. To test whether Cactus disrupts the Dorsal gradient, embryos lacking zygotic Cactus, df(cact), were compared to the control wild type, yellow-whites. All sets of embryos contained maternal Dorsal. Embryos were collected, fixed, and stained with fluorescent antibodies. They were visualized with confocal imaging to measure fluorescence of the tagged genes. Matlab was used to analyze the images. It was found that the embryos had a high population around s=0.15, but other populations could be present. The best fit values for ß, ϒ, κ, a, and delta in the modeling equation were found to be: 0.0789, 1.7258, 1.2505, 11.7598, and 0 respectively. More data is needed to make more conclusive results.

588008 Introduction to Research through Protein Expression and Genetic Engineering of a Zwitterionic Amino Acid Fusion Protein

This project served as an introduction to the research process and the scientific method with hands-on application. Primarily aimed at cadets in their freshman and sophomore years. The vehicle for this project was protein expression in E. coli. The target protein was the nerve agent bio-scavenger, organophosphate hydrolyze (OPH). Previous studies demonstrated OPH's broad substrate affinity for organophosphate compounds, such as, G-type nerve agents as well as its ability to enzymatically neutralize organophosphates (Liu, E. et al). Organophosphates are found in pesticides and nerve agents, and when exposed to the human body, inhibit the production of acetylcholinesterase causing overall action on the parasympathetic nervous system that leads to failure of neuron function (Liu, E. et al; Lerro, C. et al). Bio-scavengers like OPH have the potential to act as a protective protein. Obstacles to this include immunogenicity issues, short circulation time and low expression yield of genetically engineered fusion protein variants. Addressing these issues and building on previous studies, a genetically engineered, zwitterionic, amino acid, fusion protein was designed and initial expression results compared to the wild type OPH control.

588047 Metabolic Engineering of Yeast for Production of Biopolymers

Bacteria are capable of making exopolysaccharide biopolymers like xanthan gums and hyaluronic acid, which can serve roles in food and health industries[1]. Making the sugar nucleotides that are precursors for xanthan gum and hyaluronic acid require purifying bacterial enzymes and use of cofactors that can make this process relatively inefficient and costly[2]. The goal of this research is to work towards making a yeast platform to serve as a whole cell catalyst that can make biopolymers such as hyaluronic acid.

The first goal of this research was to overexpress the enzymes in the yeast *Pichia Pastoris* that can convert glucose into UDP-glucose, GDP-mannose, and UDP-glucuronic acid. Then, genes necessary for producing the biopolymers of interest from this increased pool of precursor sugar nucleotides can be added to the yeast DNA. Purification methods will then be used to purify the biopolymers made by the yeast.

588102 Transport of Human Milk Peptides Derived from in Vitro and In Vivo Digestion across a Model of the Human Intestinal Epithelium

Proteins in breast milk not only serve as nutrients for the infant, but also exert health beneficial properties that are necessary for the body's physiological functioning. Proteins that contain chains of peptide sequences that are necessary for body functioning, beyond nutritional value, are known as bioactive peptides. There is still little knowledge on the effect of specific bioactive peptides on the body. Therefore, the objective of our study is to analyze the transport of peptides derived from in vitro and in vivo digestion of human milk proteins across a cell model of human small intestinal epithelium layer.

Human milk was firstly under in vitro digestion and in vivo digestion in infants. The digested samples were collected and peptides were extracted from the sample. The extracted peptides were incubated with differentiated Caco-2 cell monolayer at the apical compartment of the insert. The peptides absorbed to the basolateral compartment were determined by mass spectrometry. This study will help assess Caco-2 permeability, along with which peptides will go into the bloodstream after digestion. For this study, it was hypothesized that human milk peptides released from in vitro digestion can be absorbed by human epithelium cell monolayer. Once we know more about the transport of proteins through digestion, the scientific community can further the understanding of how bioactive peptides work in the body. With the results of the experiment, there can be more research in discovering specific bioactive peptide sequences that can provide immune protection and reduce the risk of chronic diseases.

584060 Scaling Down DNA-Based Information Storage

DNA offers extremely dense and long-term information storage capabilities, enabling the ability to accommodate growing data storage needs as a potentially transformative archival medium. Recent advancements in DNA synthesis and sequencing have created the potential for storage systems of useful sizes[1,2]. Here, we focus on miniaturizing the storage of DNA files through the development of a microfluidic device capable of selective file separations[3,4]. Our design creates polydimethylsiloxane (PDMS) microfluidic reaction chambers to extract specific files using chemical handles and modified magnetic beads. We optimize the system for bead mobility across an incorporated air gap that allows for an effective physical separation of the desired file. To visually demonstrate strand retrieval, we capture 20 nucleotide (nt) fluorescently labeled primers and track them using fluorescence microscopy. We then expand to double stranded DNAs with variable lengths and show successful and specific file recovery by fragment analyzer and qPCR analyses. After file separation, we determined that the remaining solution is unaltered and can be recycled to the storage library for future use. This system provides potential physical designs for automatable and reusable DNA storage devices.

References

1. Takahashi, N. C., *et al.* Demonstration of End-to-End Automation of DNA Data Storage. *Scientific Reports.* **9,** 4998 (2019)
2. Regalado, A. The DNA data storage machine that's the size of a school bus. *Technologyreview.com, MIT Technology Review* (2018)
3. Tomek, J. K., *et al.* Driving the Scalability of DNA-Based Information Storage Systems. *ACS Synthetic Biology.* **8**, 6, 1241-1248 (2019).
4. Lin, K., et al. Dynamic DNA-based Information Storage. Biorxiv (2019).

588387 Proteins Adsorbed to Fractal Aggregates for Enhanced Stability

Pharmaceutical companies manufacture medicines to remain stable and potent at their industry standard. The loss of potency during the storage and transport phases of these manufactured products negatively influences their efficiency and safety by compromising the structure of the proteins. In order for the medicines to retain their proper functionality during storage and transport, it is recommended that they are stored in conditions between 2°C and 8 °C (36°F and 46°F). Many companies use cold chains in order to keep these products within this specific temperature range during storage and transport. However those that cannot afford or do not have the technology to consistently store medicine within this temperature range lose the functionality of their medicines. A way to preserve functionality in these high heat and intense shaking conditions is to adsorb proteins onto silica fractals. Through the use of silica nanospheres, the proteins were able to be adsorbed onto the surface of the structures. Silica, which is a compound known for its heat resistance and fairly high melting point is an ideal candidate to assist in protecting the protein structure. The protein in use, Bovine serum albumin (BSA), is known to be highly adsorbent in nature. By adsorbing onto the silica fractals, the structure of the protein will become more stable while also increasing the protein's temperature resistance.

587513 Synthesis and Characterization of Novel Phospholipid-Conjugated Topotecan Prodrugs for Lipid Vehicle Delivery

Current anticancer therapies face the innate challenge of proper targeting and dosing of chemotherapeutic agents. Low selectivity is a characteristic trait of many existing drugs, which results in unintended damage to healthy tissue alongside cancerous tissue that may cause dangerous short and long-term side effects. Our mission is to modify commonly used anticancer drugs (cytarabine, gemcitabine, topotecan, and others) to increase their selectivity through incorporation into lipid vehicles for ultrasound-mediated delivery. Currently our focus lies on topotecan—a topoisomerase inhibitor used in the treatment of solid-tumor cancers—onto which phospholipid tails were conjugated to form a prodrug that can be loaded into lipophilic environments without premature leaking. We hypothesize that topotecan is then released from the prodrug at the treatment site via non-specific esterase cleavage. 2T-T ("Two-Tailed Topotecan") prodrug was successfully synthesized by conjugating phospholipid tails to the parent drug via Steglich esterification. The resulting compound was purified with chromatographic separation, producing a yield of 24%, and characterized with 1H and 13C Nuclear Magnetic Resonance Spectroscopy (NMR). The prodrug was then loaded into liposomes and tested for biological activity via MTT cell proliferation assay. Biological effectiveness was verified by comparing 2T-T activity to the activity of parent topotecan. 2T-T showed comparable IC50 values to the parent drug at the nanomolar level, supporting our hypothesis that topotecan is released from 2T-T prodrug in its biologically active form at the cancer site. Future work includes further characterization as well as improvement of yield and biological activity. Drug modification such as this paves the way for future anticancer drugs with vastly improved safety and effectiveness.

588141 DHL Family of Novel Anti-Cancer Drugs: Synthesis, Characterization and Activities

Cancer is the second most common cause of death in the United States with nearly 600,000 deaths a year. Novel anticancer drug researchers have been looking to naturally occurring molecules in order to combat this deadly disease. In particular, 2-amino pyrroles have been identified as very important biologically active scaffolds for the development of original anticancer, antibacterial, and antifungal drugs. Using aryl-substituted pyrrolines as a starting material, a novel anticancer scaffold, DHL, was successfully synthesized and characterized through NMR spectroscopy and X-Ray Crystallography. Changes of aryl aldehydes and sulfonamide were made to synthesize different derivatives, in order to characterize starting structure activity relationships. A total of twenty derivatives now exist within the DHL family. Each derivative was tested on cancer cell lines and activity was determined through MTT Assays. The compound with the best activity has been DHL-16, showing activity of 9.72μM.

588036 Application of High-Throughput Screening Assay to Detect Glycosylated Proteins in *Escherichia coli*

Engineering of glycosylation pathways in bacterial cells has become an important tool for the development of microbe glyco-factories, conjugate vaccines, glycosylated outer membrane vesicles, and analyzing other roles glycosylated proteins play in various organisms. However, the encoding of genes that facilitate this post translational modification has been carried out in *E. coli*, an organism in which the enzyme is non-native. Thus, a robust and effective screening assay is necessary for initial detection of successfully secreted glycoproteins. A novel assay was developed for the high-throughput screening of *N*-glycosylation in *Escherichia coli*: GlycoSNAP (*glyco*sylation of *s*ecreted *N*-linked *a*cceptor *p*roteins). However, since its development, a pathway for *O*-linked glycosylation has been engineered and successfully implemented in *E. coli.O*-linked glycoproteins have a much lower specificity for sugar substrates, thus facilitating a much wider range of potential application. Here, we describe the methodology of GlycoSNAP and its variable components, as well as its potential application in regards to*O*-glycosylation. GlycoSNAP has proven to be a successful high-throughput screening methodology, and is more affordable than its common counterpart, gel electrophoresis followed by Western blot. Its efficiency and modularity can, in principle, facilitate its application in a wide range of glyco-engineered variants.

587310 Upregulation of the Glutaminase II Pathway Contributes to Glutamate Production upon Glutaminase 1 Inhibition in Pancreatic Cancer

The targeting of glutamine metabolism specifically via pharmacological inhibition of glutaminase 1 (GLS1) has been translated into clinical trials as a novel therapy for several cancers. The results, though encouraging, show room for improvement in terms of tumor reduction. Here, we demonstrate that pancreatic tumors upregulate the production of glutamate from glutamine via the glutaminase II pathway upon GLS1 inhibition. Moreover, genetic suppression of glutamine transaminase K (GTK), a key enzyme of the glutaminase II pathway, leads to the complete inhibition of pancreatic tumorigenesis in vivo; unveiling GTK as a new metabolic target for cancer therapy. These results suggest that current trials using GLS1 inhibition as a therapeutic approach targeting glutamine metabolism in cancer should take into account the upregulation of other metabolic pathways that can lead to glutamate production; one such pathway is the glutaminase II pathway via GTK.

587322 Multi-Epitopic Down-Regulating Antibodies for Cancer Immunotherapy

$CD8^+$ effector T cells are the cytotoxic force of the immune system, and their activity is carefully regulated by signaling through their cell surface receptors. The signals delivered through these receptors can be either stimulatory or suppressive; in particular, activation of immunosuppressive proteins on the T cell surface (known as immune checkpoint receptors) dampens T cell activity and leads to exhaustion or anergy. Tumor cells have evolved to exploit this immunosuppressive pathway to evade clearance by the immune system, contributing to cancer progression. To circumvent this issue, antibody drugs have been developed against immune checkpoint proteins to competitively block their activation, allowing for efficient anti-cancer T cell activity. Although this approach has demonstrated clinical success, the majority of patients do not respond to treatment, and most responses are only partial. We aim to improve upon immune checkpoint blockade by engineering novel therapeutics that act through alternative mechanisms. To do this, we engineered bispecific antibodies that simultaneously bind to 2 non-competitive epitopes on the same immune checkpoint protein (termed multi-epitopic antibodies). This design is advantageous because in addition to blocking ligand binding, the bispecific antibody can induce cross-linking of adjacent receptors to induce receptor down-regulation. We used the yeast surface display directed evolution platform and magnetic- and fluorescence-activated cell sorting to isolate novel antibodies targeting immune checkpoint proteins, which were then used as the basis for formulating multi-epitopic antibodies. Preliminary results indicate that these engineered antibodies efficiently down-regulate immune checkpoint receptors, presenting a promising new approach for cancer immunotherapy.

587546 *Examining Siglec Signaling in Alzheimer's Disease*

Alzheimer's Disease (AD) is a form of accelerated neurodegeneration and the most common form of dementia. There are currently few therapeutic options for AD patients, but the immune inflammatory response presents a potential therapeutic target once there is understanding of the underlying cellular signaling networks. We aimed to see if we could identify new targets in the immune system using an untargeted approach to signaling analysis. Using mass spectrometry to evaluate changes in protein phosphorylation of multiple mouse models of AD compared to aged matched controls, we have identified a Siglec signaling receptor that is activated on disease-associated microglial cells. We developed an expression system in BV2 cells with the wild type Siglec-F receptor, as well as inactive receptors where one or both cytoplasmic tyrosine residues have been mutated to phenylalanine. We assayed apoptosis to determine if altering functional tyrosine residues of these receptors will impact cell death and found that cells expressing wild type Siglec-F undergo significantly more apoptosis compared to those with mutated sites. We then treated BV2 cells with SHP099, an inhibitor of inflammatory transcription changes, to identify SHP-2-dependent and -independent signaling networks. This analysis identified other signaling proteins that could potentially be targeted to inhibit apoptosis. From these results, we see a potential to dissect signaling pathways activated by Siglec-F and target distinct phenotypic responses for inhibition. This approach could create a new avenue of therapeutic development for neuroinflammatory diseases.

586756 Selective Precipitation of Haptoglobin from Human Plasma Fraction IV

Haptoglobin (Hp) is the serum protein responsible for scavenging cell-free hemoglobin (Hb), the oxygen storage and transport protein of red blood cells (RBCs). During states of hemolysis, such as extensive blood transfusion, sickle-cell anemia, or Malaria, cell-free Hb is released from RBCs into the blood stream. Cell-free Hb is toxic and can lead to tissue oxidation and hypertension. Yet, when cell-free Hb is bound to Hp, these toxic effects are inhibited, making Hp a promising therapeutic for treating states of hemolysis[1]. Unfortunately, Hp is difficult to produce and expensive. Fortunately, Hp is present within Fraction IV (F-IV), a byproduct of the Cohn process used for purification of human serum albumin[2]. However, F-IV contains various additional serum proteins and insoluble material. This study sought to determine the optimal concentration of ammonium sulfate [(NH_4
selective protein precipitation. Increasing amounts of (NH_4
creating solutions ranging from 0%-70% saturation of (NH_4
centrifuged and the amount of Hp and total protein in the pellet suspension and supernatant were quantified. Total protein was quantified via the Bradford assay. Hp was quantified by reducing samples with known excess of Hb and using the spectrophotometric difference between the deoxygenated Hb-Hp complex and deoxygenated Hb to determine the Hb binding capacity (HbBC)[3]. Initially, samples had a total protein concentration and HbBC of and , respectively. Increasing amounts of (NH_4
linear decrease of the total protein. At 40% (NH_4
solution and more than 90% Hp with HbBC was also present. Increasing saturation to 50% lowered the HbBC in the solution to 2.1% and lowered total protein concentration to 50.7% in solution. Overall, the optimal concentrations for selective precipitation of Hp with (NH_4
separated half of the starting proteins in solution. Furthermore, after resuspension of precipitated Hp, bringing the solution back to 40% saturation could further remove unwanted proteins with minimal [4]. The results of this study will aid further downstream purification of Hp.

588371 Mimicking Molecular Interactions to Understand and Design Protein-RNA Regulation In Vivo

The Csr system is a model post-transcriptional regulatory network in E. coli. Two of its key components are the CsrA RNA binding protein and the CsrB small RNA (sRNA). CsrA is a global regulator protein that binds RNA. It affects the stability and translation of at least 30, but potentially hundreds of mRNAs[1]. On a phenotype-level, the CsrA protein impacts important cellular behaviors such as glucose storage and processing, synthesis of amino acids, biofilm formation, and motility[2]. The CsrB sRNA acts like a sponge and captures CsrA, which impedes CsrA's ability to bind and regulate target mRNAs[3]. Pathogenic bacteria can flourish in host organisms and can cause diseases after entering the human body. In pathogenic species the Csr system has been shown to play a critical role in the virulence gene networks of *Escherichia coli* pathogens and is conserved in other organisms[4]. *Escherichia coli* is a food-borne pathogen that colonizes the large intestine. *Escherichia coli* can cause health issues such as hemolytic uremic syndrome and bloody diarrhea[5]. Understanding the interactions between CsrA-RNA binding sites *in vivo* is significant because studies like these could potentially reveal new drug targets and guide new approaches for the interruption of pathogenic phenotypes.

1. Sowa, G. Gelderman, A. N. Leistra, A. Buvanendiran, S. Lipp, A. Pitaktong, T. Romeo, M. Baldea, and L.M. Contreras, "Integrative FourDomics approach profiles the target network of the carbon storage regulatory system," Nucleic Acids Research Journal, 45(4): 1673-1686 (2017).
2. Masse, Salvail, Desnoyers, Arguin, (2011). Small sRNAs controlling iron metabolism.
3. Vakulskas CA, Potts AH, Babitzke P, Ahmer BMM, Romeo T. (2015). Regulation of bacterial virulence by Csr (Rsm) systems. Microbiol Mol Biol Rev.
4. Christopher A. Vakulskas et al. Microbiol. Mol. Biol. Rev. 2015
5. Islam MS, KrachlerAM. Mechanosensing regulates virulence in Escherichia coli O157:H7. Gut Microbes. 2016;7(1):63–67.

588109 Designing Flavivirus Vaccines By Glycoengineering Protein Antigens

The goal of our work is to develop vaccines which provide broad protection against flaviviruses such as Zika and Dengue viruses. Zika outbreaks have been associated with a large increase in cases of microcephaly and Guillain-Barré syndrome. To date, there is no approved vaccine, antiviral drug, or other modality available to prevent or treat Zika – emphasizing the present need for a vaccine. Recent vaccine efforts have been focused on the Zika virus envelope protein (ZIKV E protein), which plays an important role in viral attachment and entry into host cells. The ZIKV E protein is composed of three spatially distinct structural domains (domain I (DI), domain II (DII), and domain III (DIII)). Previous research has shown that DIII is involved in receptor binding and elicits neutralizing antibodies. Researchers have recently characterized the epitopes for two distinct antibodies that target DIII: ZV-67 and ZV-2; only the epitope for ZV-67 was accessible in the context of the virion and only ZV-67 neutralized ZIKV *in vivo*. Thus, the goal of this study is to design an antigen that focuses the immune response on the ZV-67 epitope on DIII and effectively elicits antibodies that neutralize ZIKV. To achieve this goal, we are tuning the glycosylation of DIII. Glycans can be used as an antigen-modifier to direct the immune response away from or to a specific region of a protein. We hypothesize that shielding non-neutralizing epitopes on ZIKV DIII by glycosylation will direct the immune response towards the ZV-67 epitope to elicit more neutralizing antibodies. We also show how glycoengineering can be applied to other flaviviruses (i.e. Dengue).

584145 CRISPR Screen to Investigate Contributors to Paracellular Barrier Formation *in Vitro*

Barrier-forming cells fulfil an important role in the body by regulating paracellular transport and ultimately protecting homeostasis. To elucidate potential participants in paracellular barrier formation, we performed an arrayed CRISPR knockout screen across 8400 human genes. For the screen, we used Caco2 cells, an immortalized barrier-forming gut epithelial cell line. Our Caco2 cells constitutively express Cas-9 and are robust enough to tolerate high-throughput screening conditions. The Caco2 cells were seeded at a defined density in transfection medium into 96-well plates containing the arrayed gRNA. The cells underwent reverse transfection for 24-hours before Anti-Anti was added to the media to protect against contamination. The cells were maintained in culture for four days before we induced the barrier by switching to Neurobasal media. After six days in Neurobasal media, the cells were fixed and stained for ZO-1, a tight-junction protein that localizes to the paracellular membrane in barrier-forming cells. Each well was then imaged in a Z-stack in four locations with the ImageXpress Micro XL. The screen was performed in two replicates across 109 plates. Images from both replicates were analyzed with MATLAB code to quantify variation in barrier continuity and tortuosity. Positive hits from the screen will be confirmed and tested for barrier permeability in our human induced pluripotent stem cell (hiPSC)-derived Blood-Brain Barrier (BBB) model and a Zebrafish model.

586327 Development of RNAi Tools for Gene Knockdown in *Rhodosporidium Toruloides*

Development of RNAi Tools for Gene Knockdown in *Rhodosporidium toruloides*

August Finke [1], Anne M. Ruffing [1], James Kirby[2], and John M. Gladden[2]

1. Bioenergy and Defense Technologies, Sandia National Laboratories, Albuquerque, NM
2. Biomass Science and Conversion Technologies, Sandia National Laboratories, Emeryville, CA

Rhodosporidium toruloides is a yeast used to produce biofuels due to its fast growth, high lipid production, and ability to use different carbon sources for growth. Experiments were performed to develop and test RNA interference (RNAi) tools that can later be used to increase biofuel potential in *R. toruloides*. RNAi is a biological process that suppresses or knockdowns genes. In this study, a green fluorescent protein (GFP) – bisabolene synthase fusion gene, was randomly integrated into *R. toruloides*. GFP served to measure gene expression. An RNAi construct designed to produce double stranded RNA that knocks down GFP and bisabolene production was randomly integrated into the strain containing the GFP-bisabolene synthase fusion protein. The successful transformants were grown over several days while GFP fluorescence, cell density, and bisabolene concentration were measured to determine the level of knockdown achieved. GFP was measured using flow cytometry; optical density measurements were taken to track growth; and bisabolene production was quantified through the collection of dodecane layers and analysis by gas chromatography. The bisabolene and GFP knockdown exhibited a positive correlation, showing that the RNAi successfully knocked down the whole mRNA that coded for the GFP-bisabolene fusion gene. This RNAi technique can be applied to knockdown genes that would improve biofuel production in *R. toruloides*.

Sandia National Laboratories is a multimission laboratory managed and operated by National Technology & Engineering Solutions of Sandia, LLC, a wholly owned subsidiary of Honeywell International Inc., for the U.S. Department of Energy's National Nuclear Security Administration under contract DE-NA0003525.

583498 KRAS Digital PCR Screening Optimization for the Detection of KRAS Mutations in CTCs and Exosomes in Patients with Pancreatic Cancer

Circulating tumor cells (CTCs) and exosomes can be identified in the peripheral blood of patients with pancreatic ductal adenocarcinoma (PDAC), and can inform cancer diagnosis and treatment effect, thus avoiding an invasive primary tumor biopsy. CTCs are rare cells shed from a tumor and into the blood stream, increasing the risk of metastasis. Exosomes are secreted extracellular vesicles that carry proteins and RNA. For this investigation, samples from patients with PDAC were processed using the Labyrinth, a high-throughput microfluidic CTC isolation device, and exosomes were obtained through ultracentrifugation.

After isolation, we used digital PCR (dPCR) to screen the CTCs and exosomes for the presence of KRAS mutations, which occur in >90% of PDAC primary tumors. dPCR is an ultrasensitive technique enabling single molecule detection, critical in low input samples. Here, we present the optimization and validation of a KRAS mutation detection assay for CTCs and exosomes, which simultaneously screens for the 7 most frequent clinically detected mutations in the G12/G13 region of the KRAS gene. We compared master mixes (Bio-Rad Supermix vs TaqMan Genotyping), varied concentrations of KRAS screening assay primers and probes, tested lower concentrations of cell spike samples to mimic patient samples, and did a comparative analysis of the assay's effectiveness in CTCs and exosomes. We also report our initial findings in patient samples. Preferred results include a combination and balance of high levels of amplification as well as detectable separation between wild type (WT), mutation, and empty PCR reaction droplets. Ultimately, this assay could be used to screen high risk patients for development of PDAC and improve cancer care.

582966 Modulation of Cationic Nanocarrier Properties for Penetration of Blood-Brain-Tumor Barrier and Glioblastoma Treatment

Glioblastoma multiforme is a Grade IV malignant invasive tumor of the central nervous system that has a 3-5% 2 year survival rate. Treatment is made difficult by the inability of chemotherapeutics to effectively penetrate the blood-brain-tumor-barrier (BBTB), a highly selective interface that only allows the intercapillary transport of small (50-100 nm), lipophilic molecules; these properties are not present in modern chemotherapeutics.

Previous research in the Peppas lab has produced pH-responsive polymeric nanocarriers capable of drug complexation and targeted therapeutic delivery. To apply these nanocarriers to glioblastoma treatment, their sizes and surface charges must be controlled at the synthesis stage to fit the required parameters. The purpose of this experiment was to optimize the synthesis of these nanoparticles for glioblastoma treatment applications.

The nanoparticles were synthesized using emulsion polymerization via ARGET ATRP synthesis, a technique that allows for control over chain length and small amounts of catalyst. The organic monomers were emulsified with crosslinker, surfactants, and ATRP reagents and reduced in an oxygen-free environment. They were then purified through centrifuging and their size/charge properties were characterized through DLS and zeta potential techniques, respectively.

Varying experimental factors affected the properties of the nanoparticles. It was found that increased initiator and reducing agent concentrations resulted in a loss of swelling capabilities but increased surfactant concentration led to smaller nanoparticles. The results indicate successful synthesis of the desired nanoparticles, but further experimentation is required to confirm the properties, chemical composition, and reproducibility of these particles as well as their potential as therapeutic agents.

585981 Aggregation of Gold Nanoparticles and Virus Complexes with Osmolytes

Detection of viruses is difficult because they are some of the smallest microorganisms in the world. Typical virus detection techniques utilize expensive reagents and require highly-trained personnel. One alternative method to detecting viruses is through use of citrate capped gold nanoparticles (AuNPs) and osmolytes. When in the presence of osmolytes, viruses and AuNP complexes will aggregate due to hydrophobic interactions, increasing the hydrodynamic diameter of the AuNPs. Size changes can be detected using methods of dynamic light scattering (DLS) and a red shift in ultra violet visual spectroscopy (UV-Vis). The research conducted consists of multiple trials using 15 nm AuNPs to detect the behavior of aggregation with two model viruses.

AuNPs were synthesized by reducing gold (III) chloride trihydrate using trisodium citrate. The size of the synthesized AuNP was confirmed using dynamic light scattering (DLS), which measures size and UV-Vis absorbance, which measures the absorbance of light. Next, the virus was purified via dialysis with 1000 kDa dialysis tubing in 4.3 mM pH 7.2 phosphate buffer (PB) and size exclusion chromatography. Once purified, a 1:10 dilution was performed to obtain 5 different concentrations of virus. Additionally, 1 nM and 1 μM bovine serum albumin (BSA) and 4.3 mM PB alone were used as controls to show that only virus-AuNPs complexes were aggregated by osmolytes. AuNPs were coated by rotating for 20 hours. Once coated, a benchmark measurement was taken before aggregation with osmolytes and salt. To aggregate, the coated samples were spun down, the supernatant was removed and the samples were resuspended in 1 mL of 1 M glycine, mannitol, or sodium chloride. Finally, the aggregated samples were measured using UV-Vis and DLS.

Two viruses were used to test the ability of AuNPs to detect virus. It was hypothesized that a 1:1 ratio of AuNP diameter to virus diameter would provide the best results because it would be easier for similarly sized particles to interact with each other's geometries. Different concentrations of porcine parvovirus (PPV), diameter 18-26 nm, with 15 nm AuNP complexes showed conclusive aggregation when 1 M mannitol was added. Bovine viral diarrhea virus (BVDV), diameter 40-60 nm, and 15 nm AuNP complexes showed concentration dependent aggregation with 1 M mannitol as well. The data provides evidence that the use of osmolytes to induce aggregation of AuNP and virus complexes shows promise as a simple, inexpensive virus detection method.

586016 Improved Specificity and Sensitivity of CRISPR/Cas12a Using Toehold Modified Crrnas

Improved specificity and sensitivity of CRISPR/Cas12a using toehold modified crRNAs

The CRISPR/Cas (Clustered Regularly Interspaced Short Palindromic Repeats/CRISPR-associated) systems, including the most popular CRISPR/Cas9 system, have been used primarily for genome editing applications. CRISPR accomplishes this by utilizing both a Cas protein which acts as molecular scissors as well as a guide RNA which dictates the protein where to cut. Through simple swapping of the guide RNA, different 19-23 base sequences can be targeted for editing. Additionally, a different CRISPR/Cas system based on Cas12a, also known as Cpf1, has been used for both genome editing as well as nucleic acid detection. Cas12a has been recently used for nucleic acid detection by utilizing its single turn-over cleavage of target DNA (cis cleavage) along with a multiple turnover, collateral cleavage of non-specific single stranded DNA (trans cleavage). To quantify the amount of DNA, CRISPR/Cas12a has been combined with Förster resonance energy transfer (FRET) based nucleic acid reporters in two different studies named as DNA endonuclease-targeted CRISPR trans reporter (DETECTR) and specific high-sensitivity enzymatic reporter unlocking (SHERLOCK)[1-3]. Both methods rely on recombinase polymerase amplification (RPA) to amplify the target DNA to achieve the required sensitivity necessary for nucleic acid detection. Furthermore, they are somewhat non-specific.

To address these issues, we proposed making modifications to the 3' end of the crRNA (CRISPR RNA) that could improve the specificity and sensitivity of the DETECTR or SHERLOCK assay. The method we utilized is the addition of nucleic acids on the 3' end of the crRNA forming a toehold loop. We hypothesize that by adjusting the length of toehold sequences, the ratio of trans- to cis- cleavage can be significantly altered allowing for an increase in sensitivity and specificity. By adjusting the ratio of trans- to cis- cleavage the toehold loop can be used for either detection (by increasing trans- to cis- cleavage ratio) or specific cutting (by decreasing trans- to cis- cleavage ratio). This will have huge implications for both disease detection and *in vivo* gene editing utilizing the CRISPR/Cas12a system.

1. Chen JS, Ma E, Harrington LB, et al. CRISPR-Cas12a target binding unleashes indiscriminate single-stranded DNase activity. *Science.* 2018;360(6387):436-439.
2. Gootenberg JS, Abudayyeh OO, Kellner MJ, Joung J, Collins JJ, Zhang F. Multiplexed and portable nucleic acid detection platform with Cas13, Cas12a, and Csm6. *Science.* 2018;360(6387):439-444.
3. Kellner MJ, Koob JG, Gootenberg JS, Abudayyeh OO, Zhang F. SHERLOCK: nucleic acid detection with CRISPR nucleases. *Nat Protoc.* 2019;14(10):2986-3012.

586479 Using CRISPR Cas13a to Combat the Flu Virus at the Genetic Level: A New Approach to Flu Treatment

The influenza vaccine normally provides protection from three to four strains included in the formulation. However, since the flu viruses constantly mutate, the vaccine protection decreases because it can only protect against strains included in the vaccine but not the new ones. As a result, it needs to be reformulated each year to accommodate the new strains. The vaccine is also unable to protect against pandemic strains. As a treatment approach, antiviral medications can be administered to infected patients, but often the strains are resistant to the drugs. The long-term objective of this research is to develop a new flu medication that uses CRISPR technology to target the viral genome. Highly conserved sequences in the genome are targeted by Cas13a proteins using specially designed CRISPR RNA (crRNA). Cas13a is highly selective through the use of the crRNA, which attaches to the complementary sequence in the virus genome and designates the location the Cas13a protein should cut. As a result, the flu RNA will be cut, thereby preventing viral replication, hence treating the infection. Since the crRNA targets the highly conserved regions of influenza viruses, the treatment is expected to be broadly effective against different viral strains.

585717 A Systems Biology Model of Myeloid-Derived Suppressor Cells and Cancer Immunotherapy

Some malignant diseases can cause major changes to the immune system, which prevents the body from properly protecting itself from threats. Immuno-suppressive diseases like cancer, HIV, and AIDS disable the body's ability to fight against pathogens. This begs the question that there must be a procedure to restore the body's immune system, so it can function properly. This is why myeloid-derived suppressor cells (MDSCs) are extremely valuable to the immune system. These cells can serve many functions because they change into different types of cells depending on the situation at hand. When the immune system is activated, an MDSC can change into macrophages or dendritic cells, but when the immune system is not activated, it stays as a MDSC. Macrophages and dendritic cells are white blood cells that can sense a pathogen in the body, alert the immune system, and destroy it. Tumor Associated Macrophages or TAMs are the type of macrophages MDSCs morph into when they do differentiate. TAMs have been shown to have the potential of interacting negatively with the tumor cells by attacking it. The methodology of this research will use mathematical modeling of differential equations using MATLAB software to model the dynamics of MDSC and TAMs and their interactions with tumors. Mathematical modeling gives the flexibility of testing many different stimuli easily (quickly and cheaply) and mechanistically. I have combined two existing models and data from the literature. The first model considers MDSCs in the tumor microenvironment without any treatment and with a treatment. This model tests how each change affects the MDSC response. The second model shows how TAMs can be activated into two phenotypes to attack or ignore tumors. Validation of the combined MATLAB model will be done by comparing to data from collaborators and from published experimental studies.

586766 An Immunoglobulin G Based Steric Hindrance Assay for Protein Detection

With the imminent needs of a rapid, accurate, simple point-of-care system for global healthcare industry, electrochemical biosensors have been widely developed owing to its cost-effectiveness and simple instrumentation. However, typical electrochemical biosensors for direct analysis of proteins in human biological sample still suffer from complex biosensor fabrication, lack of general method, limited sensitivity, and matrix caused biofouling effect. To resolve these challenges, we developed a general electrochemical sensing strategy based on a steric hindrance effect on an antibody surface layer. This strategy utilizes the interaction pattern of Protein-G and immunoglobulin G (Fc and Fab regions), providing steric hindrance effect during target capturing process. The provided steric hindrance effect minimizes matrix effect caused fouling surface and altered the path of electron transfer, delivering a non-fouling and high-sensitivity detection of protein in complex matrix. Also, an enzyme based horseradish peroxidase/hydroquinone/H_2O_2 transduction system can also be applied to the system, demonstrating the versatility of this sensing strategy for general electrochemical sensor application. We demonstrated this platform through the detection of Tau protein and programming death ligand 1 with a sub-pico molar detection limit within 10 min, satisfying the clinical point-of-care requirements for rapid turnaround time and ultra-sensitivity.

588299 Characterization and Incorporation of Lipid Prodrugs into Liposomes

Targeted chemotherapeutic delivery is the process of delivering cancer drugs to diseased tissue in the body without harming healthy tissue. This project has three parts: drug synthesis, drug packaging, and drug delivery. This presentation will highlight the drug packaging portion of the project during which the synthesized prodrugs are packaged into liposomes and characterized to determine liposome size, drug incorporation, and stability. Dynamic Light Scattering (DLS) was used to determine the size of both the parent compound and the prodrug loaded liposomes. These sizes were monitored over the course of three weeks in order to determine the stability of both the parent compound and the pro drug loaded liposomes. The results from the DLS showed that the empty liposome sizes are between 100 nm and 120 nm whereas the prodrug-loaded liposomes were between 120 nm and 150 nm. These sizes stayed consistent over a three-week period. The drug incorporation was measured using UV-Visible Spectrophotometry (UV-VIS) and the results from the UV-VIS concluded that as concentration increased as the absorbance increased. The characteristic absorbance wavelengths were found at approximately 300 nm, 340 nm, and 410 nm. The outcome of these experiments confirms that the prodrugs have been incorporated into the liposomes and further research is required to determine the optimal concentration of prodrug to incorporate into the liposomes in order to maximize packing. The next step for this portion of the project is to characterize the rest of the prodrugs that have been synthesized.

586578 Fundamental Analysis of Aqueous Humor Dynamics in the Human Eye

Fluid mechanics is an essential part in numerous aspects of the composition and pathology of the human eye. In particular, glaucoma, a condition caused by an increased amount of intraocular pressure, is heavily dependent on the proper flow and drainage of aqueous humor in the eye. Currently, the most popular approach to impeding the disease is through numerous types of surgeries, such as laser or incisional surgery, which reduces intraocular pressure by creating a pathway to the Schlemm's Canal. However, minimal research has been done to improve the drugs administered to glaucoma patients nor the proficiency of the drugs being delivered to the trabecular meshwork, for example. A fundamental analysis has the potential to aid in a better comprehension of the velocity and concentration profile in the human eye to improve in the administering of drugs and validate the effectiveness of surgery on patients without damaging other functional aspects in the eye.

In this contribution, the focus is on the analysis of the mathematical models used to solve fundamental differential equations describing the velocity and concentration profile in the human eye. The Navier-Stokes equation is solved using a creeping flow assumption to obtain velocity profiles within the trabecular meshwork. The resulting velocity has been coupled with the corresponding molar species continuity equation to obtain drug concentration profiles in the same local area. A summary of these findings will be provided with significant insight on improved drug treatment and the effectiveness of surgery to glaucoma patients.

585972 Assessing Gut Transport of Methionine and Lysine in Fast-Growing Rainbow Trout

Commercial aquaculture is growing at a rate of 7% per year and uses marine-derived fish meal to feed finfish and shellfish worldwide. However, fish meal supplies are limited and the future growth of aquaculture depends on the discovery of alternative sources of protein to feed fish. One promising alternative is plant-derived fish meal, however plant proteins lack sufficient levels of methionine and lysine for fish dietary requirements. A genetically selected line of rainbow trout that grows more rapidly on a plant protein diet was compared to a non-selected line. It was hypothesized that genetically selected rainbow trout uptake lysine and methionine more efficiently in their gut via more numerous amino acid transporters as compared to non-selected fish. Amino acid uptake and total amino acids were assessed in plasma over time using [C13] and [N15] isotopically labeled methionine and lysine measured by LCMS. Amino acid transporter expression in the gut was assessed using qPCR. Results show amino acid uptake in rainbow trout selected for rapid growth on a plant-protein diet is greater than the uptake observed in non-selected fish. Quantitative PCR of amino acid transporters in the proximal intestine also show various levels of expression over time consistent with levels of amino acid uptake.

584184 Bioreactor Design for the Maturation and Organization of Engineered Cardiac Patches

Introduction: About 40,000 babies are born with a congenital heart defect (CHD) every year in the United States. Approximately 25% of these babies require surgery within their first year. The current materials available to surgeons for the reconstruction of CHDs are exclusively acellular and inert, limiting the size and location of patched repair and often lead to arrhythmias. Living engineered cardiac patches (ECPs) that integrate with the heart could reduce arrhythmias and allow for more extensive repairs. *In vivo* native cardiac tissue consists of highly organized layers and structures. We hypothesize that directional and physiological application of strain, sheer stress, and electrical stimulation can direct ECP maturation, organization, and trabeculation. In this study, we designed a bioreactor capable of presenting directional shear strain and electrical stimulation to ECPs grown *in vitro*.

Methods: A bioreactor was designed using Solidworks to fit three ECPs that measure 8 mm in diameter and 2 mm in height and two carbon electrodes. A fluid mechanics computational model was run in COMSOL Multiphysics in order to determine physiologically relevant flow and pressure profiles. The bioreactor was then 3D-printed using Formlabs' Dental SG Resin , which is an autoclavable, nontoxic, not biodegradable, nonporous, and UV-transparent material. The final design of the 3D-printed reaction chamber is compatible with the adapters, fittings, and tubing necessary for the full model.

Results: The fluid mechanics simulation of the final model showed that the bioreactor supports the targeted physiologically relevant shear stress of 10 dyne/cm^2. Throughout system operation, laminar flow is maintained as fluid flows across the chamber and perfuses through the ECPs. When in contact with the ECPs, carbon electrodes attached to an electrical source will electrically stimulate the cardiomyocytes to contract simultaneously. The bioreactor was printed using an autoclavable, nontoxic, and non-biodegradable dental resin capable of adjustable pressures, flow profiles, and electrical pacing.

Discussion and Conclusion: This study presents a bioreactor chamber that can be 3D-printed out of Dental SG Resin and is shown in computational flow models to maintain physiologic pressures, shear flow, and electrical stimulation through an ECP. Future experiments include flow testing through the device and evaluation of cell viability, proliferation, organization, tissue morphology, vascularization, contractile force, and synchronization. The bioreactor will later be incorporated into a control system that provides automated ECP differentiation and maturation using encapsulated pluripotent stem cells.

585618 Extraction of Dicarboxylic Acids from Fermentation Broth and Analysis By HPLC

Long dicarboxylic acids (DCAs), are used in the production of polymers, polyesters, and personal care products. However, the chemical process to make these compounds is unfavorable due the petroleum-based feedstock necessary to start the reactions. One alternative, renewable, and efficient method for the production of DCAs is through biotechnology. This study focused on improving methods for extraction and analysis of C18 DCAs in fermentation broth. We tested alternative methods of optimizing extraction conditions in order to reduce costs because pure C18:0 and C18:1 DCAs can sell for approximately 122 $/g. Solubility tests were performed on C18 DCA, C12 DCA, and stearic acid and it was determined that C12 DCA may be an adequate substitute for C18 DCA in testing; however using C12 DCA required changes to the analytical method in order to resolve peaks. The optimized method for analyzing C12 DCA used a decreased flow rate and an increased column oven temperature. The mobile phase was changed from an isocratic mixture of methanol, water, and acetic acid (v/v, 90:9.9:0.1) to a gradient between water with 0.1% (v/v) formic acid and acetonitrile with 0.1% (v/v) formic acid. Finally, an extraction method was developed for C12 DCA that had an average recovery yield of 92.9%. By optimizing the extraction process with C12 DCA, we obtained useful information for the process of optimizing extraction of C18 from fermentation broth.

587771 Developing a Scalable Model for Optimum Growth in Stirred Tank Bioreactors

Stirred tank bioreactors are commonly used in many laboratory and industry applications. These reactors are often utilized in chemistry, biology, and pharmacy fields to examine important parameters. The most significant parameter of a bioreactor is the volumetric mass transfer coefficient (kLA) of oxygen. The kLA of oxygen is essential to measure efficiencies and scale-ups, and is widely affected by air flow rate, impeller type, impeller speed, and temperature. The experiment was first conducted in a 1L cylindrical glass reactor with a dished bottom. In this reactor, the kLA was measured using a gas cylinder that contains nitrogen and air. To test both the concentration and temperature of nitrogen and air, a dissolved oxygen and temperature probe were used. Nitrogen was measured by waiting for the oxygen concentration to decrease down to zero. After the oxygen concentration reached zero, air was measured until the dissolved oxygen completed 100% saturation. Depending on the amount of time is takes for nitrogen to reach zero and air to reach 100% saturation, different flow rates were adjusted so that the duration would run smoothly. Various types of impellers and impeller speeds were tested to see how the amount of time is affected as well. A 3L reactor was then repeated with same process that was done with the 1L reactor.

This process was completed, and currently, the 3L bioreactor is being utilized to produce bacteria growth and cell culture. The bacteria being used in the process is Escherichia Coli due to its wide industrial use and fast-growing abilities. From using Escherichia Coli in the bioreactor with the various flow rates, impellers, and impeller speeds, we can see how fast the bacteria can absorb the oxygen and grow. To measure the concentration of the bacteria growth in the bioreactor, Optical Density (OD) measurements are recorded. This project is being tested in India as well by measuring bacteria growth and culture in larger reactors such as 5L and 7L. By doing these experiments on a smaller scale and a larger scale, we will be able to find the optimum parameters for a scalable model.

585757 Quantifying Mass Changes at a Peg Surface Complexed with Alpha-Cyclodextrin

αCyclodextrin is a unique molecule because of its hydrophobic interior cavity and hydrophilic exterior. This interior cavity can create inclusion complexes with a variety of hydrophobic guest molecules, including resveratrol, a hydrophobic molecule naturally found in red wine and grapes. This system has been well quantified using Electron Impedance Spectroscopy, Cyclic Voltammetry, and X-ray Photoelectron Spectroscopy, but not yet with electrochemical Quartz Crystal Microbalance (eQCM). eQCM provides insight into the surface coverage of the system. It quantifies the bonds between the Poly(ethyl-glycol) and αCyclodextrin by measuring the changes in mass, as a function of frequency change.

Understanding this system has implications for the field of biosensors, point-of-care tools, and any sensor that could benefit from the sensing of a guest molecule with properties similar to Resveratrol. The use of eQCM allows for mass measurements of the αCyclodextrin-PEG system. Understanding the mass changes in such a system allows for analysis of reusability and reproducibility in the context of reusable sensors.

Within a flow cell, eQCM, setup, αCyclodextrin-PEG system was modified onto a gold electrode using thiol-poly(ethyl-glycol) carboxyl-terminated (PEG-CT). A decrease in frequency (proportional to an increase in mass) would be seen upon PEG-CT attachment. After this modification, αCD was introduced to the system and attached to the PEG-CT molecules further decreasing the frequency (increasing the mass).Upon introduction of Resveratrol to the system, Resveratrol and αCD create an inclusion complex, removing the αCD from the surface, which would display a decrease in mass and increase in frequency.

587954 Evaluating Transport and Intracellular Uptake of a Protein Nanocarrier in 3-D Tumor Spheroids

The delivery of therapeutics into cells could open up many treatment options for various diseases such as cancer and neurological disorders, but these intracellular targets have been considered largely "undruggable" by small-molecule drugs that can penetrate the cell membrane. Delivery of antibodies directly to these targets would allow for more direct interaction with intracellular proteins, but intracellular antibody delivery remains an issue. To combat this, our lab has previously developed the Hex carrier, which can bind up to 3 antibodies per carrier and deliver them intracellularly in their functional form. Previously, I have shown via flow cytometry that HeLa cells treated with Hex-bound fluorescently labeled rabbit Immunoglobulin G (IgG_{Rb}) show a statistically significant increase in intracellular florescence compared to treatment with unbound IgG_{Rb}. To further expand on previous studies of uptake in 2-D cell culture systems, the evaluation was extended to a 3-D tumor spheroid model. 3-D multicellular spheroids are very common models used to evaluate nanoparticle delivery, as they better mimic actual tumors as they would appear and behave in vivo. This allows for a more realistic assessment of nanoparticle transport and uptake at the tissue level. The intracellular uptake of fluorescent IgG_{Rb} was assessed by flow cytometry to confirm that intracellular delivery is maintained in a complex, 3-D system. After being cultured and treated, the spheroids were digested with trypsin in order to break up the spheroid into individual cells for flow cytometry. Results of this analysis indicated positive fluorescence of the spheroids' cells, indicating successful intracellular delivery despite the more complex system. This successful delivery of Hex in 3-D tumor spheroids provides further evidence of the improved delivery of antibodies by the Hex carrier. Future work will involve using confocal microscopy to evaluate transport of the Hex-antibody complex across the extracellular matrix into the interior of the spheroid, providing further support for Hex as an antibody delivery vehicle.

587639 Release of Griseofulvin from Spray-Dried Nanocrystal–Amorphous Solid Dispersions (HyNASDs): Impact of Surfactant

Over the last few decades, majority of the new chemical entities (NCEs) in the drug discovery pipeline are poorly-water soluble, and their slow dissolution turns out to be the rate-limiting step during intestinal absorption. Among different approaches to enhance bioavailability, two platform approaches have gained prevalence both in the scientific literature and marketed products: nanoparticle-based dosage forms (nanocomposites) and amorphous solid dispersions (ASDs). In the nanoparticle-based dosage form, dissolution rate is improved due to the dramatic increase in the surface area of the drug crystals via size reduction down to the nanoscale. On the other hand, amorphous form of the drug can offer significant enhancement in the extent–rate of dissolution due to its higher free energy and higher apparent solubility. A major shortcoming of drug nanocomposites as compared with amorphous solid dispersions (ASDs) is their limited supersaturation capability in dissolution media.

Here, we prepared hybrid nanocrystal–amorphous solution dispersions (HyNASDs) in order to boost the drug release performance of traditional nanocomposites while elucidating the impact of a common anionic surfactant, sodium dodecyl sulfate (SDS), on drug release in dissolution tests. Wet-milled suspensions containing 1:1 to 1:5 GF:polymer mass ratios, with and without SDS, were spray-dried. Hydroxypropyl cellulose (HPC) and Soluplus (Sol) were used as suspension stabilizers and matrix-forming polymers. XRPD and DSC results suggest that nanocomposites and nanocomposites with notable amorphous GF (>10%), HyNASDs, were formed upon spray-drying. Redispersion of the spray-dried powders revealed the criticality of SDS in nanoparticle recovery from the nanocomposites/HyNASDs, which could be explained by GF wettability enhancement by SDS, as inferred from the modified Washburn experiment. Results from *in vitro* dissolution tests suggest that enhanced wettability with SDS and smaller spray-dried particle sizes led to faster GF release. For 100 mg GF dose (above thermodynamic solubility), Sol provided higher GF relative supersaturation, e.g. 250% for the HyNASDs (1:5 GF:Sol with SDS) vs. 30% for the nanocomposites (3:1 GF:Sol with SDS), than HPC (up to 50%) owing to Sol's stronger intermolecular interactions–miscibility with GF and its kinetic solubilization–recrystallization inhibition of GF. SDS provided enhanced wettability, allowing for fast supersaturation from Soluplus-based HyNASDs (~300% within 20 min), while higher Soluplus loading led to higher supersaturation. Finally, this study demonstrates how drug release from traditional nanocomposites could be boosted upon incorporating a drug-miscible, solubilizing polymer with a low GF:polymer mass ratio and an anionic surfactant.

588274 Development of Protist-Facilitated Transport for Sustainable Agriculture-Biotechnology

Protists are a diverse group of organisms that are defined as single-celled eukaryotes on the micrometer scale. Bacterivorous soil protists, such as the ones used in this study, are essential to their ecosystems by cycling nutrients for plants and increasing bacteria diversity through predation. These protists could potentially be used as transport vehicles for nano-encapsulated agrochemicals.

In this study, three soil protists collected from the rhizosphere in eastern Connecticut are tested for their ability to move through a serpentine emulated soil device. The average biomass velocity of the protists as they move through the device is correlated with their nano-encapsulated agrochemical mass transport potential.

Over 72 hours, although *Colpoda steinii* had the highest average velocity as it moved through the device, *Colpoda lucida* had the highest average biomass velocity due to its substantially larger size. *Leptopharynx costatus* had the lowest average velocity and average biomass velocity of the three. For future studies, *Colpoda lucida* is the most promising candidate.

586728 Comparison of Keit Irmadillo™ and Mettler Toledo Reactir for Pharmaceutical Process Chemistry Reaction Monitoring

In-situ attenuated total internal reflection fourier transform mid-infrared spectroscopy (ATR-FTIR) with the ATR probe placed in the reaction vessel is a well-established technique for monitoring pharmaceutical process chemistry reactions that can reveal reaction component concentration trends, such as starting material consumption, product formation, and reaction intermediates if present. The Keit IRmadillo™ is a relatively new ATR-IR spectrometer that is being marketed as a rugged instrument for monitoring reactions in liquids and slurries and is certified for use in hazardous and potentially explosive industrial environments. The Mettler Toledo ReactIR 702L is another new ATR-FTIR instrument marketed for chemical reaction monitoring, but is a more compact unit primarily intended for use in research and development laboratories.

This poster compares the IRmadillo and ReactIR 702L with the Mettler Toledo ReactIR iC10, an older ATR-FTIR instrument used extensively at GlaxoSmithKline. A major operational difference between instruments is that the ReactIR iC10 requires liquid nitrogen to cool the detector, whereas the IRmadillo and ReactIR 702L do not, which enables them to start more quickly and to run unattended indefinitely, important advantages for both lab and plant operation. However, the corresponding higher detector temperatures used by the IRmadillo and the ReactIR 702L are also expected to decrease the instrument sensitivity and increase noise, which could negatively impact reaction monitoring performance.

Results from monitoring two example reactions will be presented. The first example compared all three instruments using a pharmaceutical intermediate compound manufacturing process where the slurry reaction endpoint time can vary based on certain process parameters. To compensate for this uncertainty, ATR-FTIR is used to track the reaction intermediate concentration and determine the appropriate timepoint to sample the reaction for endpoint confirmation by HPLC analysis. The second example used the well-known homogeneous solution reaction of acetic anhydride with methanol to further evaluate the IRmadillo and ReactIR iC10 instrument sensitivity, ability to detect transient reaction species, and performance with high concentration components.

587837 Optimization of Carbon Delivery to *Nannochloropsis Oceanica* 849/10

Atmospheric carbon dioxide is a major contributor to global climate change. *Nannochloropsis oceanica* is a marine microalgae species showing viability as a large-scale carbon dioxide conversion method. Being a marine species, *N. oceanica* does not compete with agriculture for freshwater resources. This species also naturally has a high percentage of its cell weight dedicated to lipids, which can be refined into biodiesel fuel, thereby providing a renewable fuel source. In sea water, carbon dioxide is equilibrated between two forms, dissolved carbon dioxide (CO_2) and bicarbonate (HCO_3^-). Bicarbonate has previously been determined as the primary carbon source for microalgae like *N. oceanica*. To maximize carbon dioxide conversion, the algae's growth medium must have a high bicarbonate concentration, which requires a neutral to moderately basic pH. Experiments were conducted to determine the following: 1) standard growth curve of *N. oceanica*, 2) does *N. oceanica* prefer a bicarbonate source, and 3) baseline activity assays for commercially available carbonic anhydrase. Growth trials were run to determine if *N. oceanica* prefers either ammonium bicarbonate or sodium bicarbonate as a bicarbonate source. Sodium bicarbonate cultures had ammonium chloride added to maintain identical nitrogen concentrations. Total organic carbon (TOC) analysis showed that inorganic carbon (IC) concentrations, the combined concentration of carbonate and bicarbonate, decreased for cultures provided with ammonium bicarbonate. This decrease means *N. oceanica* consumed more bicarbonate. In contrast, the sodium bicarbonate/ammonium chloride cultures did not show a significant change in bicarbonate uptake. An activity assay was conducted on commercially available carbonic anhydrase for future reference on how factors such as temperature, acidity, buffered solution, and salinity affect enzymatic activity. These results are useful in determining how to improve algae-based carbon capture since carbonic anhydrase is a crucial component of the carbon concentrating mechanism in microalgae.

584728 An Electrochemical Biosensor for in-Situ Detection of Lipids By Immobilization of Cholesterol Oxidase

Atherosclerosis is the blockage of blood flow through arteries due to plaque buildup from high cholesterol. High cholesterol does not display symptoms and can only be detected through a blood test. There is a clinical need for a device able to analyze cholesterol concentration in-situ. For the development of a biosensor for cholesterol detection, an amperometric, three-electrode system was utilized. Cholesterol oxidase (ChOx) was immobilized at the working electrode through a self-assembled monolayer (SAM) of octadecanethiol (ODT) or thioglycolic acid (TGA) and electropolymerization of pyrrole. Multiple immobilization methods were assessed through amperometric measurements and cyclic voltammetry. The cyclic voltammetry experiments were run for assessing the formation of the self-assembled monolayer, by determining a change in current measured by the sensor. The amperometric measurements were conducted for cholesterol detection along with the characterization of sensitivity of the functionalized biosensor. This study looked at different enzyme immobilization methods for creating a device that can analyze cholesterol blood composition. The obtained results show that two functionalization methods yielded different results, with a fairly good liner response curve between the sensor's current signal and cholesterol concentration for the sensor with the pyrrole method. Future work includes further optimization of the functionalization process and tuning of the sensitivity, among others.

586018 In Vivo Remediation of Contaminants in the GI Tract

As a result of industrialization and urbanization, persistent organic pollutants can be found anywhere from soil to water to biological systems. High levels of contaminants in the body, like polychlorinated biphenyls (PCBs), are associated with various health risks such as hypertension, high blood pressure, and metabolic disruption. To reduce the body burden, polymeric nanocomposites were synthesized from functional monomers, including styrene, vinyl pyridine, etc. and novel crosslinkers, including quercetin multiacrylate (QMA). It was hypothesized that their aromatic properties would allow these polymers to have a high affinity for PCBs. Fourier transform infrared spectroscopy (FTIR) confirmed the presence of the components of the polymer, while wash analysis ensured a high conversation and incorporation of QMA in environments that resembled that of the gastrointestinal tract. The affinity of the polymer to the contaminant was shown in the kinetic and binding studies, which demonstrated that after 24 hours, an average of 92% of PCB-1 in solution was bound to the polymer. With further development, it is expected that these polymers will ultimately be applied to remediate the GI tract of the burden of these contaminants by preventing their absorption and aiding their expulsion.

585997 Treatment of Low Back Pain Using an Injectable Tissue Specific Scaffold

Low back pain (LBP) will affect 70-80% of the global population at some point during their lifetime. The main cause of LBP is known to be intervertebral disc degeneration, that causes spinal nerve impingement, leading to pain that may radiate throughout the body. Current treatment methods include prescription medications and invasive spinal fusion surgery which have many drawbacks and do not target the root of the problem. Therefore, this project involved using decellularized intervertebral disc tissue from pig as a scaffold to restore disc height and rejuvenate local cells to minimize LBP. This objective was accomplished through a thermally gelling collagen hydrogel, allowing for minimally invasive delivery to the core of the disc. Hydrogels containing decellularized tissue and greater than 2 mg/ml collagen formed macroscopically intact gels and thermal gelation kinetics results revealed that increasing collagen concentration led to a firmer and more viscous gel. Gelation time was also investigated to allow for optimal injectable delivery of the gel. Overall, intact gels were successfully synthesized by combining decellularized tissue supplemented with collagen hydrogels. Further work aims at improving the robustness of the gels.

584294 Modulation of Adjuvant Loading and Degradation Profiles of Biocompatible Polymeric Nanoparticles for Immune Stimulation

Poly(ethylene glycol) (PEG) hydrogels are commonly used in reparative tissue engineering and regenerative medicine due to their biocompatibility, hydrophilicity, and slow rate of degradation. Similarly, these desirable properties make PEG diacrylate (PEGDA) nanoparticles (NP) an attractive platform for delivery of therapeutic vaccines. Binding immunostimulatory ligands (adjuvants) to delivery platforms with a slow rate of degradation would encourage long-acting immune stimulation and thereby increase efficacy of vaccines. Therefore, our objective is to characterize the effect of modulus of PEGDA NP on rate of degradation and of NP surface density of CpG ODN for enhanced, sustained immunostimulatory effect. Aminated NPs for ligand attachment were synthesized by first mixing a pre-particle solution, which was emulsified in silicone oil and exposed to UV light to initiate polymerization. Pre-particle solution contained 50% or 30% wt/wt total solids in methanol consisting of 88% crosslinker [PEG diacrylate (PEGDA, Mn=700)], 10% cationic functional monomer [2-aminoethyl methacrylate], 1% photoinitiator [diphenyl(2,4,6-trimethylbenzoyl)phosphine oxide (TPO)], and 1% dye [Rhodamine B]. Particle degradation was monitored over 4 weeks using thermogravitational analysis (TGA), dynamic light scattering (DLS), scanning electron microscopy (SEM), and Rhodamine B release in pH 5, 7.4, and 10 buffers via fluorescence. A bifunctional linker, SMCC (succinimidyl 4-(N-maleimidomethyl) cyclohexane-1-carboxylate) was conjugated to NP free amines and the 5' modified thiol on CpG. CpG loading was confirmed by UV-Vis spectrophotometry at 260 nm to measure amount of unbound nucleic acid concentration in the supernatant. TGA showed linear decrease in particle mass in both 50% and 30% total solids PEGDA NP formulations of all pH buffers over 4 weeks, suggesting bulk erosion of the polymer matrix of the NP. RITC release in solution increased sharply over week 1 of buffer storage, then was maintained from weeks 1-4. While this is indicative of bulk erosion, previous studies note the high hydrophilicity and small size of RITC could result in diffusive leaching into solution rather than matrix degradation. DLS measurements of hydrodynamic diameter and zeta potential were measured to be 273 ± 4 nm and 23.9 ± 4.9, respectively. The net positive charge of unfunctionalized PEGDA-NP suggest presence of free amines on the NP surface and PEGDA NP were functionalized with CpG ODN of varied surface densities. Overall, this work demonstrated that PEGDA NP exhibit slow degradation profiles similar to those expected for PEG hydrogels. PEGDA NP therefore remain an attractive candidate for long-acting immune stimulation.

586368 Determining Pharmaceutical Powder Wettability

In this work different methodologies that are used to measure the wettability of powders are studied in detail. It has been recognized that the different manufacturing process would significantly change the wetting behavior of material. In addition, the wetting properties of raw and intermediate material directly affect the dissolution performance and bioavailability. For this reason, the characterization of wetting properties is inevitable for quality control in advanced pharmaceutical manufacturing. Several techniques have been developed to identify wettability of powders. These include the sessile droplet, capillary rise, immersion and condensation methods.

The sessile drop method (i.e. drop penetration method) is the most common method. In this method a single liquid droplet is released to a powder bed and the contact angle of the liquid is then measured on the powder bed. This method is the easiest to replicate, but it is also very sensitive to powder bed condition. For instance, air may get trapped inside the powder bed.

The capillary rise method is a layer of powder and the fluid is pushed up through it. In the immersion method the powder bed is placed under the liquid and the phase interaction between the solid and liquid is observed. In the condensation method the resting drops on the surface are analyzed. Another frequently used technique is the capillary rise method. Here, the powder will be packed into a tube and the test fluid can interact with the powder by moving up the tube. Sources of error in this method are the possible movement of the tube during testing as well as the inability that occurs due to use of non-Newtonian liquids or liquids with contact angles over 90°.

The immersion method involves taking the powder and submerging it into a pool of test fluid, while the condensation method involves making the test fluid into a saturated vapor and having the vapor condense into droplets on the powders surface for analysis. The immersion method can be difficult to replicate because the adhesive that is used to adhere the powder to the solid may lead to a wrong perimeter of the particle being recorded, that would lead to an incorrect conclusion. It is also likely that due to the liquid imbibition, the reported contact angle would be too small. Both immersion and condensation are harder to replicate and for that reason less commonly used when testing the wettability of a powder.

587425 Development of a Microfluidic Device to Study Cell-to-Cell Communication By Dynamic Sampling of Paracrine Factors

Breast cancer is the second leading cause of cancer related death in American women. Cancer cells signal to surrounding tumor stroma to produce cancer associated adipocytes, fibroblasts, and macrophages while secreted factors from stromal cells in the tumor microenvironment induce drug resistant and metastasis in cancer cells. Co-culture and conditioned media-based experiments have demonstrated that paracrine factors induce drug resistance. While multiple co-culture techniques exist, no of them are capable of dynamic sampling of extracellular media during an experiment to investigate cell-to-cell communications. To address this, a novel microfluidic device was designed, fabricated, and tested to study intercellular communications during co-culture. The device consists of three parallel channels connected by an array of 30 μm wide by 100 μm long microchannels spaced 75 μm apart to facilitate the orthogonal diffusion of biomolecules. COMSOL Multiphysics was used to model the mass transfer in the device to optimize channel length and width. The optimized device consisted of three parallel 1500 μm long with the two outermost channels for static cell culture and the center channel connected to a peristaltic pump to allow for continual perfusion and circulating of medium. The operating length of the three channels was 20 mm to allow for sufficient cell numbers for sampling of paracrine factors. Circulating media flowrates of 5, 10, and 15 μL/min were tested, with 5 μL/min providing sufficient mass transfer across the three channels with minimal shear effects on the culture cells A test device was fabricated using a combination of soft lithography and polydimethyl siloxane (PDMS) replication. The device utilized a peristaltic pump with clear, flexible tubing for constant flow of media and a three-way junction incorporating into the tubing outside of the device for dynamic sampling. Pipette-tip reservoirs were positioned at the inlets and outlets of the culture channels for media replacement. Peristaltic pump performance was characterized by running water through the tubing and timing the water collection at different pump speeds. The pipette-tip reservoirs were observed during testing to ensure media replacement without the introduction of bubbles during sampling. A careful sampling technique, which included clipping the opposite junction during sampling, was also developed to address this issue. This novel microfluidic device will be used to analyze the cell-to-cell communications between MDA-MB-231 breast cancer cells, a triple negative cell line, and adipose-derived stem cells (ASCs) to study how stromal cells enhance the proliferation and drug resistance of cancer cells.

585710 Modeling the Controlled Release of Risperidone from PLGA Microsphere Formulations

My research is focused on how to improve the predictions of the in vivo effects of controlled release of the specific drug, risperidone. Risperidone is used to treat schizophrenia, bipolar disorder, and autistic irritability. To accomplish our goal, we are using an existing mathematical model that simulates how a general drug leaves a poly(lactic-co-glycolic acid) polymer (PLGA) microsphere in vitro. We are adapting the model considering the characteristics of risperidone. So far, we have taken existing in vitro data from literature and digitized it. From there we considered the first phase of in vitro release after an initial burst to be a diffusion-controlled release regime described by Fickian diffusion with a constant diffusion coefficient. We used parameter estimation to fit the Risperidone initial burst amount and diffusion coefficients through different formulations of microspheres. Now we are looking into what happens, particularly to the second release regime of pore degradation-controlled release, when certain properties are changed, such as particle size, molecular weight, and the PLA/PGA ratios. Later we will adapt the in vitro model to an in vivo one using mass transport and pharmacokinetics of risperidone released from PLGA to create a prediction for how risperidone is delivered in the body. We will validate the final in vivo model with existing data. This approach can be compared with proposed in vitro-in vivo correlations in the literature.

585729 Effect of Feed Flow Rate and Precipitate Concentration on the Precipitation Process for the Purification of Monoclonal Antibody

Precipitation can be a highly effective step in the initial capture of monoclonal antibodies (mAb). High yield of the precipitates can be produced continuously in a tubular reactor with cross-linking and volume exclusion agents. The integration with tangential flow microfiltration for the dewatering and wash stages would allow for high removal of impurities creating a feasible fully continuous precipitation process for the purification of mAbs. However, current issues with this process are the low impurity removal rates and low yield of the final product. The purpose of this study was to investigate the effect of feed flow rate and the precipitate concentration on increasing the critical flux and product yield. The higher the critical flux, the higher the permeate flux, which would enable the membranes to remove more impurities. In our experiments, human serum Immunoglobulin G (IgG) was used as a model protein. The IgG precipitates (99.8% yield) were produced through a tubular reactor using zinc chloride and polyethylene glycol as the cross-linking and volume exclusion agent, respectively. The solutions were mixed with static mixers with a residence time of 30 seconds. Critical flux experiments were performed with a total recycle system, with the inlet feed flow rate varied to examine its effect on the critical flux. Re-dissolution experiments were performed using different IgG precipitate concentrations and buffer conditions to improve the overall product yield. Results have shown that a decrease in feed flow rate can increase the critical flux, and forming precipitates using a higher protein titer can prevent product re-dissolution and increase yield. These results provide important insights for the improved design of a precipitation – microfiltration process for initial capture/purification of a monoclonal antibody product.

584598 Critical Flux during Tangential Flow Microfiltration of Precipitated Proteins

There is considerable clinical interest in both polyclonal Immunoglobulin G (IgG) and monoclonal antibodies (mAbs) as biotherapuetics. Polyclonal IgG is currently purified from blood serum using an ethanol-based precipitation, while the initial step in the purification of most mAbs is by bind-and-elute Protein A affinity chromatography. We have been exploring the use of $ZnCl_2$ – PEG combinations for continuous precipitation of IgG / mAbs, exploiting the high product titer of current mAb products (>5 g/L). Long term operation of the system can be achieved by operating the membrane unit at a filtrate flux below the critical flux for the onset of fouling. The objective of this particular project was to explore opportunities for increasing the critical flux, including the use of co-flow to provide a more constant transmembrane pressure drop throughout the module. The precipitation of IgG was conducted in a tubular precipitation reactor with sequential inline addition of $ZnCl_2$ (a crosslinking agent) and PEG (a volume exclusion agent). High IgG yield (>99%) was obtained in the precipitate. The precipitated IgG was then concentrated and washed using hollow fiber polyethersulfone membranes. The critical flux was evaluated using a flux-stepping procedure, with the transmembrane pressure evaluated as a function of time during each constant flux interval. The addition of co-flow did not cause any measurable increase in the critical flux. However, the critical flux could be increased by using membranes with higher permeability. The critical flux was also a function of the aging of the precipitates, with fresh precipitates found to have a larger mean particle size. These results provide important insights into the development of effective tangential flow microfiltration systems for use in continuous protein purification by precipitation.

588161 Comparison of Different Methodologies for Rapid GC/MS Analysis of Volatile Organic Acids in Fermentation Media

The volatile fatty acids (VFAs) acetic, propionic and butyric acid have significant inhibition effects on activated sludge microbial consortia. Inhibition has been shown to be the highest at total VFA concentrations of 10 g/L and 20 g/L. A rapid easy to use method for the determination of VFAs was developed with no or limited sample preparation to ensure sample integrity. Liquid Chromatography (LC) has been the standard for the analysis of VFAs and requires filtration prior to analysis to remove fermentation cells and other by-products that could foul the LC column resulting in downtime and expense of column replacement. To avoid this, we have developed two methods using Gas Chromatography (GC) for a rapidly reduced sample productive alternative. The first is a Liquid-Liquid Direct Extraction with phase separation in an autosampler vial with the autosampler drawing sample from the organic phase. The other technique utilizes Solid-Phase Microextraction (SPME) technology. The sample is placed directly into a GC vial and the SPME fiber extracts and injects VFAs into GC inlet. Comparisons of the two techniques compared to traditional LC method showed the Direct Liquid-Liquid provided the lowest detection limits. The SPME detection of VFAs was shown to have slightly higher detection limits but requires no sample preparation as compared to the other techniques. Linearity of VFAs and chromatography examples will be discussed.

588313 In Vitro Model of Glioblastoma Using an Immortalized Cell Line

Glioblastoma brain cancer is the most common and deadly primary brain tumor that leaves its afflicted patients with a median survival rate of 1 year. This in vitro study seeds glioblastoma cells (U-87 MG line) into a microenvironment that mimics the brain's. By placing glioblastoma cells into a soft hydrogel environment within a microfluidics chamber that receives electrical stimulation similar to that from neural activity, a phenotypic shift is expected to be triggered. This overall effort tests the hypothesis that such an in vitro model system can induce a phenotypic shift so that the immortalized U-87 cells more closely resembles primary tumor cells. 7 wt% Gelatin Methacrylate and 50,000 cells were seeded into a 10-mm diameter glass-bottom petri dish between carbon electrodes for electrical stimulation. Cells are expected to die in tumor treating-level fields, and after 19 h of stimulation, the treated cells displayed more cell death than the non-exposed control. It was seen that the experimental setup is functional as there was an obvious reduction in the number in living cells. By adjusting the field strength, it is expected that exposure to low frequency fields typical of brain activity, in addition to the soft substratum and microfluidic flow, will significantly alter expression of mechanoreceptors without inhibiting tumor cell proliferation.

588336 Light Controllable Cell Patterning

Many different types of disease can cause damage to the heart's veins and arteries. While some can be replaced with grafts of veins and arteries from other areas of the body, it is sometimes needed to find grafts from donors. Since finding tissue matches can be difficult and have complications, such as rejection, a method that grows these veins and arteries out of a patient's stem cells would be optimal. One of the steps in growing veins out of stem cells is being able to pattern cells and achieve cell attachment in a specific location. A pattern was created on a bovine serum albumin coated slide using photobiotin, a photomask, and UV light. Cells were coated in sulfo-NHS-biotin, allowing them to attach to the patterned photobiotin. Biotinylated antibodies were also used to see if they decreased non-specific binding. Once the cells were incubated on the slides, a flow chamber was used to test the strength of their attachment. The use of antibodies was shown to help decrease unwanted attachment in addition to having stronger attachment than the sulfo-NHS-biotin coated cells. H9C2 and HUVEC cells were layered to recreate the functionality of real vein and artery tissue. This method was solely done with sulfo-NHS-biotin coated cells. The results of successfully patterning cells with strong attachment show promise for being able to one day grow a full vein or artery in a lab.

588347 Electropermeabilization of Breast Cancer Cells to Calcium Ions Using Microsecond Electric Pulses

Microsecond pulsed electric field (μsPEF) exposure causes an influx of calcium (Ca2+) that, in turn, sets off numerous signaling pathways. As a preliminary study, Ca2+ uptake into breast cancer cells is measured in response to μsPEF exposure. By first finding the threshold number of pulses, duration, and voltage to increase permeability to Ca2+, it is predicted that these same conditions can be used to initiate downstream signaling cascades. Calcium ions play important roles within the cell acting as messengers that control crucial cellular responses that affect apoptosis, muscle contraction, gene transcription, metabolism, etc. Intracellular concentrations of Ca2+ are very small compared to the extracellular matrix. This small concentration is essential for the signaling and communication processes to exist within the cell. When the cell is exposed to the μsPEF, pores form in the membranes of the cell causing a rise in concentration of Ca2+ within the cell, interfering with calcium homeostasis. This rise activates calcium - ATPase which normally transfers Ca2+. It also activates an absorption of Ca2+ into the mitochondria. Because of this gradient change, the mitochondria will collapse. With a lack of ATP being produced within the cell, cell death is triggered. By observing clear Ca2+ uptake, it is shown that Ca2+ dependent paths can be manipulated. In turn, Ca2+ reliant cellular activities will be disrupted. Presently, we are exploring downstream responses to μsPEF with breast cancer cells.

587948 Development of Wearable Cortisol Biosensor for Medical Application

Anxiety and depressive disorders alone affect over 400 million people and costs of over 1.5 trillion dollars per year. Although many people are affected, the psychiatric field still suffers from minimal understanding. There is a need to directly observe the essential biological nature of these disorders and move into objective, readily available indicators that can guide care.

Cortisol, the body's main stress hormone, is one of the most interesting biomarkers for many psychological disorders. There is a correlation between cortisol levels and several neurological disorders such as major depression, bipolar disorder, and anxiety. Cortisol levels can be measured using sweat, saliva, or blood. The current state of the art for measuring cortisol levels is phlebotomy, which uses blood samples, is invasive, and not applicable to tracking a patient continuously.

The ability to continuously monitor cortisol levels could potentially give great insight into biological nature of these disorders and make way for a form of a diagnostic biological test which is currently nonexistent. It has been shown that the cortisol levels in sweat have a good correlation with those levels in blood. We believe that body sweat is a valuable source of biological information that can be used readily and monitored continuously.

We chose commercial available metal oxide silicon field effect transistors (MOSFET) with an extended gate design as our sensor due to their high production speed, mechanical compatibility with skin, and low cost. Using competitive immunoassay, cortisol, a small molecule without charge removes the antibody from the surface of the extended gate displaying the large signal change necessary for our sensor. We are integrating this biosensing method into the extended gate MOSFET design due to the need to miniature the device for wearable application. Since one possible rout for sensing with extended gate structure is detecting the charges of the biomolecules, the research presented analyzes the effects of the buffer solutions Debye length, the scale which traveling charge carriers screen out electric field in conductors on the drift of the signal. The research presented will also detail the signal changes of the sensor as different MOSFETS with capacitances ranging from picofarad (pF) to nanofarad (nF) are paired with extended gate varying in surface area. By changing the surface area we can increase and decrease the total capacitance of the extended gate since the inherit capacitance is defined by the chemical functionalization itself.

586802 Microfluidic Synthesis and Purification of Protein Nanoparticles

MICROFLUIDIC SYNTHESIS AND PURIFICATION OF PROTEIN NANOPARTICLES

Nanoparticles have revolutionized the field of colloidal science and have been applied for a myriad of purposes in medicine, coatings, and even electronics. Many such applications, especially in medicine, necessitate precise control over parameters including size, polydispersity, and chemical content. Protein nanoparticle systems, like Abraxane®, are successful drug carriers due to their biocompatibility and biodegradability. However, nanoparticle synthesis procedures are limited in their scalability due to the critical interplay between particle size and solution conditions. Furthermore, medical applications necessitate the removal of trace chemicals and cytotoxic impurities – a challenge for nanoparticle suspensions. Here we present a scalable microfluidic process that addresses the need for high-throughput continuous production of homogenous, size-tunable, purified protein nanoparticles.

Our platform takes advantage of laminar flow to enable the reproducible synthesis of monodisperse protein nanoparticles. Nanoparticle synthesis occurs via protein desolvation and controlled aggregation in the presence of ethanol, followed by cross-linking with glutaraldehyde. Laminar flow facilitates the precise control over protein, solvent, and cross-linker concentrations within the device, allowing for tuning of nanoparticle size. We validated the efficacy of our platform with the synthesis of bovine serum albumin (BSA) nanoparticles.

Our results contrast with reports on the bench-scale synthesis of the ~50 nm BSA nanoparticles necessary to avoid an immune response. Such efforts have traditionally suffered from increased polydispersity. The particles produced by our platform demonstrate superior stability and size tunability. Our system provides a 33% reduction of required solvent, and an 80% reduction of crosslinking agent compared to benchtop BSA nanoparticle synthesis. Furthermore, our platform can produce 16.2 mg/hr of crosslinked protein nanoparticles, compared with 11.3 mg/hr for benchtop synthesis.

Purification is also an important challenge in nanoparticle synthesis because the removal of small molecule cytotoxic impurities typically requires time-intensive centrifugation processes. The H-filter in our platform enables purification of nanoparticles in a continuous manner by exploiting differences in diffusion times between small molecules and nanoparticles. Therefore, our platform can remove excess solvent and crosslinker, eliminating additional purification steps.

The present process introduces significant improvements to current protein nanoparticle synthesis processes, potentially enabling new applications in drug delivery. The approach is general and future work includes extrapolation into a plethora of other protein and polymer-based nanoparticle systems. Additionally, the planar chip design allows for the process to be easily scaled-out for a variety of academic and industrial applications.

588144 Dispersion and Coating of Titanium Nitride Nanoparticles for Photothermal Therapy

Photothermal therapy is a minimally invasive cancer therapy strategy that damages targeted tumors by hyperthermia. Photothermal nanoparticles absorbs noninvasive near-infrared (NIR) light to elevate the temperature to ablate the targeted tumor cells. Titanium nitride (TiN) nanoparticles offer several advantages over conventionally used photothermal agent gold nanoparticles, such as strong absorption of light within the biological window (700-1000 nm), lower cost and biocompatibility. However, the main challenge to employ TiN nanoparticles for photothermal therapy is their colloidal instability in aqueous solution. They make aggregates larger than 200 nm in solution which is not suitable for delivering to the targeted cell. The goal of the project was to modify the surface of the TiN nanoparticles using polyethyelene glycol (PEG-40 Stearate) and liposomes to enhance their dispersibility in solution. Ultrasonic probe sonication was used to disperse the nanoparticles in deionized water and in 1x PBS (Phosphate Buffered Saline, PH 7.4) solution followed by centrifugation to remove large aggregates. The size of the nanoparticles was characterized using dynamic light scattering and the surface coating was confirmed using Fourier-transform infrared spectroscopy. The mean hydrodynamic diameter achieved for both liposomes coated and PEG coated nanoparticles in water were less than 200 nm. Additionally, liposome coating increases the colloidal stability of the nanoparticle's dispersion in the aqueous solution. Unfortunately, neither the PEG-40 stearate or liposome coatings could prevent the aggregation of nanoparticle in PBS. Further research would be required to find suitable surface coating to effectively disperse the particles in a salt solution.

586795 Screening Powder and Binder Combinations for 3D Printing Personalized Medication

Screening Powder and Binder Combinations for 3D Printing Personalized Medication

Leila Sorrells[1,2], Shing-Yun Chang[2], Kavin Kowsari[2], Karthik Nagapudi[3], Koyel Sen[2], Bodhisattwa Chaudhuri[2], Anson Ma[2]

[1] University of South Florida; [2] University of Connecticut; [3] Genentech, Inc.

Binder jetting printing is a form of 3D printing that has recently become the preferred additive method for creation of drug tablets due to its precise printing and control of drug release. It operates using inkjet print heads and a moving print bed where prints are created by repeatedly layering a powder, binding the powder with fluid, and rolling another layer of powder on top.

This research focuses on using disintegration and hardness testing to determine the best combinations of powders and binders to use as a matrix for holding an active drug component for 3D printing drug tablets. Eight polymer-based binders were tested against fifteen powder mixtures created from varying concentrations of lactose monohydrate (LM), calcium sulfate hemihydrate (CSH), and sodium croscarmellose (SC). The goal was to find a combination or combinations that would reach a hardness of roughly 15 kgf and a disintegration time less than 15 minutes in 36 °C water.

For all binder types, increasing CSH percentage leads to an increase in hardness. However, the binder that consistently created samples that best aligned with the hardness and disintegration goals was Kollidon, which can be introduced in the powder or in the liquid form as an aqueous solution. Samples prepared from the same percentage of Kollidon, regardless of the method of introduction, have similar properties.

588272 Smart Self-Oxygenating Tissues for Organ Engineering

The engineering of organ sized tissues has the potential to address one of the greatest clinical challenges: the continuous and evergrowing donor organ shortage. However, large engineered tissues do not remain viable upon implantation as the limitation of oxygen diffusion causes massive anoxia-induced cell death. To overcome this, many researchers have attempted to expedite angiogenesis i.e., via implantation near vascular-rich tissues, co-implantation of endothelial cells, and release of vascular endothelial growth factors. These approaches still need improvement to prevent the imminent anoxia-induced cell death following implantation. Consequently, the local release or generation of oxygen has been considered by using oxygen-rich fluids such as perfluorocarbons. Although these approaches could indeed supply implants with oxygen for several hours, implant neovascularization typically requires several weeks. Consequently, despite numerous attempts, there currently still is no effective strategy to bridge the implanted tissues' prevascular phase.

This work addresses this crucial challenge by pioneering a unique solution that allows for the controlled generation of oxygen for an indefinite period, which will sustain the survival of large 3D tissues until the neovasculature is formed (Scheme 1). Our innovative solution is based on the transformation of the most abundant moleculewater) into the most needed substance (oxygen). To accomplish this, we hypothesize that a biodegradable and biocompatible solid-state electrochemical hydrolysis cell inside the tissue scaffold.

The proposed technology will thus combine concepts of flexible electronics, catalysis, and 3D bioprinting to build the world's first smart self-oxygenating tissues (SSOTs). The water will be converted to oxygen inside the BioGel via water-splitting using the 3D printed CP/GH/SNP electrode as the catalyst. This will be printed by an extrusion 3D printing technique to realize functional, implantable, and tether-free SSOTs. The ability to sustain life of the implantable SSOTs will be evaluated *in vitro* and *in vivo* to demonstrate the ability of our approach to sustain the survival of organ sized tissues.

When successful, this technology will allow the unprecedented feat of producing oxygen for an indefinite time in a highly controlled manner within 3D tissues. We hypothesize that this technology can prevent the detrimental effects of anoxia or poor vascularization that hinder the *in vivo* use of organ-sized tissues. This yet unrealized feat represents a key stepping-stone towards the development of viable clinically sized engineered organs.

587270 Development of an Osteochondral Implant Using Microencapsulated Mesenchymal Stem Cells

Osteoarthritis remains a poorly treatable outcome of traumatic cartilage injury as cartilage is unable to self-repair. Transplantation of an engineered, osteochondral tissue plug has been proposed as a better way to achieve repair of cartilage defects with implants that can readily integrate with the subchondral bone. Composite materials composed of chitosan and glycosaminoglycans (GAGs) have been shown to support rapid invasion and integration with host tissues post-implantation and are a promising platform for assembling such an osteochondral implant. In this study we explored the use of bone marrow-derived mesenchymal stem cells (MSCs) within glycosaminoglycan-chitosan capsules as a platform for generating spherical cartilage and bone tissue "modules" for use in assembling an osteochondral construct. Rat bmMSCs were suspended (1×10^7 cells/ml) in a solution of 1.5 wt% carboxymethylcellulose (CMC), 4 wt% chondroitin 4-sulfate (CS), and 1 mg/ml collagen type I. The suspension was then electrosprayed to generate uniform droplets that were captured into a stirred and grounded chitosan solution (0.6 wt%). Electrostatic interactions between negatively charged GAGs and positively charged chitosan results in the formation of hollow microcapsules (300-500 µm in diameter) entrapping MSCs at high density. The encapsulated MSCs were cultured under static conditions in either osteogenic or chondrogenic induction medium for either 28 days (osteogenic) or 42 days (chondrogenic). Between 28 and 42 days osteogenically induced capsules were maintained in MSC growth medium. Capsule cultures were sampled at intervals and analyzed to evaluate cell/biomaterial morphology, DNA content, and extracellular matrix composition. After loose capsule culture, osteogenic and chondrogenic capsules were used to generate osteochondral constructs as follows. Cultured capsules were first briefly washed with 0.1 wt% chitosan followed by a wash with 0.1 wt% glycidyl-methacrylated polygalacturonic acid (GMA-PGA) to deposit a photocrosslinkable polyelectrolyte film. A transparent cylindrical mold was then loaded with a 4 mm layer of coated osteogenic capsules followed by a 4 mm layer of coated chondrogenic capsules. The mold was then immersed in PBS and capsule fusion was induced by photocrosslinking with 400 nM light using LAP as the photoinitiator. The fused constructs were then cultured for an additional 15 days in MSC growth medium. Results to date demonstrated the feasibility of the culture, coating and photocrosslinking method of osteochondral assembly using long-term cultured GAG-chitosan capsules. In addition, GMA-PGA was found to be an efficient, photocrosslinkable surface coating for generating fused capsule constructs.

587282 Investigating Health Effects of Aerosol Particles on Single-Cells in a High-Throughput Air-Liquid Interface Platform

Air pollution and its detrimental health effects have been an increasingly alarming concern for the world. Cardiovascular diseases, chronic respiratory diseases, and different types of cancers can all be linked to air pollution effects. Furthermore, over 8.9 million people die per year from the direct effects of ambient particulate matter. One conventional method for studying the health effects of aerosol particles is to collect particulate matter (PM) from the atmosphere, resuspend the PM in cell culture media, add the PM/media to cell cultures, and assess single cell oxidative stress. However, this method does not resemble the *in vivo* conditions of alveolar macrophages. Alveolar macrophages are found in the alveoli of the lungs where the oxygen and carbon dioxide exchange between blood and air takes place and are a key cell type that is affected by PM. In this project, we aim to create a microfluidic air-liquid interface environment that mimics how alveolar macrophages are exposed to PM in order to study effects of PM at a single-cell level. Utilizing a microfluidic platform provides uniform microenvironment and enables high-throughput and high content single-cell analysis for hundreds of cells. We adapted a previously developed microfluidic single-cell analysis technology and developed a method for exposing cells to an air-liquid interface. We then characterized cells by microscopy and LIVE/DEAD staining to assess if cells remain viable during air exposure. Following this validation of the air-liquid interface platform, we will use it to study the single cell responses of intracellular reactive oxygen species in exposure to aerosol particles.

584092 Exploring a Missing Mechanism That Links Redox Signaling to the Actin Cytoskeleton

In the cell, reactive oxygen species (ROS) and the actin cytoskeleton work together to control a large range of activities, such as immune response, neural growth, formation of blood vessels, and spreading of cancer cells. However, our understanding of how ROS crosstalks to actin is very limited which has hindered the development of intervening agents. My project is to determine the biochemical mechanism by which a major generator of ROS molecules named NOX and a central regulator of the actin cytoskeleton named WAVE interact with each other. This interaction may serve as a key linkage between ROS and actin in the cell. To understand this mechanism, we use recombinant protein production and various biochemical assays to determine if and how a protein named p47phox can serve as a bridge to bring NOX and WAVE together. We design various mutant proteins to map the regions responsible for p47phox binding to NOX and WAVE, respectively. We also test if p47phox binding to the WAVE Regulatory Complex (WRC) to promote formation of the actin cytoskeleton. This knowledge will help in the understanding of how the cell coordinates ROS signaling and the actin cytoskeleton to drive various processes and has the potential to reveal new targets for therapeutic development.

585240 Grape Ripening inside and Outside the Fruit Region

Grape growth and wine production are dependent on location, temperature, and time. The time required for grapes to ripen also varies by species of grape and location of the grape cluster on the vine. There are two main regions on a grape vine, the fruit region (FR), and outside the fruit region (OFR). The FR is the lower-middle area of the vine where the majority of the fruit grows. The region OFR consists of the upper part of the vine where there is less fruit and more foliage. Throughout many harvest periods, Fero Vineyards & Winery has determined that grapes in the OFR ripen at a slower rate than the fruit in the FR. This is due to the amount of direct sunlight and leaf exposure. FR grapes receive a moderate to high amount of direct sunlight because the leaves are pruned in this area and are not covering the fruit. The OFR grapes receive less direct sunlight because they are covered by leaves. Direct sunlight has photosynthetic effects on ripening rates of grapes through its effects on the vine's leaves. This is because leaves provide the grapes with sugar to make them sweeter. Thus grapes outside of the fruit region do not receive the benefits of sunlight which affects their rate of ripening. Wine production proceeds almost exclusively from FR grapes, presenting an opportunity to reduce waste and increase production by determining when OFR grapes can be harvested.

While general characteristics of wine grape behavior is well known, regional variations in climate and sunlight require any model of grape ripening to be based primarily on local data. This research project focuses on the time dependence of ripeness for Pinot Noir, Lemberger, Riesling, and Gruner Veltliner grapes both FR and OFR. Ripeness is determined by the balance of brix (sugar content), ph, and titratable acidity. This data will aid Fero Vineyards & Winery in harvest time estimates and possible economic benefits from creating new OFR-based products.

Sources:

Zoecklein, B W, and B H Gump. *Evaluation of Grape Maturity and the Factors Impacting Maturity*. www.apps.fst.vt.edu/extension/enology/downloads/wm_issues/Fruit%20Maturity/Evaluation%20of%20Grape%20Maturity%20and%20the%20Factors%20Impacting%20Maturity.pdf
.

587234 Understanding the Relationship between Citrus Fiber Suspension Rheology and the Mesoscale Arrangement of the Fibers

The pectin depleted citrus peel is a side-stream from the pectin industry that is usually discarded. An excellent way to avoid wasting this part of the fruit is to turn it into citrus fibers by means of high pressure homogenization and drying for applications in the food industry. When mixed with water, these fibers form a viscous suspension that can be used as texturizer in food products Even so, it was observed that, after dehydrating and rehydrating a citrus fiber suspension, the viscosity of the suspension decreased significantly. This study aimed to understand why the decrease in viscosity occurs by analyzing the relationship between citrus fiber structure and rheology using viscometry, and sum frequency generation (SFG) vibrational spectroscopy. The effect of acidity and basicity on the viscosity of suspensions were investigated as well, and viscosity decreased as pH decreased. SFG indicated that the mesoscale arrangement of citrus fibers changes as the viscosities change in suspensions. From the findings of this study, the packing of the particles in mesoscale changed as fibers went through the dehydration process.

587764 Extraction of Biologically Produced Indigo Dyes

Indigo is a chemical that is used to dye clothing such as Levi Blue Jeans. The original method to produce dyes was extracting dyes from plants. However, this method took more than two weeks to ferment making it inefficient when the blue jean industry became popular. The current methods for producing clothing dyes are synthetic processes containing toxic and harmful steps. In order to reduce the health and environmental hazards, a biologically produced dye is being investigated as an alternative. In a previous study, it was found that a mutation in the enzyme, toluene ortho-monoxygenase, (TOM) of the bacteria *Burkholderia cepacia* would cause a change in color in rich media. The mutation A113V in the TOM enzyme results in the production of indigo. Using *E. coli* TG1, extraction and reduction methods are being investigated to develop a process for the production of indigo for use as a clothing dye. *E. coli* TG1 is grown for 48 hours with Isopropyl ß-D-1-thiogalactopyranoside (IPTG) added at the 24-hour mark. The IPTG activates the production of indigo. Once cells are grown, the cells are pelleted and frozen to rupture the cells walls. Currently, chloroform is being used, but different solvents are being investigated for indigo extraction. Other solvents being studied include dimethyl sulfoxide (DMSO), tetrahydrofuran, and transcutol. Testing various solvents will help determine a chemical that decreases health and environmental hazards. After extraction, the dye is insoluble in water and must be converted into the leuco-form by a reduction mechanism. The leuco-form is water soluble and will adhere to the clothing. The current industrial method uses sodium dithionite which cannot be recycled, so other methods are being investigated such as biological and electrochemical methods. After the clothing is dyed, air oxidation converts the dye back to its original form, fixing it in the clothing.

588124 Functionalization of Feed Spacers for Listeria Remediation in the Dairy Industry

Listeria Monocytogenes (LM) is a harmful foodborne pathogen that has the potential to contaminate industrial dairy processes. Pasteurization cannot completely prevent downstream recontamination, thus investigation into further sanitation methods is warranted. Quaternary ammonium compounds are of interest because the associated permanent positive charge promotes adhesion of negatively-charged bacterial membranes, along with subsequent leakage of cytoplasmic contents. Quaternary ammonium functional groups were integrated into the feed spacer material used in spiral wound membrane elements to treat LM recontamination. Unfortunately, European industrial regulations governing food processing may render the use of quaternary ammonium compounds impractical. Therefore, additional effort was devoted to the integration of phenolic compounds in the feed spacer material. Phenolic compounds induce enzyme deactivation and can contribute to degradation of bacterial cell walls. The immobilization of phenolic compounds on feed spacer material could offer an antimicrobial alternative to quaternary ammonium compounds when processing dairy products for the European market.

In this investigation, polypropylene feed spacer mesh was modified to produce an antimicrobial material. The feed spacer was modified by three methods. In the first method, vinylbenzyl chloride was reacted with hexadecylamine and photopolymerized in the presence of the polypropylene mesh. The presence of the amine was assessed by Raman spectroscopy and dye analysis. Bacterial toxicity was assessed via biotesting with a surrogate bacterial strain, *Lactobacillus plantarum*. The second and third methods for feed spacer modification were designed to immobilize phenolic functional groups on the polypropylene material. Photopolymerization was performed in both instances, for a tyrosine-vinylbenzyl chloride mixture in the second method and a eugenol-butanol mixture in the third method. The presence of phenolic functional groups was assessed by aromatic ring-based dye characterization, and quantitative analysis was performed on the dye results via imaging analysis software.

588150 The Effect of Milling on Protein Enrichment of Yellow Pea during Electrostatic Separation

The standardized processes implemented for the separation of agro-material components (i.e. protein, starch and dietary fibers) are wet fractionation methods that involve the use of concentrated acids and/or alkali as well as energy-intensive dehydration step. In our Bioprocess Engineering Research Laboratory at Howard University, the focus is on developing a novel tribo-electrostatic separation process to effectively produce a protein-rich fraction of an agro-material in a water- and chemical-free environment. Our custom-built tribo-electrostatic separator encompasses of a fluidized bed, polytetrafluoroethylene (PTFE) tribo-charging tube, and a rectangular separation chamber supplied with two copper-plate electrodes. For this study, yellow pea flours milled using two different milling types (i.e. pin and Ferkar milling systems) at two different milling intensities were first fluidized under the influence of pressurized air and then traveled through PTFE tribo-charging tube where protein particles were charged positively and attracted by the negatively-charged electrode plate. Our results highlight the importance of the milling types and intensities along with the air flow rates and electric field strengths for effective enrichment of yellow pea protein particles at their native states. Using the optimal milling and operating conditions, the original yellow pea protein content is increased by about 60% in the final collected protein fractions.

588188 High Temperature Measurements of Multicomponent Foods

Low-water-activity and low-moisture foods are generally considered safe due to the inability of microorganisms to grow under such conditions. Recently, there have been a surprising number of recalls on low-water-activity foods due to pathogens such as *Salmonella agona*. To understand this issue, we investigated the dependence of water activity on *S. agona* inactivation in multicomponent (carbohydrate, protein, fat) food systems. Initial results indicated that a high-fat multicomponent food requires a longer time for complete *S. agona* inactivation. We conducted a follow-on study to confirm water activity at high temperatures and the starting moisture contents of each multicomponent food system using a series of moisture content and water activity measurements. To date, the initial and current data have been consistent across the studies. These results imply that further studies are required to understand how different types of carbohydrates, proteins, and fats will differ while undergoing *S. agona* inactivation.

588306 Land of Hopportunity: Characterization of New Mexico-Grown Hops

Essential to the craft of brewing are hops: the flowers of *Humulus lupulus* which provide beer with bitterness, flavor and aroma. Containing a unique combination of bittering acids (a and ß), essential oils, and polyphenols, each hop variety contributes to distinguishing brews. Recent research on hops has focused on the effects of growing location on the characteristics of the hops and the effects on the finished beers. Over the last two years, the NMSBrew team at New Mexico State University have been developing a technical assistance program to provide hop characterization analyses and best practices information to New Mexico hop farmers. This season, hop characterization focused on sensory evaluation, UV-Vis spectroscopy measurement of a and ß acids, hop storage index based on moisture and temperature, and steam distillation of essential oils. Relatively new to the picture is *Humulus lupulus neomexicanus,* a variety of hops native to New Mexico with unique (but not always pleasant) aroma and flavor profiles, and relatively high levels of polyphenols. While some *neomexicanus* cultivars are available commercially, much more work is needed to understand the potential of this crop in the Southwest to support the local brewing industry.

587042 HA-CNT Nanofibers with Anti-Inflammatory Microspheres for Enhanced Nerve Regeneration

This research aims to develop a material to promote nerve growth following spinal cord injury (SCI). The proposed design is to develop a conductive nanofiber and microsphere composite for electrical stimulation induce drug delivery to improve SCI regeneration. The main focus of this project is to fabricate the material and evaluate the drug release profile through the microspheres.

Microspheres are made from gelatin with an oil, water mixture and crosslinked with genipin (GMS). GMS were loaded with methylprednisolone, which is used in treating inflammation in SCI. Nanofibers were electrospun methacrylated hyaluronic acid (MeHA) with carbon nanotubes (HA-CNT). GMS were incorporated into the HA-CNT electrospinning solution. Currently, three different concentrations of GMS were tested, 0, 30, 60 mg GMS per ml of electrospinning solution. Nanofibers were stimulated with 200 mV/mm bi-phasic square AC wave. A sample from each well is taken at increasing time intervals. These were measured for absorbance at 243nm to determine the concentration of methylprednisolone released in order to determine the methylprednisolone release profile.

GMS HA-CNT was spun with both 3ml and 5ml solutions. With the 30mg/ml solution, the microspheres were well dispersed, however 60mg/ml formed some large structures (figure 1). Clumping of the GMS is an issue due to uneven local methylprednisolone release and delaying the release of trapped GMS's in the center of the clumps. Breaking up the microspheres and mixing may become an issue in the future for this concentration of GMS to HA-CNT

The GMS HA-CNT was electrospun successfully in three different concentrations of microspheres. 5ml of solution will be used due to better coating for the electrical stimulation. Initial electrical stimulation studies showed drug release following stimulation. We are currently testing the drug release for electrical pulses which may allow us to partially release the drug and further control release dynamics. Future plans for this research will be include growing neurites and astrocytes on the material and determine the effect of drug release on cell behavior.

Figure 1. SEM image of the MeHA-CNT GMS

588294 Formulation of Magnetic Particle Imaging Tracers Using Flash Nanoprecipitation

Magnetic particle imaging (MPI) is a novel and rapidly developing *in vivo* imaging method, which boasts significant advantages over traditional imaging methods (e.g. fluorescence or x-ray imaging). These advantages include the lack of ionizing radiation and tissue depth attenuation. MPI image quality depends on magnetic nanoparticle tracers, of which there is a large variety, including ferrocarbotran nanoparticles, magneto-liposomes, micelles, and individual PEG coated iron oxide nanoparticles (IONP). Here, we propose a new MPI tracer, which uses IONP clusters encapsulated by Flash Nanoprecipitation. Flash Nanoprecipitation is a size-tunable process which encapsulates hydrophobic agents within amphiphilic block copolymer membranes. This process is typically used in the realm of drug delivery to achieve controlled drug release rates, high drug loading (lower doses), and cell targeting (modifying block copolymer). By using FNP to encapsulate IONPs, however, we create magnetically controlled nanocarriers (MCNCs) comprised of IONP clusters with a block copolymer membrane. These MCNCs can range from 50 nm to150 nm in diameter. We show that clustering and encapsulating IONP into MCNCs has a negligible effect on MPI signal, and that MCNC tracers have comparable MPI resolution to unencapsulated IONP. By showing that FNP-encapsulated nanoparticles maintain their MPI properties, we pioneer the development of new MPI tracers that have greater capacity for tissue diffusion, efficient dosing, and cell targeting capability.

585868 Production of Dicarboxylic Acids By Biotransformation of Oleic Acid Using *Candida Viswanathii*

Dicarboxylic acids (DCAs) are wildly used in biopolymer production and personal care products. Among various DCAs, long-chain dicarboxylic acids (LCDCAs) are used mostly in production of biodegradable polyesters and polyamides such as nylon. Production of DCA through biotransformation has drawn great interest from researchers recently for sustainability reasons. Previous studies have demonstrated that the yeast *Candida viswanathii* can potentially be used for production of dicarboxylic acids from a variety of different fatty acids, some of which are found in distillers corn oil (DCO), a byproduct of dry grind ethanol production.

In this work, small scale fermentations of *Candida viswanathii* under various controlled conditions were conducted. As one of the major components in DCO, oleic acid was chosen as a model feedstock. Conversions with both the free acid and the methyl ester were explored. Products extracted from fermentation broth were analyzed with a previously developed high-performance liquid chromatography (HPLC) method. Initial results indicate the consumption rate of oleic acid was faster than that of methyl oleate. Among components extracted from fermentation broth, DCAs were detected at lower levels than expected based on feedstock consumption. Improvements to the extraction protocol are being explored to ensure high recovery of DCA. In addition, interfering medium components imposed difficulty in quantifying concentrations of DCA with the existing HPLC method. Adjustment of column temperature in the HPLC method reduced such signal interference.

587369 Using Zebrafish to Identifying the Role of Wisp1 in Melanoma Metastasis

Wnt family members play an important role in many developmental processes in humans and other animals, they also have been linked to tumorigenesis. A gene named WISP1(Wnt inducting signaling pathway1), is a member of this family, it's expression is increased in melanoma(a deadly type of skin cancer) and is associated with reduced overall survival of patients diagnosed with primary melanoma. The goal is to find the role WISP1 plays in melanoma metastasis. Zebrafish was the model selected for this purpose,because they have melanocytes that can become malignant and develop melanoma through specific mutations. Zebrafish also have two homologs of the gene WISP1 in their genome. Here, we knocked down each homolog of WISP1 separately in wildtype zebrafish, using MOs (Morpholino oligonucleotides) and compared their phenotype to other control groups to understand what role WISP1 plays in Zebrafish morphology and then perform a dose control experiment to ensure that the resulting phenotype is a result of the knockdown, and not any other off-target effects. Results so far indicate that knockdown groups exhibit a distinguishable phenotype compared to control groups. After determining the ideal dose to knockdown the gene of interest, we would then incorporate the hWISP1 gene in the minicoopr,introduce this modified plasmid in Casper zebrafish and test for changes in melanoma development. The minicoopr plasmid would rescue melanocytes via re-expression of mitf while introducing specific mutations that will promote melanoma development

585881 Testing Methyltransferase Inhibitors to Therapeutically Target Glioblastoma Cancer Cells

Glioblastoma is an extremely aggressive and deadly cancer where patients have limited therapeutic treatment options. Currently, most patients must resort to chemotherapy or surgery in order to treat glioblastoma. Therefore, developing a viable therapeutic treatment option for patients suffering from glioblastoma would represent a major advancement. The process of protein methylation has been an understudied cellular process despite its importance, and the Graham Lab has developed novel proteomic techniques utilizing mass spectrometry to study peptide methylation.

Due to the proximity of the *MTAP* gene (5-methylthioadenosine phosphorylase) to the tumor suppressor *CDKN2A*, *MTAP* is deleted in about 50% of glioblastomas. When *MTAP* is deleted, there is a buildup in the concentration of MTA (methylthioadenosine), which acts as a natural inhibitor for PRMT5, a methyltransferase enzyme responsible for cellular protein methylation, further sensitizing the cell to drug inhibition of PRMT5. Data from the Graham Lab indicates that PRMT1 and PRMT5 inhibition have a synergistic effect. Furthermore, this trend should only selectively target glioblastoma cells that have *MTAP* deleted, as normal cells with MTAP should be able to withstand drug-mediated inhibition.

We hypothesize that when both PRMT1 and PRMT5 are inhibited within a cancer cell deleted for *MTAP*, the cell will die because its methyltransferase enzymes will no longer be able to methylate essential proteins. To test this hypothesis, we will culture *MTAP*-deleted cancer cell lines engineered to overexpress either luciferase (negative control) or MTAP (reconstitution of MTAP). We will treat the cells with different concentrations of PRMT1 and PRMT5 drug inhibitors and analyze cell growth. We will then lyse the cells and use both High pH SCX (Strong Cation Exchange) and Sequential IAP (Immunoaffinity Purification) to enrich for methylated peptides and investigate the samples using mass spectrometry. Both SCX and IAP are powerful proteomic enrichment techniques that can be used to study methylated forms of arginine. Using the mass spectrometry results, we will identify differential methylation sites to determine the impact of global arginine methylation.

We expect to pinpoint a specific concentration of PRMT1/PRMT5 inhibitors that will selectively target cells deleted for MTAP while keeping normal cells alive. We expect for the PRMT1 and PRMT5 inhibitors to display a synergistic effect. Additionally, we expect to learn more about how methyltransferase inhibitors impacts cellular survival. What we learn about protein methylation has the potential to advance our knowledge on cellular processes, and similar methods can be applied to therapeutically target different diseases.

585619 AKT but Not MYC Promotes Reactive Oxygen Species-Mediated Cell Death in Oxidative Culture

Oncogenes can generate metabolic vulnerabilities in cancer cells. Here, we tested how AKT and MYC affect the ability of cells to shift between respiration and glycolysis. Using immortalized mammary epithelial cells, we discovered that constitutively active AKT but not MYC induced cell death in galactose culture, where cells must rely on oxidative phosphorylation for energy generation. However, the negative effects of AKT were short-lived, and AKT-expressing cells recommenced growth after ~15 days in galactose. To identify the mechanisms regulating AKT-mediated cell death, we used metabolomics and found that AKT cells dying in galactose upregulated glutathione metabolism. Next, using proteomics, we discovered that AKT-expressing cells dying in galactose upregulated nonsense-mediated mRNA decay, a marker of sensitivity to oxidative stress. We measured levels of reactive oxygen species (ROS) and discovered that galactose induced ROS in cells expressing AKT but not MYC. Additionally, ROS were required for the galactose-induced death of AKT-expressing cells. We then tested whether these findings could be replicated in breast cancer cell lines with constitutively active AKT signaling. Indeed, we found that galactose induced rapid cell death in breast cancer cell lines and that ROS were required for galactose-induced cell death. Our collective results demonstrate that AKT but not MYC induces a metabolic vulnerability in cancer cells, namely the restricted flexibility to use oxidative phosphorylation.

587416 Non-Standard Amino Acid Incorporation in Proteins for Biofilm Surface Attachment

Biofilms are consortiums of microbes in which cells stick together or on a surface embedded within a slimy extracellular matrix composed of extracellular polymeric substances mainly polysaccharides, proteins, lipids and DNA. The biofilms formed on living or non-living surfaces prevalent in hospital settings are generally by pathogenic microorganisms such as Staphylococcus spp, Escherichia coli, or Pseudomonas aeruginosa leading to numerous health infections. Microbes form biofilm in response to various factors, which include nutritional cues, or in some cases, by exposure of planktonic cells to sub-inhibitory concentrations of antibiotics. A cell switches to the biofilm mode of growth undergoes a phenotypic shift in behavior in which large suites of genes are differentially regulated and gradually develop resistance to conventional anti-microbial drugs. Therefore, in this study we are exploring the interaction of probiotic and pathogenic bacterial species to make a monolayer probiotic biofilm on any surface to prevent pathogenic biofilm attachment and formation. Usage of probiotic bacteria might lead to a possible alternative solution for inhibiting and killing pathogenic microorganisms rather than developing a new array of antibiotics.

585067 Qualitative Phytochemical Analysis of a Tropical Plant in Search for a Cancer Treatment

According to the CDC approximately 38.4% of men and women will be diagnosed with cancer at some point during their lifetimes (based on 2013–2015 data). My own aunt lost her second battle with cancer a little over two years ago, her devotion to Moringa Oleifera is the inspiration for this project. In Ethno-Medicine arena, Moringa Oleifera is known as the "miracle" tree. The overall goal of this research project is to extract cancer fighting (antioxidant) properties (if any).The preliminary results after phase I of this project reveal the presence of saponin and glycosides/glycolipids in some of the samples tested. Saponin is just one of the important naturally occurring antioxidants thought to be in Moringa. Saponins are considered to have strong antitumor effect on many cancer cells. "Several saponins inhibit tumor cell growth by cell cycle arrest and apoptosis with IC50 values up to 0.2 mM...found beneficial targeted on inhibition of tumor angiogenesis by suppressing its inducer in the endothelial cells of blood vessels, and then on prevention of adhering, invasion, and metastasis of tumor cells," (Man, 2010). Further experiments are needed to analyze the rest of the samples and identify the most promising anti-cancerous properties. Once the anti-cancerous properties are identified, conduct In-Vivo and In-Vitro anti—cancer assays.

585016 Rheological Study of Natural Starch Biomaterial for Building Biomimetic Immune Tissue

Mimicking lymph tissue is challenging, yet crucial for understanding cell-mediated immune responses, and translating into clinical immunotherapy treatments. The 3D bioprinting technology is emerging for creating in-vivo like scaffolds as extracellular matrices, allowing spatial arrangement and 3D tissue culture of cells. However, existing bioinks still rely on crosslinkers and chemical modifications to produce printable bioinks, particularly for most biocompatible collagen materials. The chemical crosslink and modifications can introduce toxicity and unwanted cell-matrix interactions. Thus, we proposed the natural starch material to blend with collagen as a novel, non-toxic 3D bioink. We systematically investigated the rheological behavior of starch under different concentrations and gelation heating temperatures. We also studied the mixture ratio with collagen for developing the 3D printability. The optimized rheology is correlated with 3D printing shape fidelity, which leads to a new, non-chemically crosslinked collagen bioink. We also evaluated the biocompatibility of this bioink using MTT assay. The cell viability with our developed bioink scaffold showed good consistency with biocompatible collagen materials. This study introduces a new class of 3D bioink which could advance 3D tissue engineering and facilitate fast growing of immunotherapy treatment strategy.

587424 Elucidating Protein Remodeling Mechanism By Molecular Chaperones

HtpG is a molecular chaperone protein found in *Escherichia coli*. This protein belongs to a family of ATP-dependent molecular chaperones and couples conformational changes from ATP hydrolysis to protein remodeling; HtpG assists other proteins in reaching their native conformation, preventing aggregation, and mediating targeted unfolding and disassembly. In this study mutants of HtpG defective in ATP binding or substrate binding were constructed. HtpG mutants E34A and W467R were made by site directed mutagenesis. HtpG wild type and mutant proteins were overexpressed in E. coli, collected, and purified using Fast Protein Liquid Chromatography. The purified proteins were tested for ATPase activity, stimulation of ATPase activity by substrate proteins, and prevention of aggregation assays. Both mutants studied, HtpG E34A and W467R, were defective in chaperone activity and prevention of aggregation. Comparing these mutants with their wild-type form, they exhibited decreased rates of ATP hydrolysis and were not stimulated by model substrate proteins such as rhodanese and citrate synthase. Also, in the prevention of aggregation assay using urea- denatured Rhodanese, HtpG wild-type and ATP and substrate defective mutants were unable to prevent aggregation.

586001 Comparison of Mathematical Solution Approaches for a Simple Theoretical Dialysis Model

Emerging new disciplines and technologies have made it possible to move in the direction of smaller and wearable artificial organs. This is the case of alternative treatments for renal insufficiency which may include dialysis. Miniaturization of traditional methods, such as hemodialysis and peritoneal dialysis, has open the doors for the development of portable and implantable devices. On the design aspect, considerable attention is been devoted to analyze the fundamental principles governing the intended down scaling for this type of applications. Typically, the analysis involves the solution of non-trivial differential equations followed by simulation of different performance scenarios. The approach, predictive in essence, makes experimental work more effective and less intense as optimal device designing can be achieved using less resources.

In this contribution, the focus is on the analysis of the mathematical methods utilized to solve the fundamental differential equations describing the dialysis phenomenon. These methods may have an impact on the prediction of performance and therefore the potential design of a device. For the typical dialysis model, two mathematical methods apply to the solution of the resulting differential equations: variable separation and the asymptotic solution. In this study, the molar species continuity equation is solved analytically using these two methods and the prediction results are compared with the numerical values. A summary of these results will be provided with important insights into the development and design of wearable artificial kidneys.

585889 Effects of Altered Quorum Sensing Regulatory Networks on *Pseudomonas Aeruginosa* pathogenesis

*Pseudomonas aeruginosa*is an opportunistic bacterial pathogen that commonly causes chronic lung infections in cystic fibrosis (CF) patients. People with CF acquire *P. aeruginosa*from their environment, after which the pathogen undergoes genetic and phenotypic changes to adapt to the lung. Many of the adaptations occur in quorum sensing (QS) systems, which regulate cell-to-cell communication that aids in coordination of group behaviors within a bacterial population. *P. aeruginosa*expresses a wide variety of QS-regulated phenotypes that can be used as prognostic markers of infectious stage within each patient. This study examines the link between the accumulation of QS genetic mutations and phenotypic expression in *P. aeruginosa* laboratory strains and clinical isolates. We utilized several plate-based and colorimetric assays to quantify the production of pyocyanin, rhamnolipids, and protease from the lab strain PA14, its QS gene deletion standards, and paired clinical early- and late-stage chronic infection isolates across 16 patients. Exoproduct production of each isolate was compared to the mean production of pooled isolates to classify high producing (QS-sufficient) and low producing (QS-deficient) isolates. We also sequenced important QS genes in these clinical isolates to identify mutations in order to link changes between genotype and phenotype. Future research of the QS regulatory networks will identify whether reversion of genotype will result in corresponding phenotypic changes in QS-deficient chronic infection isolates.

587952 Modeling the Amphiphilic Behavior of Quorum Sensing Signaling Molecules and Their Interactions with Lipid Bilayers

Quorum sensing is a bacterial communication process used to coordinate group behavior, such as the formation of antibiotic-resistant biofilms. Quorum sensing is controlled by the secretion and sensing a group of amphiphilic quorum sensing signaling molecules (QSSMs). In general, Gram-negative bacteria mainly use *N*-acyl homoserine lactones (AHLs), consisting of a polar homoserine lactone head group and a nonpolar aliphatic tail, as QSSMs. The urgency and severity of antibiotic resistance has led to interest in understanding the behavior of QSSMs in order to develop strategies to regulate bacterial communication. In this work, we developed a new protocol based on the conductor-like screening model for realistic solvation (COSMO-RS), in conjunction with molecular dynamics (MD) simulations and quantum chemistry calculations, to investigate the amphiphilic behavior of AHLs by studying their self-assembly and their interactions with lipid bilayers. Atomistic MD simulations can determine the aggregation free energy and free energy of partitioning an AHL into a lipid bilayer with a large computational cost, whereas the new protocol incorporating COSMO-RS theory is at least 300 times faster than the MD simulation protocol with minor error. The results from the new protocol indicate that AHL head groups containing an additional 3-oxo group inhibit aggregation compared to their counterparts with only a single 3-oxo group in the head moiety, which is in agreement with the MD simulation results. The COSMO-RS results also indicate that longer aliphatic (more hydrophobic) tails facilitates aggregation and partitioning into a lipid bilayer, again in agreement with the MD simulation results. Comparison of the free energies for aggregation and partitioning indicates that aggregation is generally preferred, suggesting that aggregation may stabilize QSSMs in solution. Our faster COMSO-RS-theory-based protocol could be used as a high-throughput screening method for QSSM amphiphilic behavior and help guide the design of new AHL delivery techniques based on their membrane transport for better controlling the bacterial communication.

586007 Estimation of Subcutaneous Electroosmotic Velocities in a 2-D Iontophoretic System: An Area Averaging Approach

Although the technique is well known since approximately 3,000 BC, transdermal drug delivery became a common practice starting only five decades ago. Initially used to treat skin disorders, today's transdermal delivery is an alternative treatment for patients with vascular limitations, epilepsy, advance pregnancy, heart condition, artificial joints, and lumps among others. Iontophoresis, a modern version of this old technique, uses electrical field gradients over the skin to more accurately and in less time deliver drug to a specific area of the body. The process facilitates the transport of species by two mechanisms, electrophoresis and electro-osmosis. Depending on the order of magnitude of the electrical field gradient one of these two mechanisms can be dominant of the procedure. Investigating the best or most effective conditions at which electro-osmosis has an improvement impact in drug delivery is the driving force for the present effort. This is the continuation and follow up of a previous contribution presented on the topic[1].

In this contribution, the mathematical governing expressions capturing the fluid dynamic of the system, coupled with the electrical field, is solved by applying the method of spatial averaging to the bulk movement of the transporting fluid. In addition, explicit analytical expressions are derived for the local and average electrostatic potential, and the local and average electroosmotic velocity. Illustrations of how a 2D electrical field can aid in increasing drug delivery with the target system. Practical conclusions and potential applications with less disruptive treatment protocol are discussed.

1. Analysis of a 2D Iontophoretic System Using an Area Averaging Approach, ***AIChE Annual Meeting***, Pittsburgh, PA Oct 28 - Nov 2, 2018. Alisa J. Kidwell1, Mario Oyanader2 and Steffano Oyanader.

588060 Antimicrobial Activity of a Synthetic Polythiophene Oligomer

Cationic polythiophenes are used as antimicrobial compounds due to their ability to kill bacteria via the production of reactive oxygen species from the absorption of visible light. They are also able to damage negatively charged cell envelopes by selectively associating with them. This study demonstrates the ability of imidazolium- functionalized poly(3-hexylthiophene) (P3HT) to inactivate Gram-negative Escherichia coli under photolysis and dark conditions. Cell plating is used as a method to quantify cell death. The compound shows high levels of killing at 10 (μg/mL) polymer concentration for the Escherichia coli species after photoactivation as well as high levels of dark inactivation.

587780 Application of Imaging Fluorescence Correlation Spectroscopy to Model Cell Membranes during Immunoreceptor Signal Transduction

The goal of this project is to understand the dynamic organization of the Rat Basophilic Leukemia (RBL) cell plasma membrane in its resting state. This is the first step in evaluating the extent of membrane remodeling that occurs during and after immunoreceptor signal transduction. We employed Imaging Fluorescence Correlation Spectroscopy (Imaging FCS) for this purpose. Imaging FCS is used to quantitatively measure diffusion modes of fluorescently labeled plasma membrane proteins. The changes of diffusion properties will be monitored as a function of physical and biochemical properties such as surface charge, post-translational modification, characteristics of transmembrane regions and availability of regulatory domains within the protein quaternary structures of membrane proteins. The observed dependence of membrane diffusion on these physico-chemical properties will eventually lead to a comprehensive understanding of the resting state of the RBL cell membrane.

588152 Investigation of Bacteriophage Encoded Toxins through Gene Analysis and Development of a Cell-Free Protein Production System

Prophage are viral sequences that have integrated into bacterial genomic DNA. Under certain conditions, these phage can be induced to replicate independently within the cell which allows the phage to disperse and infect other cells, thereby facilitating the horizontal spread of virulence and antibiotic resistance traits. We investigated the presence of a related group of prophage incorporated into the genome of *Xanthomonas* bacteria, a genus of common bacterial plant pathogens. We have designated this group of bacteriophage V-phage after the initially observed 'Valentino' and also for their potential role in virulence. We assessed the scope of bacterial strains possessing the phage in their genome and examined the conditions under which the prophage containing zot would enter the lytic cycle to begin actively replicating.

Our aim is to characterize the specific proteins encoded for by the prophage, specifically focusing on the Xap toxin, which could be contributing to virulence in *Xanthomonas* and may be related to the zonula occludens toxin (zot) commonly produced by pathogenic strains of *Vibrio cholerae*. Toxins frequently impair bacterial growth during characterization experiments. For this reason, we are developing a cell-free approach to toxin production. This method, along with inducible cell-based systems, can be used to produce, purify, and study bacteriophage toxins such as Xap. Information regarding Xap and the group of V-phage could aid in understanding the role that bacteriophages and related molecules play as reservoirs of virulence factors in both plant and mammalian pathogens. Furthermore, future work to characterize Xanthomonas gene expression and related phage could aid in the creation of target-specific antimicrobials for common plant pathogens.

585283 Sublimation Process of Ibuprofeniconamide Eutectic

Drug eutectic can effectively improve drug activity and bioavailability, especially for the clinical effect of oral drugs with poor water solubility.Ibuprofen is a kind of antipyretic, analgesic drug, but its water solubility is poor, its clinical effect is not good. Ibuprofen-nicotinamide eutectic was successfully prepared by sublimation crystallization technique, which solved the problem of poor water solubility of ibuprofen and achieved good clinical effect.

In this paper, the experimental temperature of ibuprofen-nicotine eutectic prepared by sublimation crystallization technique is 85°C, the experimental temperature of ibuprofen placed in low temperature region is 75°C, and the total pressure is a standard atmospheric pressure. The mass flow rate of carrier gas is 30 ml/min. This experiment also used cooling method and grinding method as auxiliary means, combined with optical microscope, infrared spectrogram, XRD and other characterization means, proved that ibuprofen-nicotinamide eutectic was successfully prepared by sublimation crystallization technique in this experiment, and it was proved that ibuprofen-nicotinamide eutectic was successfully prepared by sublimation crystallization technique. And it has high yield and purity. It is proved again that sublimation crystallization technology has great scientific research and commercial value.

584448 Effective Modeling of the Coal Combustion in Fluidized Bed Reactor Using the Aspen Plus.

Despite numerous attempts to replace fossil fuels with biomass for the energy production to reduce the environmental pollution, coal remains one of the main sources of energy contributing to the 25% of the world's energy generation. The need to study the coal combustion process remains crucial to develop its usage to an efficient level. For this purpose, several models of the coal combustion process were studied and developed both experimentally and computationally using different types of coal and different types of reactors like the fixed and fluidized bed. However, these works are insufficient to design a general model of the coal combustion process due to their specificity in process and design. Therefore, it is important to design the process simulations for the coal combustion that reproduce the real behavior of the process. This became possible with the recent introduction of new tools by Aspen Plus software that proposes a new built-in operation model for the design of the fluidized bed processes. Herein, we design and study the coal combustion process applying the novel tool at different operational conditions, i.e. temperatures and coal/air flow rates. The simulation explores the hydrodynamics and kinetics of the combustion process. To validate the performance of the simulations, the fuel combustion process was performed on a pilot-scale fluidized bed combustor using the Ekibastuz coal, a type of low-grade Kazakhstan coal. Obtained experimental results were used for the validation and fine-tune the Aspen Plus model. Eventually, it was noticed that both models are in a close agreement representing accurate results for the gas compositions, including SO_x and NO_x measurements, and the hydrodynamics of the reactor. Moreover, to examine the temperature and feed flow rate effect on the process efficiency and flue-gas compositions several sensitivity analyses were performed. Taken together, work assist in the development of an efficient and accurate model for the coal and biomass co-combustion processes designed using the fluidized bed reactor.

585665 Production of Energy from Flare (Zero Flaring)

Chemical waste is one of the most hazardous issues of the industry. So, when the chemicals are burned, they emit harmful gases in the environment and increase the pollution around the world. The statistics show around 150 billion cubic meters of the flaring gas is vented to the air each year. Also, flaring gases take around 30% of the total yearly gas. So, flare gas is a device which is used to burn waste chemical in industries. The research shows the different types of waste chemical in each industry. So, increase in oil and gas industries equals to an increase in the amount of waste gas which burns per hour. As a result, there is an increase in air pollution.

In order to overcome this problem. A plant should be designed which produces electricity using a fairly new technology known as Fuel Cells. This method is highly efficient, cost effective and clean. There is no greenhouse gases production as there is no burning of gas involved therefore, no air pollution.

After extensive research and study in this field, I initially designed the project by adopting research into the appropriate type of cell to this project until I got satisfactory results. The results showed that the percentage of energy production increased by changing the factors affecting the cell. On the other hand, I studied the positives of the used equipment and compared it to the rest of the equipment. Fuel cell has Quiet Operation, Scalability, Small area required etc. Therefore, the fuel cell has the lesser cost while compared to gas turbine and other type of equipment to produce electricity.

Also, I did research for the amount of production form fuel cell which is around 4.5 MW per one ton of gas. So, the amount of gas assumed in this study equals to 10.5 ton/hr from flare system related and verified with the data from ORPIC. After long calculations, the amount of electricity produced from this process will be around 47.25 MW per hour. That means the cost of electricity per day is around 103080.20 USD and the cost of electricity per year is around 37108871.81 OR.

586376 Cold Flow Properties of Biodiesel Blended with Model Diesel Solutions

Biodiesel is widely becoming a viable alternative to petrodiesel due to its carbon neutrality and the sustainable future it offers. However, the use of biodiesel is inhibited by its inferior cold flow properties, specifically, its higher cloud point. This issue is problematic in colder climates, because the biodiesel will solidify at temperatures of around 0 to -10°C, approximately 20-40°C higher than petrodiesel. While the cloud point of biodiesel can be depressed by varying the fatty acid methyl ester (FAME) composition, it is important to consider other viable strategies for cloud point depression. One such method is to mix biodiesel with petrodiesel, creating blends with favorable cold flow properties.

This work considered blends of biodiesel and petrodiesel to understand the effect blending has on cold flow properties. Model diesel solutions were created using decane, dodecane, hexadecane and mesitylene to investigate the influence of different diesel components on important cold flow properties. These model diesel solutions were mixed with biodiesel to create B0, B25, B50, and B100 fuels, the phase transitions of which were studied using differential scanning calorimetry. The experimental cloud points of blends were used to consider the effect of the biodiesel fraction as well as the components in the model diesel solution. Additionally, the heats of crystallization were measured to determine the heats of mixing corresponding to each blend. Furthermore, this work used ideal solution theory to predict the cloud point, and consequently study how the blends deviate from ideal solution behavior. The difference from ideality was considered based on the biodiesel fraction and the model diesel solution components.

584109 Development of Falling Thin Film Liquid Phase Electrolyzer for CO_2 Reduction

Anthropogenic CO_2 emissions are on the rise despite rapid advances in renewable energy. This discrepancy can be attributed to the difficulty in replacing established technologies, which are heavily reliant on fossil fuels. Electrochemical CO_2 reduction offers a promising solution to reduce industrial CO_2 emissions by enabling the production of carbon-negative and profitable industrial chemicals.

For electrochemical reduction to be feasible, the electrolyzer needs to achieve several figures of merit: low cost, high current density, high energy efficiency and stability. Liquid phase electrolyzers (LPEs) confer an advantage over gas phase electrolyzers (GPEs) by operating at higher current densities, which may lower capital costs. Additionally, LPEs can produce a wide variety of value-added carbon products such as plastic monomers provided that a selective and active catalyst can be discovered; GPE products are restricted to C_{1-3} products with a low boiling point which typically also have low market value.

The primary barrier to industrially scaling up LPEs is the occurrence of liquid flooding of gas channels that impairs the mass transport of CO_2 to the catalyst. However, the chlor-alkali industry has developed falling film electrolyzers to circumvent the pressure differential at the root of the flooding problem. This technology is considered the state-of-the-art in large scale electrolyzers. In a falling film configuration, the electrolyte runs from the top of the cell down by action of gravity, the small liquid volume reduces the hydrostatic pressure, eliminating differences in pressure over the height of the electrode. Drawing from this established configuration, we created an electrolyzer prototype using 3D printing that does not suffer from flooding.

Using 3D printing, we also created a proprietary percolator design which increases electrolyte mixing while optimizing the ionic conductivity. The design ensures continuous wetting of the entire electrode surface with minimal blocking of catalyst surface area. The architecture of the cell prevented fluid flooding of the catalyst, reduced cell resistances resulting in enhanced current densities; thereby indicating that this design is advantageous for scaling to industrial dimensions for CO_2 electrolysis.

586529 Subsurface Modelling of Deep Direct-Use (DDU) Geothermal on the West Virginia University Campus - Morgantown, WV

Subsurface modelling of Deep Direct-Use (DDU) geothermal on the West Virginia University Campus - Morgantown, WV

Kevin B. Donnelly[1], Nagasree Garapati[1], Yingqi Zhang[2], Pierre Jeanne[2]

[1] *Department of Chemical and Biomedical Engineering, Benjamin M. Statler College of Engineering and Mineral Resources, West Virginia University, Morgantown, WV 26506.*

[2] *Lawrence Berkeley National Laboratory, Berkeley, 1 Cyclotron Rd, Berkeley, CA 94720*

The north-central region of West Virginia hosts a unique geological opportunity for the development of deep direct-use (DDU) geothermal energy. Being in the midst of this region, West Virginia University's (WVU) Morgantown campus could potentially supply its heat from DDU geothermal in an effort to increase renewable energy on campus. This work features the results of simulations done for a subsurface reservoir using TOUGH2/EOS1, a program specifically developed to simulate coupled transport of water and heat in geothermal reservoirs. Reservoir behavior (pressure, temperature) over 60 years of production time is analyzed. In order to achieve a more accurate pressure profile at well bores, a radial mesh is used around injection and production well bores in the simulations. Reinjection temperatures and flow rates were varied for different well configurations to analyze their effects on the reservoir productivity. The reservoir behavior is analyzed through temperature and pressure contour maps created at various time periods within the 60-year timeframe. Longevity of the reservoir decreased with increased mass flowrate and lower reinjection temperatures. Comparing single production wells to multiple production wells showed that at the same total production flow rate, single production wells experienced faster thermal breakthroughs and higher-pressure drop than the multiple production wells. These results from numerical simulation help in designing the optimum well configuration.

Acknowledgment: This material is based upon work supported by the U.S. DOE's Office of EERE under the GTO, under Award Number DE-EE0008105.

587137 Hydrogen Rich Syngas Production through Biomass - Flare Gas Synergistic Co-Processing on Carbon Nanofiber (CNF) Based Catalyst

Researchers at West Virginia University have proposed a new reaction pathway focused towards process development of synergistic biomass – flare gas coprocessing to produce hydrogen rich syngas. The novelty of such a process is in-situ co-processing of waste natural gas (flare gas) and biomass along with CO_2 utilization to produce syngas with H_2:CO = 2. Methane activated biomass gasification produces hydrogen rich syngas with >80 mole% H_2 in product gas. With addition of 1% CO_2 in the gas feed along with 5% CH_4, synergistic methane – biomass – CO_2 reaction system produces controllable ratio of H_2 and CO based on the requirement downstream. New catalyst was developed with carbon nanofiber (CNF) support impregnated with iron (Fe) and molybdenum (Mo) for the natural gas – biomass gasification system. Fe-Mo/CNF catalyst offered dual functionality as Fe active sites caused CH_4, CO_2 activation while Mo is highly active for deoxygenation of biomass oxygenates through strong oxygen binding. The deoxygenation of oxygen rich biomass provides steam adsorbates for steam reforming and dry reforming of CH_4, CO_2 which produces hydrogen rich syngas. FeMo/CNF catalyst regeneration and reuse in multiple cycles was also investigated at temperatures 800, 850, and 900°C. After 15 cycles of FeMo/CNF being subjected to reuse without external regeneration, it was observed that hydrogen rich syngas could be produced consistently with H_2:CO ratio between 1.3 to 3.5. Interestingly, the CNF based catalyst undergoes self-regeneration after every 4 to 5 cycles evident from periodic rise and fall in the H_2:CO ratio. Although the self-regeneration mechanism is not completely understood, it is evident from TEM imaging and XRD patterns of CNF catalyst subjected to 1, 5, 10, and 15 cycles. Multiple reuse of CNF without external regeneration causes shielding of nanofibers and nanotubes by coke but the metallic active sites were clearly visible for all the CNF samples under TEM. Carbon nanotube and fiber structures were visible and similar in the CNF catalyst samples subjected to 1 and 10 reuses. This indicates the catalyst undergoes self-regeneration to some extent. Self-regeneration of the catalyst may occur due to 3D coke formation leading to nanotube formation on the Fe active sites, as reported by Wang et al and Robinson et al. Hydrogen rich syngas production could be achieved by co-processing flare gas with biomass on-site along with CO_2 utilization on inexpensive CNF based catalyst synthesized in-house.

587090 Design Integration of Gasification and Reforming Technologies for Syngas Production

This study is focused on the production of syngas using gasification and dry methane reforming (DMR) technologies. The two processes are compared as a standalone design for a syngas production capacity of 10,000 kmol/h with H_2 to CO ratio of 0.88. The two processes are then integrated to develop a novel design that can reduce the CO_2 emissions to a near-zero level with reasonable amount of energy requirement. In this study, Aspen Plus® has been employed as a process simulation tool. The results reveal that the DMR process consumes 3.4 times less energy compared to a gasification plant of same capacity. The result also shows that the integration of gasification and DMR design off-sets the negative aspects of gasification technology enabling it to utilize conventional solid fossil fuel with low emission levels.

585077 The Effect of Non-Alkanes on the Gelation and Thermodynamic Properties of Petroleum-like Solutions

Wax deposition in deep sea oil pipelines is an important and costly issue in the petroleum industry. The wax that forms in these pipelines is composed primarily of *n*-alkanes. However, there are many other components within the wax that can alter its crystallization and agglomeration properties. This work focused on the role of aromatics on the crystallization properties of the model waxy crudes. These waxes were composed of one or two *n*-alkanes, either C_{28}, C_{32}, or C_{36}, mixed with either a single ringed aromatic durene or the polycyclic aromatic compound phenanthrene in a dodecane (C_{12}) solvent. Differential Scanning Calorimetry (DSC) showed that the cloud points of these model waxes were not affected by the addition of aromatic molecules at low concentrations, but at high concentrations phenanthrene caused an increase in cloud point because the phenanthrene molecules would crystallize out of solution first.

585895 Computational Fluid Dynamics Analysis of Polymer Elasticity and Pore Morphology on Residual Oil Saturation

Abstract

Polymer flooding is a widely used Enhanced Oil Recovery (EOR) method that uses polymer solutions to increase sweep efficiency by decreasing the water/oil mobility ratio. Viscoelastic polymers have shown potential in reducing residual oil recovery, primarily due to the additional forces exerted such as normal stress forces, which help overcome capillary forces.

In this work Computational fluid dynamics (CFD) simulations were used to analyze the flow of viscoelastic polymers both within a single pore throat and a network of pores. Simulations were carried out in two pore geometries, namely a pore centered geometry and a constriction geometry similar to that used by Afsharpoor et al.[1], for viscoelastic fluids with viscosities of up to 50 cP and relaxation times of up to 0.2 s. The relative effect of the various forces, specifically the normal forces, on the fluid's ability to mobilize the residual oil within the pore is investigated.

Moreover, CFD was used to analyze the performance of the viscoelastic fluids in 5 different porous media geometries with different pore heterogeneities, pore connectivity, pore shape and tortuosity, and the performance of the viscoelastic polymer properties on the enhanced oil recovery, breakthrough time and fluid trapping within the 5 different porous media geometries was investigated.

References:

[1] A. Afsharpoor, M. T. Balhoff, R. Bonnecaze, and C. Huh, "CFD modeling of the effect of polymer elasticity on residual oil saturation at the pore-scale," *Journal of Petroleum Science and Engineering,* vol. 94-95, pp. 79-88, 2012/09/01/ 2012.

583598 The Study on the Hydro-Liquefaction Reaction of Lipid-Extraced Microalgae (LEM)

The Study on the Hydro-liquefaction Reaction of Lipid-extraced Microalgae (LEM)

LEM are mainly composed of protein and polysaccharides. In order to make full use of the organic matter in microalgae, and improve the application channel and comprehensive value of microalgae, the hydro-liquefaction reaction behavior of LEM was studied.

The influences of solvent types and reaction conditions including reaction temperature, reaction time, solid-liquid ratio and initial hydrogen pressure on the yield of liquid-phase products were investigated, and found that the conversion of LEM hydro-liquefaction reaction increases with the increase of solvent polarity, and ethanol is the most suitable solvent. The yield of the liquid-phase products can be increased by increasing reaction temperature, reaction time and ethanol dosage. The optimum conditions for LEM hydro-liquefaction reaction were reaction temperature 290 °C, reaction time 30 min, initial pressure of hydrogen 3 MPa, and solid/liquid ratio 1:10 (w/v). Under the optimum condition, the yield of the liquid-phase products was 65.72%.

The effects of the type and amount of ionic liquids (ILs) on the yield of the liquid-phase products from LEM hydro-liquefaction reaction were studied. It was found that when the amount of [Bminm] Cl was 1.0wt%, the highest liquid-phase products yield was obtained.

By comparing the characterization of liquid products prepared in different catalytic systems, it was found that ILs/nickel-based catalysts could improve the liquefaction reaction degree and calorific value of LEM. However, the liquefied products of LEM hydro-liquefaction reaction are not suitable for preparing high calorific value biofuel oil because of the low H/C ratio, high N/C ratio and low calorific value.

584426 Hydrothermal Liquefaction of Micro-Algae to Produce Liquid Biofuels

Algae biofuel is a carbon neutral alternative to fossil fuels, which are the main source of carbon dioxide emissions. Biocrude oil can be extracted from algae through hydrothermal liquefaction (HTL), but the produced product contains heteroatoms such as oxygen, sulfur, and nitrogen. However, the biocrude oil can be upgraded into higher quality oil using heterogenous catalysts. The green algae, *Micractinium Immerum*, was hydrothermally liquefied at 300 °C. The crude oil was then separated from other fractions using dichloromethane as the solvent. After HTL, the biocrude from *Micractinium* was upgraded with catalysts MO_2C, HZSM-5, and Pt/C at 350 and 400°C using hydrogen at 1000 psi. The upgraded oil's elemental composition was measured using an elemental analyser. The products were analyzed for high heating value using a bomb calorimeter. Although all the three catalysts decreased the heteroatoms in the crude oil at these process conditions, the HZSM-5 catalyst performed the most effectively, while the Pt/C catalyst yielded the highest heating value. By improving the energy density, algal-derived fuel may become a competitive alternative to fossil fuels and ultimately reduce additional carbon emissions.

587828 Co-Solvent Hydrothermal Liquefaction of Wastewater Algae

Decreasing supplies of non-renewable fuels create demand for a new and renewable source of transportation fuels. Algal biomass harvested from wastewater treatment plants has the potential to fulfill this need due to its abundance and suitability for conversion into fuels. Hydrothermal liquefaction (HTL), a thermochemical conversion process in which biomass is heated in a pressurized liquid, can be used to generate products with energy contents higher than that of the original feedstock. HTL of algae produces liquid bio-crude oil, which can be upgraded into a liquid fuel, as well as residual char and aqueous phase which can be used for nutrient recovery and reuse.

Alcohol co-solvents have been shown to increase the yield and quality of bio-crude oil from HTL of algae. Here, ethanol, isopropanol, and glycerol were tested as HTL co-solvents using a 1.8 L Parr reactor at different operating temperatures and residence times. Glycerol produced the highest yield of desired products compared to the other alcohols. Crude glycerol, which contains residual methanol and fatty acids, is a byproduct biodiesel production. Using crude glycerol in HTL of algae creates a value-added use for two low-value process streams. Co-solvent HTL products were characterized using bomb calorimetry, CHNS elemental analysis, and Fourier transform mass spectroscopy.

583733 Development of an Inexpensive Fine Particle Stabilized Aqueous Nitrogen Foam System for Enhanced Oil Recovery

Development of water flooding is the pivotal method in the process of utilization of global oil and gas resources, which is studied by more and more engineers and scientists all over the world. However, with the development of water flooding, it is inevitable that oilfields will enter the extra high water cut stage. Currently, the oil recovery rate by water flooding is generally less than 40% on a global scale. Therefore, how to continue improving oil recovery in the extra high water cut stage is an urgent problem to be solved. Especially with the fluctuation of oil prices in recent years, exploring inexpensive and high-efficient oil displacement system is a key measure to overcome this problem. It is worth noting that foam, as an efficient oil displacement system, has been gradually accepted on a global scale. However, this technology is limited by the cost of the foam stabilizer and its foam stabilizing ability, which is difficult to be promoted on a large scale.

In this work, the composite clay and fine particles which are from coal combustion are compounded and used as an inexpensive foam enhanced system. The experimental results show that fine particles can significantly improve the stability of foam, and can make the bubbles stabilized in the porous medium for a long time, which overcomes the Oswald effect between the bubbles to some extent and weakens the separation of gas and liquid. Besides the addition of particles also reduces the volume of the bubbles, which makes the bubbles more rigid, thereby increasing the apparent viscosity of the foam and improving the sealing ability of the foam in the formation. Therefore, fluid mobility capacity can be effectively controlled by special foam in the extra high water cut stage. Moreover, the sweep efficiency of the oil-displacing agent around the foam is remarkably enhanced, meanwhile the addition of the particles also enhances the washing efficiency of the foam as the oil-displacing agent.

585940 Natural Gas Sweetening By Membrane Cascade Using a Standardized Model

Natural gas prices are at a low due to high availability which has driven several companies to consider using C1 chemistry to generate petroleum products and petrochemicals, particularly from stranded gas reserves that are typically flared. [1] Processes which aim to take advantage of these resources must be modular, but have similar requirements to centralized processing plants in terms of required feedstock and product purities. Membrane cascades, which use a series of interconnected, permi-selective membrane modules to separate fluid mixtures, have been proven for removing contaminants from natural gas, mainly CO_2. [2] However, most commercial process simulators lack the models needed to simulate gas separation membranes. As part of the RAPID Center for Process Modeling's effort to provide models and recommendations to the engineering community, a "membrane separation toolkit" has been developed to standardize simple membrane modeling. A test case involving the purification of 25 MMSCFD of CH_4 laden with CO_2 as the primary impurity using a standard cellulose acetate membrane was developed to demonstrate the use of the toolkit. The cascade was optimized with respect to the stage operating pressures, temperatures, and cuts to produce pipeline quality natural gas containing ≤ 3 % CO_2 by minimizing OPEX and maximizing the produced volume of salable gas.

References:

[1] H. Fleisch. [White Paper], World Bank-Global Gas Flaring Reduction Partnership, 2015.

[2] R.R. Zolandz and G.K. Fleming, "Applications" in *Membrane Handbook*, ed. W.S.W. Ho and K.K. Sirkar, 78 - 94. New York, NY: Van Nostrand Reinhold, 1992.

588282 Development of a Simplified Combustion Kinetic Model for Autothermal Biomass Pyrolysis

A simplified combustion kinetic model was developed to assess the autothermal operation of a biomass fast pyrolysis reactor. For that, single- and two-step combustion mechanisms of Westbrook and Dryer (1981) [Ref 1] were reviewed, coded using Aspen Plus and then validated with the detailed kinetic model provided by Chemkin for hydrocarbons and alcohols. Artificial intelligence tools were further used to predict the values of important kinetic rate constants associated with the simplified model for fuel chemicals generated in the biomass fast pyrolysis reactor. The two-step combustion mechanism model qualitatively describes fuel combustion patterns for hydrocarbons once the fuel is ignited, although it does not simulate chemical ignition delay during which the fuel undergoes radical decomposition before actually starting the oxidation reaction. Machine Learning algorithms prove quite satisfactory in correlating the kinetic parameters for the combustion of alkanes but fail to assertively calculate them for oxygenated molecules due to the lack of such data points in training set. Future work involves the refinement of the two-step combustion mechanism for oxygenated fuels and the improvement of the machine learning features to achieve robust prediction for the rate constants associated with oxygenated molecules.

References

Ref 1: CK Westbrook and FL Dryer, "Simplified Reaction Mechanisms for the Oxidation of Hydrocarbon Fuels in Flames," Combustion Science and Technology (1981) 27:31-43.

588139 Thermal Conductivity of Ionic Liquids and Hydrofluorocarbons: 1-Hexyl-3-Methyl-Imidazolium Bis(trifluoromethylsulfonyl)Amide and 1,1,1,2-Tetrafluoroethane

Novel vapor absorption refrigeration processes utilizing ionic liquids (ILs) and the hydrofluorocarbon refrigerant gas, 1,1,1,2-tetrafluoroethane (R-134a) have been recently investigated. The low volatility of ionic liquids allows them to be used as ideal solvents in such processes with no need for bulky rectifying columns to purify the refrigerant of residual solvent. Employing the use of ILs in vapor absorption refrigeration cycles paves way for more efficient energy usage. Transport properties of such mixtures are needed for further design processes. The IL, 1-hexyl-3-methyl-imidazolium bis(trifluoromethylsulfonyl)amide ([HMim][Tf_2N]) was investigated as a model solvent. In this study, liquid thermal conductivities of [HMim][Tf_2N] saturated with R-134a at 25 °C, 50 °C, and 75 °C with pressures ranging from 0.7 bar to 18.9 bar were measured. Herein, a transient hot wire method was used to determine that the values of the thermal conductivity of pure [HMim][Tf_2N] were in good agreement with those reported in the literature. The ionic liquid and R-134a mixture values show a trend of decreasing thermal conductivities with increased refrigerant solubility compared to those values obtained for pure [HMim][Tf_2N]. Trends in thermal conductivity change based on composition of R-134a in [HMim][Tf_2N] have provided compelling evidence that the effects of [HMim][Tf_2N] on thermal conductivity are greater than those of R-134a.

586559 Renewable Power Generation through Reverse Electrodialysis

Renewable Power Generation through Reverse Electrodialysis

Abstract:

The rapidly growing energy demand along with the extensive use of the limited fossil fuels and the consequent climate change issues drives the search for alternative energy sources. Recent estimates suggest that approximately 40% of global electricity demands can be met by capturing the potential energy contained in the mixing of seawater and freshwater at coasts, by utilizing the salinity gradient, separated by a semi-permeable membrane to derive a potential difference. The chemical potential difference between salt and freshwater generates power which is around 2.3 MJ/1 m^3 of river water. The gross power potential of this unconventional and unrevealed energy source is estimated to be 2.4-2.6 TW when the average discharges of all rivers were taken into account.

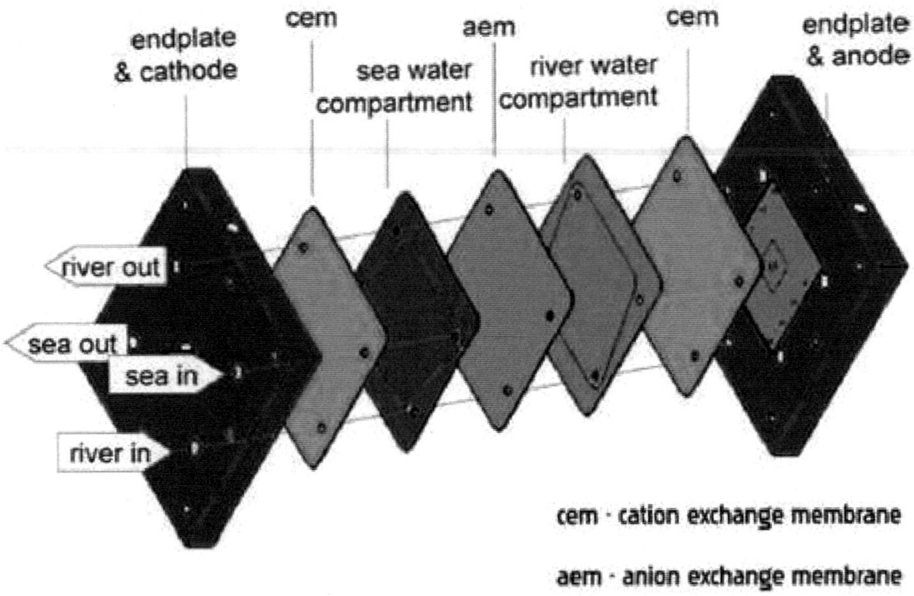

Salinity Gradient Power (SGP) is environmentally sound as well as a sustainable method of power generation from the reversible mixing of two kinds of water with different salt concentrations. The Gibbs free energy that is released when saltwater and freshwater partially mix, can be captured in a process called Reverse Electrodialysis (RED). In RED, a series of anion exchange membranes (AEM) and cation exchange membranes (CEM) are alternately stacked between two electrodes, thereby creating adjacent low concentration compartments (LCCs) and high concentration compartments (HCC), which are fed by two aqueous salt solutions of different salinity. The salinity difference on either side of the membranes initiates selective transport of ions across ion-exchange membranes towards the electrodes. The ionic current in the cells is converted into electricity by redox reactions at the electrodes. Literature data show that the obtained power density increases progressively with higher salinity gradients, despite lower membrane permselectivity, as a result of reduced electrical resistance at higher concentrations. Significant improvements in power and current generation were obtained by using brush current collectors. The large surface area of these brushes reduced electrode resistances by increasing charge transfer to/from the carbon particles.

In the present work, the effect of the spacer (which determines the distance between the stack membranes) thickness on power density was evaluated. Results revealed that lowering the spacer thickness from 500μm to 200μm improved the power density up to 32%. Moreover, the effect of flow velocity was also evaluated and critical velocity was found to be ~2.5 cm/sec.

585398 Electrochemical and Thermal Analysis of an Ionic Liquid Based Solid Polymer Electrolyte

Lithium Ion Batteries (LIBs) have emerged as the leading technology in energy storage, with potential applications ranging from consumer electronics to hybrid vehicles and storage of energy produced by photovoltaics. State-of-the-art LIBs utilize a liquid electrolyte that poses issues such as flammability, potential leakage from battery casings, and dendrite growth when used with lithium metal. Solid polymer electrolytes (SPEs) show promise in replacing liquid electrolytes, as they offer decreased flammability and the potential for dendrite prevention, enabling the use of lithium metal. However, SPEs are often hindered by low ionic conductivities and poor contact with electrodes. In this work, a SPE is proposed that consists of a high molecular weight polyethylene oxide (PEO) matrix with an ionic liquid and lithium bis(trifluoromethanesulfonyl) imide (LiTFSI) as plasticizers. A range of different compositions of SPE were fabricated and their electrochemical and thermal properties were characterized using cyclic voltammetry, electrical impedance spectroscopy, and differential scanning calorimetry. Our experiments demonstrated a high lithium conductivity and wide electrochemical and thermal stability windows for the electrolyte. Cycling cells, constructed using lithium metal and a composite Lithium Iron Phosphate (LFP) electrode, demonstrated high coulombic efficiency and minimal capacity fade through cycling at both elevated temperature and room temperature. As a whole, the SPE compositions with a higher ratio of ionic liquid to polymer showed higher conductivity and improved cycling performance. These electrolytes with high ionic liquid content show promise as safe, stable, and conductive alternatives to liquid electrolytes for lithium ion batteries and are also commercially relevant due to their ability to cycle at a wide range of temperatures with commercially available electrodes.

588113 Permselective Separators for Grid Storage Alkaline Zn/MnO_2 Batteries

Increasing awareness of climate change and the dangers of fossil fuels have spurred a need for faster integration of renewable energy sources onto the electrical grid. Rechargeable alkaline Zn/MnO_2 batteries have emerged as a candidate to fill this need as they are low-cost, possess a high energy density, are environmentally inert, and can be readily mass produced [1]. A layered Bi-birnessite oxide with Cu^{2+} intercalation has recently been developed that stabilizes the MnO_2 structure and allows continuous cycling at near full 2e- capacity [2]. However, one of the technical hurdles is overcoming limited cycle life caused by the formation of irreversible and resistive $ZnMn_2O_4$ and ZnO species on the cathode when zincate ($[Zn(OH)_4]^{2-}$) diffuses across the battery separators [3].

In this presentation, we demonstrate the use of new permselective polymeric separators that can be used in high depth of discharge Zn/MnO_2 batteries. Several separators were screened for their zincate blocking abilities using our recently developed anodic stripping voltammetry methodology [4]. The polymeric separators were also tested for hydroxide diffusion where it is shown that they demonstrate selectivity for the monovalent hydroxide anion. A primary discharge of the full $2e^-$ capacity of MnO_2 (617 mA h g^{-1}) was used to further test the ionic selectivity of the separators where the 2nd e^- discharge plateau, indicative of the absence of zinc, was observed only for Zn/MnO_2 batteries that utilized the selective polymeric separators.

This work was supported by the Office of Electricity Delivery and Energy Reliability and the Laboratory Directed Research and Development program at Sandia National Laboratories. Sandia National Laboratories is a multi-mission laboratory managed and operated by National Technology and Engineering Solutions of Sandia, LLC, a wholly owned subsidiary of Honeywell International, Inc., for the U. S. Department of Energy's National Nuclear Security Administration under contract DE-NA0003525. Dr. Imre Gyuk, Energy Storage Program Manager, Office of Electricity Delivery and Energy Reliability is thanked for his financial support of the project. The views expressed in this article do not necessarily represent the views of the U. S. Department of Energy or the U. S. Government.

References:

[1] N. Ingale, et al., *J. Power Sources* 276 (2015) 7-18.

[2] G.G. Yadav, et al., Nat. Commun. 8 (2017) 14424.

[3] J. Gallaway, et al., *J. Power Sources* 321 (2016) 135-142.

[4] J. Duay, et al., *Electroanalysis* 29 (2017) 2261-2267.

583824 Bipolar Plate Design for an Alkaline Direct Ethanol Fuel Cell

The growing energy demand and all the alliances between governments to combat global warming and climate change make that most of the research works focused on taking advantage of renewable energy fonts or lest contaminant technologies. One those technologies are the Fuel Cells; thus, they are electrochemical devices where the chemical energy of reactants can be transformed on to electricity without going through a thermal process and can operate with some fuels that allow to name It as no or low contaminant technology.

Usually, it is normal to find advances in this technology in connection with Hydrogen as Fuel, since it is the simplest fuel on the world and the product of the reaction is water. Because of the low production of Hydrogen in Colombia, Ethanol is regarded as the best option to implement fuel cell technology as a solution for the non-interconnected zones throughout the territory.

A fuel cell is, in essence, an open continuous electrochemical reactor, so it is necessary to transport the reactants inside the cell and carry out of the cell the products and reactants in excess. In essence that is the main function of bipolar plates and in fuel cells stacks the bipolar plates guarantee the electrical conductivity to close the electrical circuit.

It is well known that the electrochemical reactors are heterogeneous and during the normal performance the cell will work under diffusion control and it will demand good mass transfer coefficients. That is because a well-designed bipolar plate it is necessary to obtain the best performance of a cell. This work aims to combine CFD simulation and Multiphysics simulation to taste different designs with guidance of previous works and result of general channel designs as interdigitated or serpentine and choose the perfect one for ethanol as a liquid solution and for oxygen from the air.

586859 Improved Surface Passivation of Halide Perovskite Quantum Dots Using 5-AVA

Shifting the world from fossil fuels to renewable energy sources is becoming increasingly vital, and requires efficient and low cost renewable electricity generation. One emerging class of solar cell which could drive this change is halide perovskite solar cells, a solution-processable semiconductor material. State-of-the-art perovskite solar cells have surpassed 25% efficiency, but further efficiency, and perhaps more importantly stability, gains appear to be tied to defects at the material's surface. Halide perovskite quantum dots (QDs) are an elegant vehicle for studying surface properties due to their high surface to volume ratio. In this study, three different halide perovskite quantum dots (~7 - 9 nm in diameter) were synthesized: $CsPbI_3$, $CsPbBr_3$, and $FAPbI_3$ along with alloys of the form $Cs_{1-x}FA_xPbI_3$ and $CsPbI_{3(1-x)}Br_{3x}$. In a typical synthesis, the QDs are capped by amine and carboxylate ligands which are quite labile and form a dynamic QD surface. These as-synthesized quantum dots were treated with the iodide salt of the bifunctional ligand molecule 5-aminovaleric acid (5-AVA) to better passivate the QD surface. $CsPbI_{3(1-x)}Br_{3x}$ resulted in a quantum yield increase by 300% following 5-AVA treatment while other QD alloys had little to no change. These experiments provide insight into the complex surface chemistry of varying halide perovskite materials. Ultimately, improved surface passivation of QDs could lead to a better understanding of the surface properties of halide perovskites in general and thus improve optoelectronic devices.

584755 Fabrication and Evaluation of Lightweight Proton-Exchange Membrane Fuel Cells

The objective of this study was to create lightweight proton-exchange membrane fuel cells (PEMFCs) by replacing traditional bi-polar plates with combinations of polycarbonate (PC) flow channels and metallic mesh current collectors. The concept was based on previous work in literature, which demonstrated the effectiveness of using fuel flow channels fabricated from thermally imprinted PC film. While the cited literature uses a heated hydrostatic pressing technique, this study demonstrates that a similar thermal imprinting method using a hydraulic press, heating blocks, and a two-piece die assembly can also produce effective flow channels.

This study involved the testing of four PEMFCs: one commercially available cell used to collect control data and the remaining three made by replacing components of the control cell with lab-fabricated ones. After cell testing, qualitative conclusions were made based on the performance of each cell. Fuel cell performance is explained with the aid of measured polarization curves. It was concluded that a serpentine flow channel design operates more effectively than other unchanneled designs, and because a silver mesh current collector far outperforming current collectors made with stainless steel, it was affirmed that current collector electrical conductivity plays a major role in PEMFC power output capacity. Also noted in the study was a progressive decrease in performance across tests of the control cell, speculated to be a result of membrane-electrolyte assembly (MEA) degradation due to extensive use. This observation along with the those which led to the previously stated conclusions open the door for further testing and research into cell performance as it relates to cell design.

587493 Optimization of Graphene/Mn_3O_4 active Layer and Gel Polymer Electrolyte of All-Printed Supercapacitor for Space Applications

Energy storage is essential for space exploration, communication, and operation of electrical systems beyond Earth's orbit. Due to the inherent variability of renewables, reliable energy storagedevicesare necessary to make future renewable energy sources viable. The most common form of energy storage is the battery, because of its high energy density, but the battery lacks the superior specific power as well as high cyclic durability required for fast energy transfer as in a capacitor. A supercapacitor is an energy storage device which implements the best of batteries and capacitors, ultimately identifying a device with high specific energy and specific power, and robust cyclic durability.

A key project of space energy storage is In Space Manufacturing (ISM). ISM furthers advancement of space technology by allowing for complete fabrication of devices in space, reducing the cost and energy required to transport materials to target locations such as the International Space Station, Moon or Mars. Supercapacitor technology in space is important for triboelectric energy harvesting and solar harvesting; variability of wind and solar sources requires storage in a reliable device.

The crucial features determining performance of a supercapacitor are the active layer and the electrolyte layer. This study focuses on optimizing not only these essential layers, but also the extrusion printing technique. Success if determined by analyzing cyclic durability and printing stability. The studied supercapacitor utilizes a silver current collector layer, a graphene active layer, and a lithium-chloride gel polymer electrolyte. This research can be implemented further with analysis on radiation effects of the 3-D printed supercapacitor and future optimization of the active layer including research into non-binder inks.

588159 Carbon Nanotube/Platinum Nanoparticle Composite Inks for Enhanced Oxygen Reduction Reaction Catalysis for PEM Fuel Cells

Carbon Nanotube/Platinum Nanoparticle Composite Inks for Enhanced Oxygen Reduction Reaction Catalysis for PEM Fuel Cells

Delaney A. Marbach[1], Aaron N. Romero[1], Dan Rabbia[1], Hugh Mahr[1], Deryn Chu[2], F. John Burpo[1]*, Enoch A. Nagelli[1*]

[1]Department of Chemistry & Life Science

United States Military Academy, West Point, New York 10996

[2]U.S. Army Combat Capabilities Development Command, Army Research Laboratory

2800 Powder Mill Rd., Adelphi, MD 20783-1107

Hierarchically porous and high aspect ratio platinum nanoparticles were synthesized using electrostatic assembly of a combination of square planar $[PtCl_4]^{2-}$ with $[Pt(NH_3)_4]^{2+}$ in aqueous solution. The subsequent chemical reduction of these platinum salt crystals using sodium borohydride forms high-surface platinum micro- and nanotubes. We demonstrate an all aqueous and scalable approach to synthesizing high aspect ratio platinum nanoparticles with tunable aspect-ratios and surface-area to combine with conductive carbon binders for electrocatalyst inks for proton exchange membrane (PEM) fuel cells. We combined our high aspect ratio platinum nanoparticles with composite with various carbon allotropes (graphite, carbon nanotubes, and graphene) to form conductive inks for electrocatalysis of the oxygen reduction reaction (ORR) at the cathode of a PEM fuel cell. Our resultant inks are comparable to conventional platinum black-based inks for fuel cells catalysts and would be a more cost-effective alternative. The catalyst inks were characterized using scanning electron microscopy (SEM), energy dispersive X-ray (EDX), cyclic voltammetry (CV), and linear sweep voltammetry (LSV).

585797 High Energy Density Flexible Aluminum-Air Fuel Battery Performance

This article mainly talks as a new type of energy storage system, metal-air battery has great development potential. At present, there are two main problems in the development of aluminum -air battery: one is the corrosion of aluminum anode to produce hydrogen, and the other is the low ORR activity of air cathode. The most important factor determining the activity of air electrode is the ORR activity of cathode catalyst. The starting point of this paper is to improve the activity of cathode catalysts.

Compared with traditional noble metal platinum-based catalysts, perovskite catalysts have low cost and high catalytic activity, thus becoming a research hotspot. The innovation of this paper is to replace La element at a position of $LaMnO_3$ perovskite oxide with appropriate Sr element, so as to improve its catalytic activity. In the experiment, in order to improve the conductivity of perovskite catalyst, a certain proportion of conductive carbon black was added. Finally, using hydrophilic conductive carbon cloth as catalyst substrate, flexible electrode was prepared, and flexible aluminum air battery was assembled by combining aluminum anode and flexible solid electrolyte membrane. The electrochemical activity of prepared cathode catalyst was tested, and the results showed that the catalytic activity of 50%LSM30 is the best, which means the molar content of strontium is 30% and the mass ratio of the carbon black is 50%.After assembling the flexible aluminum-air battery, the battery performance was tested, and the electrochemical performance and mechanical performance were good.

587263 Investigating the Microstructural Effects of Electrode Compression on Redox Flow Battery Performance

Redox flow batteries (RFBs) promote the decarbonization of the electricity sector by enabling the storage of intermittent renewable energy over mid to long-duration timescales (hours-days). However, further cost reduction is necessary for ubiquitous adoption. Electrodes are particularly important components of the RFB system, and this study examines the effect of electrode compression on performance. Freudenberg paper and Avcarb cloth, common RFB electrodes with differing microstructures, were chosen for this study. We found that cloth electrodes elicited the highest surface area and electrochemical performance at 25% compression, whereas the carbon paper electrodes showed optimal performances at 10% compression. Additionally, the permeability of the cloth electrode increased at 10% compression compared to uncompressed material, but permeability decreased significantly at higher percentage compressions. We hypothesize that there exists a trade-off between porosity, permeability, and accessible surface area of the electrodes; this trade-off results in maximized performance at moderate compression of 25% and 10% for cloth and paper, respectively. Elucidating the relationship between compression, microstructure, and performance enables targeted electrode design that will optimize the performance and reduce costs of RFBs, promoting their utilization at the grid scale.

584154 Design of Pumped Hydro Energy Storage System for Smart Grid Application

While countries are pushing their goals of producing only renewable (green) energy, engineers need to solve the challenge that arises by relying on such resources - irregularity. Renewable energy sources need plenty of storage to ensure that the grid can function in their absence. This is where the idea of a pumped hydro energy storage (PHS) system saves the night!

A pumped hydro system works like an energy reserve; when the demand for electricity is not met by the traditional generators or renewables, the PHS system powers up and generates electricity whereas when the demand is low and the production is high, for example during the day or during windy periods, the turbine utilizes the excess energy to pump water up an elevation and store it as potential energy.

The objective of this project was to design a PHS system through the development of a model and optimization of parameters to maximize net present value. Traditional grids do not have the capability of getting feedback from the consumers. In such cases, Smart grids bring the feedback element into the equation. This is utilized to predict the fluctuations in the prices of electricity and channel the excess energy that is otherwise lost in the grid. Additionally, a demand response study provides day-ahead prices of electricity for an interval of time. The characteristics of the smart grid in conjunction with the day-ahead electricity prices are used by the PHS model to generate an operation schedule for the plant that will maximize profit.

The model repurposes abandoned quarries as reservoirs. Using the concepts of linear programming and receding horizon (model predictive control), an estimate for reservoir capacity and profit from revenue was obtained. An economic analysis was performed on this model to approximate an initial capital investment in this project. Losses were calculated to simulate the plant operation which was utilized in the code. Overall, the investigation suggests the feasibility of a PHS in a quarry system.

587836 Waste Plastic Generation Profile of Kampala, Uganda

Pollution is a significant global challenge. Municipal solid waste accumulates on land due to many reasons. In some rural areas, waste management solutions are nonexistent, meaning waste piles around residential locations. One sustainable solution to this problem considers the complete elimination of plastic from the ecosystem, unlike many other current methods. While methods such as recycling are great for reusing plastic already produced, it can only be used so much before it ends back in the waste pile. Instead, by creating another purpose for this waste plastic, there will be a lot less plastic that ends up in the environment and buried underneath the ground. This solution is termed as Trash to Tank (3T), a process that converts waste plastic types (polystyrene, polypropylene, low-density polyethylene, and high-density polyethylene) into Plastic Derived Fuel Oil (PDFO) via pyrolysis. Pyrolysis is a method of thermal decomposition that occurs in the absence of oxygen while burning. The fuel produced is similar in composition to diesel and kerosene.

Uganda is a suitable country for the use of 3T method because it lacks sufficient capital and infrastructure to manage all of its waste. In order to optimize the utilization of 3T process, reduce waste plastic accumulation on land and in the natural waterways of Uganda (River Nile and Lake Victoria), a waste plastic generation profile must be developed and analyzed. This study accomplishes this for a small subset of Uganda, Kampala, the capital city. Through a partnership with Kampala Capital City Authority (KCCA), a local waste management municipality, existent waste plastic generation was determined for the city. This information was correlated with available data on population density and geographical analysis of income level via ArcGIS (a spatial analysis tool) to understand the impact of population and income on waste generation. The results of this study reveal the interplay of these factors for waste plastic management in Kampala and similar sub-Saharan African cities.

587456 Experimental Evaluation of the Temperature Coefficient of Quinones' Redox Reactions

Due to their low cost, high abundance, and extensive tunability, organic redox couples, such as quinones, have become of high interest. Moreover, their wide range of redox potentials have recently become highly sought after in thermogalvanics, which describes the generation of a proportional electric potential when a temperature gradient is applied to electrochemical redox couples. With applications in low-powered electronics, quinones as thermogalvanics have shown great potential in waste heat harvesting. The main objectives of this study were to manufacture a prototype setup that would record the hot and cold temperature, and the open-circuit voltage of an electrochemical reaction, leading to the evaluation of the non-isothermal temperature coefficient (α) of the reaction; and to determine the α value of different quinone redox reactions at different pH levels in aqueous solution. The redox reaction between p-benzoquinone and p-hydroquinone was successfully studied in unbuffered and buffered acidic media with $\partial E/\partial T$ ranging from -0.32 to -1.02 mV/K. Although data showed hysteresis, at a pH of 10, $\partial E/\partial T$ was -2.07 mV/K, surpassing the ferri/ferrocyanide α value of -1.4 mV/K, which is considered the highest temperature coefficient recorded. Experimental values demonstrated a linear fit as per the Nernst equation. Moreover, electrochemical oxidation of p-hydroquinone above neutral pH resulted in irreversible reactions producing anionic species of p-benzoquinone.

587007 Microemulsions in Acidic and Alkaline Environments As Electrolytes for Energy Storage

Redox Flow Battery (RFB) technology can store a large amount of energy as chemical energy that can later be discharged on demand. Intermittent renewable energy sources, such as solar or wind power, are becoming more common, thus the need for state-of-the-art batteries is increasing. A main component of RFBs is the electrolyte solution that is used to store energy. The most common type of RFB is the vanadium RFB (VRFB), in which vanadium serves as the charge carrier. Disadvantages of VRFBs are the high cost of vanadium, crossover of vanadium, and corrosion from sulfuric acid. In this work, microemulsions are explored as the electrolyte solution in RFBs. Microemulsions use surfactants to stabilize two immiscible liquids, thus increasing the number of redox pairs that can be incorporated into the electrolyte solution by solubilizing hydrophobic charge carriers. In this work, the microemulsion system used consists of water, Tween-20 (surfactant), 1-butanol (cosurfactant), toluene (oil phase), ferrocene (charge carrier) and various acids and bases (supporting electrolyte). We used cyclic voltammetry to study mass transport and reversibility parameters of the microemulsion system while varying the water to surfactant ratio and pH. The results obtained from the acidic and alkaline systems have CVs that show reversible behavior. Interestingly, the basic system exhibits a classic EC reaction at low scan rates likely producing dihydroxyferrocene. Unsurprisingly, at higher concentrations of acid or base the ester bond in the Tween is hydrolyzed, thus destroying the microemulsion system. While at intermediate pH values, however, the microemulsions are stable for extended periods of time.

587866 The Effects of pH and Complexation on Metal Cation Uptake in Nanoparticle Organic Hybrid Material (NOHM) Systems

The potential use of nanoparticle organic hybrid materials (NOHM) as electrolytes in redox flow batteries is being explored. Redox flow batteries present the advantage of providing large-scale energy storage, beyond the capacity of more common battery types. The most common active material used is currently Vanadium, which is expensive in a pure form, has a corrosive environment, and decreases battery capacity over time. NOHMs are composite particles formed by polymers grafted (either ionically or covalently) onto surface-modified inorganic nanoparticles. Such structure has great potential to capture and carry nonpolar electroactive species. A NOHM comprised of a silica core ionically grafted with polyethylenimine (PEI) polymers is focused on in this study, we refer to this as NOHM-I-PEI. Copper is introduced as a probe to evaluate the effect NOHM structures have on metal cation uptake. Cyclic voltammetry (CV) was used to investigate the electrochemical behavior of the copper species as the concentration of NOHM increases at a fixed $Cu(NO_3)_2$ concentration. NOHM-I-PEI showed inhibition of the oxidation and reduction peaks as the NOHM concentration increased, suggesting the formation of the cuprammonium complex. UV-Vis measurements also suggested the presence of this complex. pH measurements taken for each solution shoed a drastic change along the range of solutions tested. The pH of the solutions was controlled via Britton-Robinson buffers and CVs were ran with this set of solutions. Results from testing these solutions allow for better isolation of the NOHM's effect from the effect of pH. It is observed that peak intensity decreases as the pH increases. At high pH (>6) complete inhibition is observed, while at lower pH, two cathodic peaks can be seen. Based on these results, it is hypothesized that the bound copper is not available for oxidation/reduction at high pH. Whereas at lower pH, the observed variation in intensity can be related to complexation as a function of the nitrogen to copper ratio. The presence of two cathodic peaks can be explained by the stabilization of the cuprous ion due to complexation.

584795 Identifying Optimal Conditions for Aluminum Catalyzed Conversion of Glucose to HMF in NMP

HMF is a desirable chemical that can be produced from renewable and lignocellulosic materials and which has several useful applications, including biofuels and polymer synthesis. Aluminum catalyst $L^{Ph}AlMe_2$ (**2e**) was explored as a method of glucose isomerization to fructose and subsequent fructose dehydration to HMF in N-methyl-2-pyrrolidone (NMP). Reaction temperature and the addition of co-catalysts were studied to identify reaction conditions that maximize glucose conversion to HMF. Glucose and catalyst **2e** were reacted in NMP and analyzed via HPLC to study glucose to fructose isomerization. A reaction temperature of 120 °C provided the highest yield of glucose to fructose in NMP (yielding an 80% fructose selectivity), however fructose was found to degrade in the presence of catalyst **2e** in NMP at all temperatures. Co-catalysts P_2O_5 and Nickel (II) Chloride were added to the reactions to induce HMF production in NMP. P_2O_5 is a strong dehydrating agent and has been previously shown to improve fructose to HMF conversion. Starting with glucose in NMP, P_2O_5 led to glucose conversion to unwanted by-products and no fructose or HMF was formed. Nickel (II) Chloride was considered as a co-catalyst because the halide anions created when in NMP allow fructose to be easily converted to HMF. Nickel (II) Chloride was found to give a glucose conversion of 55% along with an HMF selectivity of 24%. This reaction should be further optimized to effectively utilize Nickel (II) Chloride in NMP to improve the conversion of glucose to HMF. Further improvements in this reaction increasing selectivity of HMF, combined with harvesting glucose from cellulose, would allow industrial scale production of HMF from lignocellulose.

586639 Evaluating the Impacts of Extended Coordination to the Uranyl Cation Using a Tetracarboxylic Acid Ligand

Two novel crystal structures containing tetrahydrofuran-2,3,4,5-tetracarboxylic acid (THFTCA) and the uranyl (VI) cation were synthesized and characterized with single crystal X-ray diffraction and Raman spectroscopy. THFTCA is important in separating uranium from other actinides and is used in the disposal process of radioactive waste. This study analyzed how the coordination environment of the uranyl cation impacted its interactions with THFTCA. Two uranyl structures were isolated. The first, a simple monomer with piperazine acting as a charge balancing counter-cation. The second a framework with two unique uranyl environments and a sodium dimer. Within the uranyl structures, an increase in complexity was observed when a non-organic counter-cation was introduced, which contributed to the formation of a 3D framework. Shifts in the Raman spectra were analyzed to determine vibrational differences that resulted from the coordination of THFTCA to the uranyl cation.

585407 Novel Electrochemical Technique to Track Rapid Decay of Redox Active Species

The availability, abundance, and energy density of fossil resources have driven progress over the last century leading to dramatic improvements in quality of life worldwide. However, to avert catastrophic climate change, there is an increasingly urgent need to reduce carbon emissions without stifling economic growth. Deep decarbonization necessitates mass deployment of renewable technologies, such as wind and solar photovoltaic, to displace existing fossil generation. Though increasingly inexpensive, the inherent difficulties in converting the intermittent output of renewable installations into reliable, dispatchable power sources represents an important constraint and motivates research into storage technologies to buffer the intermittency without significantly increasing the cost of electricity. Redox flow batteries (RFBs) hold promise for long duration electrochemical energy storage by employing permanganate, a low cost redox active species. However, this species auto discharges to manganite and oxygen, introducing a potential flammability concern. Due to the rapid decay kinetics, typical ultramicroelectrode (UME) cyclic voltammetry cannot be employed and so a novel electrochemical technique is developed and validated to track rapidly decaying electrochemical species. In this work, a reducing potential is held in a permanganate electrolyte and the limiting current is tracked continuously. From current, a concentration can be determined and fit numerically to a general rate equation. The decay order was found to be close to second order with the rate constant near 40 M^{-1} s^{-1}, in close agreement with experimental observations. These findings are corroborated with UV-vis spectrometry.

585988 A Biomaterials Approach to Kidney Engineering: Structural and Chemical Modification of Silk Fibroin Scaffolds for *in Vitro* Tubulogenesis

About 14 percent of the general population suffers from chronic kidney disease (CKD). Although organ transplants and dialysis provide treatment for loss of kidney function, these are unsustainable solutions. A promising avenue lies in tissue engineering, particularly in the development of biomaterials that mimic the extracellular matrix (ECM) through the introduction of chemical markers and integration within intracellular signaling pathways. *In vivo*, Jagged1 (Jag1), a surface adhesion morphogen, contributes to proximal tubule cell fate and nephron patterning via Notch signaling within the ECM. However, Jagged1 shows very little activity in solution, and therefore must be bound in order to effectively signal to cells.

In this work, both physical and chemical cues are introduced to silk fibroin (SF) scaffolds. A novel fabrication methodology was developed to create directionally aligned pores to promote tubule structure formation. Sponges were tested with renal proximal tubule epithelial cells (RPTECs). Compared to random pore scaffolds, the directionally aligned sponges possessed greater cell densities, with possible lumen formation occurring after two weeks in culture and two weeks under perfusion. For the binding of Jagged1, a diazonium coupling reaction was utilized to introduce sulfonic acid moieties of varying concentrations to free tyrosine residues within the directional sponges. The development of these modified tissue scaffolds for improved tubulogenesis can provide the groundwork for the development of a tissue engineered solution to chronic kidney failure.

585608 Development of Bio-Inspired Nanopatterned Surface on Polymer Thin Films for Anti-Reflective and Anti-Bacterial Coatings

Serious infectious diseases from contact with pathogenic microorganisms in our daily lives as well as the fast evolution of drug-resistant bacteria present the urgent need to develop a new functional surface that transcends antibiotics. Recently, physical methods emulating natural, bactericidal surfaces with nanopatterns have been employed in the efforts to combat bacterial adhesion by triggering mechanical rupture of the bacterial membrane. Such biomimetic nanostructures that exhibit antibacterial properties have shown promise in hard materials used for orthopedic implants such as titanium and stainless steel. Yet, little is known about the surface engineering of soft materials to attain bactericidal properties, despite numerous applications of polymers that need prevention of bacterial attachment. We aim to develop antibacterial structures on polymeric surfaces of which conventional benefits from intrinsic materials' properties are particularly desirable. The main goal of the current study is to develop the biomimetic moth-eye structure on polymer surfaces with both antibiotic and antireflective properties. To achieve this goal, first, we will fabricate the moth-eye mimetic nanopillar-patterned structures on silicon wafers (hard molds) with differing pillar geometries and spacing. By counting colony-forming units (CFU) and live/dead cell assays of gram-negative *E. coli* adhered on the engineered surfaces, we will find the optimal nano-surface geometry to maximize antibacterial properties. Then, we will obtain the moth-eye mimetic structure on poly(urethane acrylate) that presents a high refractive index by the soft molding pattern transfer technique using a fluoropolymer mold with a very low surface energy. The detailed investigation on the correlation between the surface properties and the antibacterial performance will be conducted to engineer polymeric surfaces for target applications. Development of nanopatterns on polymer thin films to induce bactericidal and antireflective properties will offer a wide variety of polymer coatings for future application in monitors, phone displays, and other optical devices.

585773 FTIR-Based Method for Measuring Diffusion in Organogels

In the past decade, various efforts to innovate drug delivery have been developed. The diffusion of nanocarriers in hydrogels has been a common approach. This research studies the diffusion of nanocarriers in organogels, and the results can help similar studies involving both organogels or hydrogels in the future. The main objective of this research project is to develop and validate a method to characterize nanocarrier diffusion in block copolymer organogels. The main driving force of diffusion is the concentration gradient which, in this case, is the difference in the amount of the nanocarrier in the gel and the outside solution. Experiments consisted of three disk-shaped gels each containing 0-1 wt% of a selected nanocarrier. The amount of nanocarrier was measured using quantitative Fourier-transform infrared spectroscopy (FTIR) in combination with gravimetric measurements to track gel swelling—both as a function of time. The data was then analyzed by normalizing the peak corresponding to the nanocarrier using the polymer peaks in order to see how the nanocarrier content dropped over time. After running various experiments with a wide range of polymer concentrations and two unique diffusing species, trends on how diffusion can be manipulated were established. For example, higher polymer concentration translated to a slower diffusion rate than low concentrations. Furthermore, unimer diffusion species diffused much faster than nanocarriers (*i.e.*, reverse micelles). Ultimately, the proposed FTIR-based method to track diffusion of species in organogels was valid regardless of what polymer concentration or type of penetrant was used. With this information, we can help further the advancement of controlling the rate at which drugs are delivered throughout the body.

587580 Colloidal Gelation of Silica Suspensions with Varying Salt and Surfactant Concentrations

Colloidal gelation can be useful to prevent sedimentation of complex fluid products such as oral medicines or paints. Colloidal particles remain suspended in a liquid when interparticle repulsive forces are stronger than the attractive forces, especially at low volume fractions. When attractive forces dominate, colloidal particles may form an "attractive gel". For aqueous suspensions of charged colloidal particles, electrostatic repulsive forces can be weakened by counterions. Similarly, agents can be added to strengthen attractive forces. For ionic surfactants at concentrations below the critical micelle concentration (CMC), there are no micelles, and they act as simple electrolytes. It is hypothesized that above the CMC of sodium dodecyl sulfate (SDS) surfactants (CMC = 8 mM), micelles will introduce attractive depletion forces between particles, altering gelation. Consequently, when compared to the baseline system of a simple salt, one may expect colloidal suspensions to gel at lower concentrations of SDS than of NaCl. Using negatively charged silica in water as a model colloidal system, the goal is to understand how gelation phase behavior differs in the presence of increasing concentrations of simple NaCl electrolyte versus SDS. The phase behavior of Ludox colloidal silica was characterized according to transitions from a fluid to an attractive gel phase. Purified suspensions were prepared at a variety of silica concentrations. Suspensions were monitored over time to record the onset gelation time, which was characterized qualitatively based on flow characteristics of the sample.

Thus far, a phase diagram mapping the effects of added salt and SDS on the silica suspensions has been prepared. The phase boundary between gel and fluid phases for SDS samples was observed at two to four times higher concentrations as compared to the phase boundary for NaCl samples for all the silica concentrations studied. In addition, gelation was faster for samples containing NaCl as compared to those containing equivalent concentrations of SDS. Whereas all NaCl added to solution contributes to screening of electrostatic repulsions, SDS micelles do not contribute to screening. Thus, only a fraction of sodium counterions for SDS added above the CMC contribute to screening. The results indicate that attractive depletion forces induced by SDS micelles were not sufficient to compensate for the weakened ionic strength enhancement caused by SDS addition compared to NaCl. The ionic strength effects dominate the gelation process.

587660 Polymeric Microsphere Fabrication for the Controlled Release of Therapeutics

Degradable polymeric microspheres (MS) were fabricated to encapsulate therapeutics, brimonidine tartrate (BT) and mesenchymal stem cell-conditioned media (MSC-CM), with an interest in the controlled release drug delivery for the treatment of glaucoma and chronic tympanic membrane (ear drum) perforations. MSC-CM MS and BT MS were fabricated using poly(lactic-*co*-glycolic acid) using single emulsion and double emulsion techniques with polyvinyl alcohol (PVA) used as a surfactant. PVA from different lots and different vendors were used to try to determine their effect on MS degradation. Microspheres were suspended in phosphate buffered saline (PBS) and rotated in an incubator at 37°C, with drug release samples taken daily over the course of 2-4 weeks until the MS degraded. Spectrophotometric analysis, with a colorimetric assay used for the MSC-CM MS releasates, of the total protein concentration in the MSC-CM MS and BT MS releasates showed a stable plateau release preceded by a burst of drug release. Scratch assays were performed *in vitro* to determine the wound healing potential of human dermal keratinocytes that are in the presence of MSC-CM. ImageJ analysis of images of the scratches taken at 6, 12, and 24 hours indicated that the presence of MSC-CM caused an increase in wound closure.

587821 Hydrogen Peroxide Releasing Biomaterials for Vascularization in Bone Tissue Regeneration

In the United States, 5-10% of bone fractures result in nonunion. Traditional solutions such as autografts and allografts have significant side effects, including donor site morbidity and disease transmission, respectively. Although new regenerative engineering treatments have been developed, their high cost and limited scalability have prevented their widespread clinical adoption. An additional limitation of bone tissue regeneration is inherent tissue complexity as vascular and neural networks must be regenerated in addition to osteoblasts. The Biomodulatory Materials Engineering Laboratory is working to develop a cheaper and more widely applicable option for large volume bone repair using novel biomaterials. This project focuses specifically on using the simple signaling molecule hydrogen peroxide to induce the differentiation of endothelial cells from mesenchymal stem cells for bone tissue vascularization applications.

First, the therapeutic window for the differentiation of mesenchymal stem cells into endothelial cells using hydrogen peroxide was determined by exposing the cells to various concentrations of hydrogen peroxide for one, three, and seven days. Proliferation and differentiation of the cells was determined for each timepoint using DNA/ATP assays and fluorescent microscopy, respectively. A novel hydrogen-peroxide-releasing biomaterial was then synthesized by modifying glutamic acid with hydrogen peroxide to form a peroxy acid (i.e., perglutamic acid).

Future research will focus on leveraging perglutamic acid as a monomer for various degradable polymers (e.e, polyanhydrides, polyesters, polyamides), for which cell studies can be conducted guided by the already established therapeutic window. The effects of these new biomaterials on off-target cells will also be studied. After biomaterials for osteogenesis, angiogenesis, and neurogenesis have each been synthesized, they can be integrated for large volume bone repair through tissue regeneration.

587139 Rheology of Binary Thermo-Reversible Colloidal Gels

In this work we investigate how the mechanical properties of thermo-reversible colloidal gels composed of binary mixtures of octadecyl silica nanoparticles in decalin change as a result of varying size ratios, attraction strength and concentration. We test the feasibility of theoretical models including those pertinent for composite materials and those within the Naïve Mode Coupling Theory framework to develop universal scaling criteria to predict elastic moduli.

586614 The Influence of Relative Humidity on the Thermal Stability of Biocomposite Films

Biomaterials have been a topic of interest in material science, bioengineering, and medical fields because of its abundance in nature, environmental friendliness and compatibility in living systems. In this study, the relationship between relative humidity and the compositions of biocomposites films is explored. The introduction of water into the system leads to plasticization, or the process of changing the thermal and mechanical properties of a polymer such as thermal stability, crystallinity, and flexibility because the water molecules may interrupt intermolecular forces between the molecules present in the film. Previous research has shown that the introduction of water to such systems lowers glass transition temperature and increases the stiffness or crystallinity of the films. The biocomposites made up of *Bombyx mori* silk and microcrystalline cellulose were regenerated using 1-ethyl-3-methylimidazolium chloride and ethanol and hydrogen peroxide as coagulation agents. The thermal, mechanical, and structural properties of these films were tested using techniques that include thermogravimetric analysis (TGA), differential scanning calorimetry (DSC), and Fourier transform infrared spectroscopy (FTIR). The relative humidity tests were only run with DSC and conductivity because the relative humidity could be controlled under those procedures.

586764 Colloidal Synthesis and Characterization of Metal Phosphide Nanoparticles for Optoelectronic Applications

Our growing energy usage, stressed by the apparent effects of climate change, has created an urgent need for sustainable sources of lighting and electricity. Successful development of such technology relies on our discovery of scalable photonic materials that can compete with current traditional group IV (Si, Ge) and III-V (GaN, GaAs) semiconductors, but have low toxicity and processing costs. Comprising of abundant, non-toxic, recyclable elements, nanoscale metal phosphide systems have garnered interest for such applications. To date, however, development of nanoscale metal phosphides has been deterred because of issues like the toxicity of metal and phosphorus precursors, poor stability under atmospheric conditions, and low quantum yield. In this work, we present reliable colloidal routes to group II and group IV phosphide (II – Zn, IV – Sn, Ge) nanocrystals from various metal and phosphorus precursors, some environmentally benign. We present optimization of synthesis protocols to allow for tunability of nanocrystal size and size dispersity. Structure and optoelectronic property relationships are explored via x-ray diffraction, transmission electron microscopy, and x-ray absorption spectroscopy. Finally, we show preliminary results on implementation of these materials in functional devices.

588045 Van Der Waals Adhesion Force for a Particle Interacting with a Rough Surface

The small gravitational forces characteristic in dust systems facilitate opposing dynamics: airborne suspension and adhesion to surfaces. To characterize the extent of particle adhesion, we investigate the van der Waals force between particles and rough surfaces using well-defined sinusoidal surfaces. Exact numerical solutions are carried out by creating an atomistic model and explicitly summing atom pair interactions between the particle and surface. In the case that the particle size is larger than the length scale for surface roughness, we derive an analytic model that incorporates force contributions from all asperities on the surface. Our analytic model accurately predicts the van der Waals force, in contrast to previous models which underestimate adhesion by not accounting for the effects of all surface asperities.

583171 Dynamics of Graphene Sheets at a Water-Vapor Interface: A Molecular Dynamics Study

Understanding the dynamics of particles at fluid-fluid interfaces has attracted considerable research interest over the past several decades as fluid interfaces create an environment where monolayer thin films can be assembled through a 'bottom-up' approach. For example, fluid-fluid interfaces have been used to manufacture highly transparent and electrically conductive thin films of graphene flakes as it has been shown that graphene flakes are thermodynamically favorable at the interface between two immiscible fluids. However, the dynamics of film formation and the interactions between graphene flakes are currently not understood. Furthermore, it has proven difficult to isolate and experimentally probe the dynamics of pristine monolayer graphene flakes at fluid-fluid interfaces. This study aims to address this gap by using computational simulations, which have been shown to reliably estimate real-world experimental scenarios. Molecular dynamics simulations were employed to investigate the lateral interactions and stacking dynamics of mono- and few-layer graphene flakes at a vapor-water interface. Adaptive biasing force (ABF) simulations were used to render potential of mean force profiles and identify different stacking pathways. Additionally, distance versus time profiles were generated for two interacting monolayer flakes and the results were fit according to a power law relation to understand how the interaction pair potential scaled with center-to-center separation distance. Ultimately, this study offers a new perspective into the investigation of interactions between interfacially-bound particles.

585865 Stability and Kinetics of Phase Transformations of Amorphous Dispersions

Many pharmaceuticals are composed of two main components: active pharmaceutical ingredients (API) and excipients. APIs are the active ingredients, while excipients are the inactive ingredients. Specifically, in pills/tablets, an API's characteristics will greatly influence the potency or bioavailability (the ability of a compound to dissolve in the blood to be used by the body). The more bioavailable a drug is, the less is required to achieve a desired effect. Moreover, the structure has a major impact on the bioavailability. The excipient can help to control the structure of the API because whether it is amorphous, crystalline, or some intermediate can influence solubility. Generally, amorphous particles are more soluble and crystalline particles are less soluble. However, crystalline structures tend to be more stable than amorphous structures, and, consequentially, sometimes crystallize fully or partially over time. The polymer excipient helps prevent or slow down this since polymers themselves do not crystalize, so a homogeneous mixture of API and excipient would be expected to have little or no crystallization (Myerson, 2002).

The Vibrating Orifice Aerosol Generator (VOAG) helps form amorphous dispersions. These dispersions comprise of a polymer excipient, polyvinylpyrrolidone (PVP), and a model API (a dicarboxylic acid). These are chosen since PVP is a common excipient in many pharmaceuticals and dicarboxylic acids resemble many APIs.

Solutions of various weight percents of dicarboxylic acid to PVP are dissolved in isopropanol (IPA). The VOAG works similarly to a spray dryer (commonly used in the pharmaceutical manufacturing): a solution is fed into the machine where it is sprayed, allowing individual drops to dry independently creating small particles. One main difference between the VOAG and spray dryer is that the spray dryer operates at higher temperatures, whereas the VOAG uses ambient temperatures with lower flow rates and improved droplet dispersion to allow for drying. Also, the VOAG produces a narrow droplet distribution which allows the VOAG to create particles of similar size to eliminate size dependency. We form amorphous particles and analyze their dynamic stability. Using powder x-ray diffraction (pXRD), samples are monitored regularly to determine if the particles begin to crystallize, how long this takes, and percent of API crystallization. Standards of known crystallinity are made to compare to the data to determine the percent crystallinity in the sample. This is done using partial least squares regression, peak area, and peak height.

References

Myerson, A. "Handbook of Industrial Crystallization." Butterworth-Heinemann. 2002.

584844 Computational Studies of the Self-Assembly of Tapered Brush-Coil Polymers in Solution

Materials formed by the assembly of block copolymers have many applications in environmental cleanup, drug delivery, electronics, optics, etc. In the past, block copolymers with linear architecture have been studied extensively using experiments and simulations. Moving to non-linear architectures, such as star, bottlebrush, and brush-coil, has the potential to advance assembly studies by producing new trends in structure and thermodynamics not seen in linear systems. One specific architecture of interest to us is a tapered brush-coil where the solvophobic polymer block has a bottlebrush architecture of varying density and gradient of side chain length, and the solvophilic polymer block has a linear architecture. In this poster, I will present my recent work, focused on the assembly in tapered brush-coil polymer solutions using a coarse-grained molecular dynamics simulation. For varying grafting density of side chain, gradient, and the direction of gradient of side chain length, I compare the assembled micelle size, aggregation number, and concentration profiles. At low grafting densities, there are observable differences in micelle size between a positive and negative gradient. At high grafting densities the micelle sizes are statistically similar for both gradient directions; however, the chains pack differently within the micelles. This packing is done in a way that does not drastically alter the concentration profile of the micellar core between grafting densities. This implies that while the bulk properties of the micelles may be similar, their internal composition and properties are completely different.

584147 Development of IL Based Low-Temperature Electrolyte for MET Siesmometer Utilizing Molecular Interactions

Room temperature ionic liquids (IL), cation-anion pairs that remain liquid at room temperature, have proven to function effectively as electrolytes in a variety of applications, including energy storage devices and sensors. Several unique traits allow for IL to broaden the operating temperature window of the electrolyte. Among these are its wide liquidus range, low volatility, and good thermal and electrochemical stability. The strategy of incorporating other solvents or salts with ILs is broadly applied for enhancements on critical physicochemical properties of an electrolyte. In particular, IL based systems show tremendous potential as electrolytes for the molecular electronic transducer (MET) seismometer to conduct seismology studies under harsh environments. In this work, we present a system consisting of IL, organic solvents, and salt, specifically designed for MET sensing technology at ultra-low temperatures. We discovered that through the formation of hydrogen bonds between the imidazolium cation and the organic solvents, the coulombic attractions between cations and anions are weakened, as evidenced by NMR spectrums. Furthermore, the molar ratio of components in the system has been tuned to optimize the molecular interactions, where neither organic solvent caused undesired first-order phase transitions. This tuned electrolyte boasted a glass transition temperature (T_g) at -119 °C, marking the lower temperature working limit of the system. Additionally, experimental results confirmed improvements in fluidity and ionic conductivity over the temperature range of 25 to -75 °C when compared to a previously developed aqueous based electrolyte system. Furthermore, the electrochemical stability of the proposed electrolyte is validated to be compatible with the MET sensing technology. Knowledge of the presented work may support the operation of MET seismometers and provide insights for future developments of low-temperature IL based electrolytes.

588317 An Improved Characterization of Topological Features in Molecular Simulations Using Persistent Homology.

We apply a mathematical method known as persistent homology to analyze the geometrical features of mesophases produced from molecular dynamics simulations. Specifically, we study ionic liquid/hydrocarbon mixtures. Ionic liquids (ILs) are salts with unique properties, which have made them ideally suited for many applications. Their ability to dissolve different, sometimes amphiphilic, compounds is likely tied to the formation of segregated nanoscale domains. These domains, or aggregates, are created due to the separation of the polar and nonpolar regions in the ionic liquid. Determining the shape of these domains is very important in understanding the behavior of the system.

Persistent homology uses a process known as filtration that incorporates a geometric measurement into topological invariants. By using a filtration parameter, simplicial complexes, such as triangles, are generated. The lifespans of the invariants are then recorded in a barcode. Such barcodes can be interpreted to determine the geometry of the object. Persistent homology was tested on objects with well-defined shapes, providing an understanding of the detailed barcode resulting from the analysis, such as the interpretation of the lifetime of lines in the barcode and the different dimensions. Comparisons between the shapes of these mesophases was done via persistent homology using the bottleneck distance to determine how similar two topologies are from one another. The smaller the value of the bottleneck distance, the more similarity exists between the two shapes. Additionally, persistent homology can be used to analyze different properties, such as thickness and diameter, of 2D and 3D shapes to see how each property affects the barcode. Such tests gave an insight on how persistent homology works, allowing us to use it on systems of ionic liquids to quantitatively characterize their topology.

Multiple simulations containing the mixtures of lipid-inspired ILs with either hexane or cyclohexane were conducted at various mole fractions and analyzed. The barcodes from each simulation showed that increasing the concentration of IL in the system increases the number of aggregates in both 2D and 3D. Furthermore, most holes generally appear to have the same size, suggesting a characteristic size that the IL aggregates like to adopt. Future studies will compare these simulated IL aggregates with a library of idealized shapes to identify its characteristic shape. Moreover, work will be done to extend the persistent homology code to accurately deal with topologies extending across periodic boundaries.

587383 Flow-Induced Crystallization of Fatty Acid Methyl Esters

Determining effects of concentration and shear-flow on the crystallization temperature of pure Fatty Acid Methyl Esters (FAMEs) and binary mixtures of FAMEs could lead to methods of altering cloud point of biodiesel fuels. The widespread implementation of biodiesel fuels is hindered partially by the cloud point of their FAMEs while in fuel or pipelines. Three FAMEs were studied, methyl myristate, methyl palmitate, and methyl stearate; pure and binary mixtures were prepared at 0/100, 20/80, 40/60, 60/40, 80/20 and 100/0 weight percent. Crystallization temperatures were observed as part of a phase study for the pure and binary FAME mixtures under static conditions. Then, rheology studies were conducted to determine the flow-induced crystallization temperature. The hypothesis of the project was that increasing shear rate increases crystallization temperature. Binary mixtures have shown to reduce the crystallization temperature from that of the pure components under static conditions and were expected to show even lower crystallization temperatures under flow. Crystallization temperature under flow has not consistently depressed, rather it has followed the trend taken by the highest concentration FAME.

587765 Thermochromic Cholesteryl Liquid Crystals Characterization with Polarized Light Microscopy

Cholesteric liquid crystals have temperature responsive optical properties due to anisotropic molecular orientation that is dependent on temperature. Specifically, the twisted nematic phase creates a helical structure that reflects visible light; the pitch of the helical structure and the wavelength (i.e. color) of reflected light both depend on temperature. Such liquid crystals have applications for thermal mapping by cameras. This introduces a need to evaluate the behavior of liquid crystals through visual analysis. In this work, a method to determine the thermochromic properties of liquid crystals was developed by using polarized light microscopy (PLM). To examine the liquid crystals, rectangular capillary tubes were filled with different formulations, comprised of cholesteryl benzoate, cholesteryl pelargonate, and cholesteryl oleyl carbonate, of liquid crystals. The composition of the ternary mixture was varied to tune the thermochromic properties of the liquid crystal. The rectangular capillary tubes were sealed at one end, filled with liquid crystal; a thermocouple was introduced, and the capillary was sealed at the other end. After heating to approximately 50°C, the samples were imaged using polarized light microscopy as they cooled to ambient conditions. To observe the color change due to the change in the molecular orientation of the liquid crystals, imaging was performed at 10x magnification, the minimal field diaphragm, and about 50% aperture. To avoid saturation, images were captured with auto exposure, auto white balancing at 1 frame per second. The digital images in standard red, green, blend (sRGB) were transformed to L*a*b* color space Performing frame-by-frame analysis, observed changes in color were related to quantifiable changes in CIE L*a*b* values. Depending on the composition, the blue-start to red-end varied from of 23.0 to 25.4 °C, 25.3 to 30.3 °C, and 36.9 to 42.6 °C. Specifically, color transition temperatures between 23.0 and 42.6 °C were obtained by varying the fraction of cholesteryl pelargonate in the formulation. The analysis method developed was validated with UV reflectance; the color transition temperatures measured by PLM and UV were comparable. Interestingly, the polarized light microscopy method allows for more detailed characterization of color transition temperatures than achieved by UV spectroscopy.

587207 CFD and PBM Modeling of Particle Aggregation in Mixing Tanks

Latex emulsions are systems containing a discrete phase of latex particles dispersed in a continuous aqueous phase and are widely used in industrial mixing processes. In the mixing of latex emulsions, small particles aggregate into larger, more uniform particles to establish a particle size distribution (PSD) required for quality standards. Particle aggregation and colloidal interactions are governed by the Derjaguin, Landau, Verwey, Overbeek (DLVO) theory. Perikinetic aggregation due to Brownian motion is the dominant form of aggregation for very small particles in latex emulsions. Computational Fluid Dynamics (CFD) and Population Balance Modeling (PBM) are two techniques that are used to develop simulations and computational models of the mixing and particle aggregation involved in emulsion mixing processes. CFD is used to model the fluid and flow properties associated with the mixing, whereas PBM uses population balance equations to track the PSD of the latex particles. This project aims to understand the complexities associated with colloidal dispersions and emulsion aggregation in mixing tanks and the issues involved with scaling-up these mixing processes. By examining (1) the velocity profiles, (2) the change of PSD over time, and (3) the specific areas of the tank at which particles accumulate, two tanks of different sizes can be compared. Using ANSYS Fluent, CFD-PBM simulations of latex particle aggregation in water were run in a Pilot Plant tank and a larger Manufacturing tank using a user-defined function based in DLVO theory to calculate particle aggregation rates. The simulations illustrate a lack of mixing in areas at the top and near the wall of the tank in addition to slower and less uniform particle aggregation patterns in the Manufacturing tank compared to the Pilot Plant tank.

585312 Permeability of Cured Composite Skins Produced Using Co-Cure over Honeycomb Core

Honeycomb sandwich structures made of fiber-reinforced polymer facesheets, film adhesive, and core can be produced using co-cure, during which the facesheets and adhesive are processed simultaneously. Frequently, honeycomb skins (facesheets and adhesive) must be impermeable during service to protect the core from environmental degradation or maintain internal pressure. However, skins can be rendered permeable by process-induced defects such as porosity. This paper describes an investigation of relationships between process conditions during co-cure and skin permeability. Half-sandwich structures composed of a four-ply prepreg laminate, a layer of film adhesive, and a honeycomb core were fabricated using a custom designed lab-scale fixture and controlled thermal and pressure conditions. Autoclave, bag, and core pressures were varied to evaluate hypotheses about mechanisms leading to permeability. Parts were cured with vacuum drawn in the bag and core (resulting in porosity), with super-ambient bag and core pressures (leading to minimal porosity), or with several constant pressure gradients between bag and core (leading to possible continuous flow between these regions). The resulting skins were tested for permeability within a second custom-designed fixture that allowed accurate, repeatable gas flow rate measurements across the skins. Gas flow pathways were also located using a water layer. The micro-structure of the skins was analyzed using light microscopy of polished sections to assess the presence, amount, location, and distribution of porosity and other defects. Results showed that porosity is related to permeability, but that gas flow through skins is possible even without high void contents. Pressure differentials led to complex variations in permeability, showing that variations in gas flow rate between cured skins is at least partly stochastic. Overall, the study contributes to a fundamental understanding of processing and manufacturing science for composites, and can provide useful guidelines for improving manufacturing processes and ensuring part quality.

583647 Rheology & Thermal Stability of Supercritical CO_2 foams at Different Temperatures and SDS Concentrations

The use of nanoparticles (NP) combined with surfactants has been a novel practice in the generation of CO_2 foams for hydraulic fracturing. This combination reduces water consumption and formation damage, while improving apparent viscosity and stability of the generated foam. This study evaluates the synergistic effects between nanoparticles and surfactants in CO_2 foams and how surfactant concentration affects the rheology and stability of the CO_2 foam. To assess the rheology and stability of CO_2 foam, nanoparticle dispersion with surfactant is co-injected with high pressure CO_2 at 1500 psi in the CO_2 accumulator and 1300 psi at the system outlet through a bead pack at different temperatures. The generated foam flows through a capillary tube and the pressure drop along the tube is used to calculate the foam apparent viscosity. The heights and texture of the generated foam in the sapphire tube is recorded over time to assess foam stability. To assess the effects of surfactant concentration, the temperature in the chamber is fixed and the concentration of surfactant is changed. To assess the effects of temperature, the surfactant concentration is fixed, and the temperature is changed. From the results, the mixture of nanoparticles and surfactant offered stronger and more viscous foams as compared to foams generated with nanoparticles alone. This is attributed to the synergistic effects between nanoparticles and surfactants, which is believed to form a steric barrier comprised of nanoparticles and enhanced with surfactants at the water/CO_2 interface. The structural integrity reduces the film drainage, the interfacial tension, and film coalescence, which contribute to higher foam performance. The combination of surfactant with nanoparticles in foam generation can offer enhanced foam stability and foam apparent viscosity, which helps maintain foam properties and provide better proppant transport under harsh reservoir conditions.

584155 Immobilization of Laccase on Plantain Fibers for Dye Degradation

Immobilization process is a commonly used method involving a support used as a structural component and a catalytic component to maximize the properties of both elements and increase the application range. This is a common technique for generation of heterogeneous catalysts. Characteristics of a competitive support for catalyst are affinity, availability and low-cost. In this study, fibers extracted from pseudostem plantain (waste produced by plantain harvests) were used as support for commercial laccase immobilization by adsorption process. Mechanical and enzymatic pretreatments were made to modificate morphology of fiber surface or its superficial area. Laccase adsorption was evaluated through remaining enzymatic activity on fiber, and characterization assays as SEM, FTIR and electrokinetic properties were studied to examine the structural modifications on material surface. Process parameters used during the immobilization (Temperature, contact time and pH value) were analyzed in terms of indigo carmine dye degradation. The change of absorbance was evaluated by UV-Vis. The results showed that a 98 % of degradation was observed after 2 h using immobilized laccase at basic pH values and 5°C of storage temperature. Thus, the immobilization of commercial laccase on plantain fibers is an available strategy for production of low-cost, and functional catalyst for textile industries.

587476 Architected Porous Media for Metered Liquid Transfer Flexographic Printing

The development of printed and unconventional electronic devices to meet application-specific needs requires innovation in printing technologies. Flexography (flexo), a rubber-stamping method developed in the 19th century, has proven scalable (meters per second) for graphic arts but is underdeveloped for printed electronics, particularly transistor and transparent electrode applications, due to limited feature resolution (>50 μm).

Although much work has gone into understanding the structural and fluid mechanics of the ink transfer processes, little attention has been paid to the actual stamp, typically fabricated via polymer replication of a hard master. The elastomeric stamp is top-side inked and compressed on a substrate, a process that has inherent limits for materials transfer and results in uneven pixel quality (poor feature resolution) due to compression-induced spreading. However, consider a porous stamp that undergoes precise deformation such as negative Poisson's ratio (NPR) during this process. Here, ink transfer could be a metered process with the pore-space being the reservoir and the mechanical deformation being the "metering pump". Moreover, a stamp that exhibits a slight NPR may allow for controlled expulsion and sharper transfer foot print (minimal line-edge roughness). Only recently has it been feasible to produce such engineered structures at high resolution, for example, using multiphoton-induced, direct laser writing (DLW). Here we investigate how precisely architected (e.g., NPR and a structured pore size distribution), 3D porous media can control the fluid saturation/capillary pressure characteristics upon mechanical compression to enable high fidelity/metered material transfer for high speed printing. In this study, we systematically investigate the effects of porosity, pore size distribution and microstructure compression on flexographic ink transfer using iterative arrays of high-resolution (<1 μm feature size) DLW structures to converge on optimized forms. Fluid dynamics simulations of defined poroelastic media provide further insight into metering ink transfer during compression. Overall, this work illustrates the design flexibility and precision control of micro- features/fluid dynamics enabled using form-fabrication of flexographic forms.

587489 An Assessment of the Biofilm-Prevention Efficacy of a Silver Carboxylate Eluting Titanium Dioxide and Polydimethylsiloxane Matrix on Spinal Implants

Healthcare acquired infections (HAI) cause an increased burden and excess length of hospital stays (LOS), which accounts for approximately 90% of total costs. [1] Moreover, infection rates in biomedical implants can be as high as 4% for some devices and can cause up to $50,000 to fix. [2] These infections specifically may involve the spread of a family of gram-negative bacteria, Enterobacteriaceae (CPE), including *Serratia marcescens*, and is becoming a global public health problem associated with healthcare facilities. [3] Human infections by this pathogen were not recognized until the latter half of the 20th century. [4] *S. marcescens* is a gram-negative bacterium commonly associated with chronic orthopedic wounds, specifically in the spine. These infections may be comprised of biofilms, which are communities of microorganisms that are attached to a surface and are resistant to antibiotics, resulting in recurrent Surgical Site Infections (SSIs). [5] However, in recent years the use of silver as an antimicrobial agent has received attention for its ability to combat biofilm formation through its long-lasting release, broad antibacterial spectrum, and low incidence of antibiotic resistance. [6] In response to these concerns, we have developed a coating composed of a titanium dioxide and polydimethylsiloxane (PDMS) matrix doped with silver carboxylate that allows for the controlled release of silver, acting as an antibiotic-independent antimicrobial by preventing biofilm formation.

This project involved the assessment of the antimicrobial capabilities of various composition ratios of the matrix and silver carboxylate on three spinal implant materials. Stainless steel, titanium, and polyetheretherketone (PEEK) were chosen for this study due to their varying topographies, but also widespread use and effectiveness as surgical implants. The coating does not affect the mean roughness of the implant and does not interfere with osseointegration. The layering is thin, on the order of 10-100 nm, and can be applied incrementally to create a desired thickness. Multiple trials showed that microbial adherence and percent area coverage decreased with increasing silver dosage. Across all implants, the best performance was found using a 95% TiO_2 and 5% PDMS matrix, with a 10x concentration of silver carboxylate. Scanning electron microscopy and confocal microscopy analysis showed the highest CFU per area for the stainless steel, resulting in a higher adherence rate for *S. marcescens* on stainless steel than the other two materials. Future direction of this study will constitute of an in-depth characterization and improvement of the coating itself.

586028 Morphology and Density of Organic Particles

My primary research goal is to determine the impact of initial droplet concentration and the fraction of the concentration that is small organic molecule versus polymer on the resulting particle morphology (both internal and external) with special attention paid to the particle density. It is hypothesized that a larger initial droplet concentration will lead to a particle shell earlier in the droplet evaporation process leading to a potentially hollower particle, while the particle density dependence with regards to the fraction of solids that is small molecule organic vs. polymer may be determined by the relative solubility of each molecule. Testing will be done using succinic acid as the model small organic molecule, polyvinylpyrrolidone(PVP) as the polymer of interest with methanol as the solvent.

In order to accomplish these goals, a range of solutions was prepared with respect to both the initial solids concentration, as well as the relative amount of succinic acid and PVP. Thus, for this work, I focus on solution concentrations of 23.90, 47.79, and 95.58 g/L. This was done by the droplet composition of a solvent, and polymer, and a small molecule with an 11.5% limit of small molecule. I am working with succinic acid and PVP. For each of those values of total solids concentrations, I also considered a variety of fractions of succinic acid and PVP. The initial focus for these was on a range of values where the resulting particles are fully amorphous.

Another focus is the internal morphology of the particles with the PVP as the polymer of interest with methanol as the solvent. In the past, researchers have used Araldite 502, DDSA, and DMP-30 to make an Araldite 502 epoxy resin for embedding, but due to the epoxy resins slightly malleable nature the particles appeared to be warped by the microtome blade, which deformed the spherical shape and internal structure of each particle. My goal is to find a more refined embedding technique which has no large change in temperature and cannot have a malleable nature. This past summer I worked with Benzil, Bisphenol-A based Bismaleimide resin, and Cashew Nut Shell Liquid as a resin to embed the particles into. I am going to further my research with the Cashew Nut Shell Liquid because of the long alkyl chains provide some flexibility because of the internal plasticization effect, which lead to a softer resin.

587772 Optimizing Lead Perovskite Formation in High Humidity

Perovskite solar cells (PSCs) have gained popularity recently due to their high efficiencies and low production costs. PSCs are often fabricated in a humidity controlled environment, because water degrades the perovskite layer. This project investigates the synthesis of PSCs in high humidity (>50%). In this study PSCs were synthesized using different perovskite and antisolvent combinations to achieve the best performance. The best outcome was achieved using a combination of 2.4% 1,3-Dimethyl-2-imidazolidinone (DMI). in the perovskite and diethyl ether as the antisolvent. This approach yielded an efficiency of approximately 4% (without a Hole transporting layer(HTL)).

586765 Rate Measurement and Extraction for Thermally Activated Delayed Fluorescence Materials

In recent years, an increasing amount of researches has been focusing on organic light-emitting diodes (OLEDs), which have unprecedented advancements in display and lighting technologies. Organic electronic materials generate lights through a process called electroluminescence in which singlet and triplet excitons are formed in the first step. Singlet excitons relax to ground state and produce light while the triplet excitons dissipate their energies to the surroundings as heat. Several strategies such as thermally activated delayed fluorescence (TADF) have been proposed to convert up to 100% of the excitons in the device into the light. In this study, a type of TADF material, 4Cz-IPN is used to demonstrate the delay fluorescence property. A combination of optical, electronic and cooling systems is built to study the temperature dependence of TADF materials and to extract the rates in the TADF process. These rates can be used to calculate the energy gap between triplet and singlet excited states, and thus giving guidance of TADF material design. This system can be applied to various topics such as the influence of gold particles doped in the TADF materials, and the host-dependence of TADF materials.

586499 Enhancement of Biocide Mitigation of Carbon Steel Biocorrosion By an Oilfield Biofilm Using a Nature-Inspired 14-Mer Anti-Biofiom Peptide

Microbiologically influenced corrosion (MIC) is a threat to hydrotest, which is a critical step in confirming whether a pipeline can be commissioned. Seawater is frequently used in hydrotesting large pipelines. In this biofilm prevention study, an oilfield biofilm consortium was grown in an enriched artificial seawater followed by anaerobic incubation at 37 °C for up to 60 days. The combination of 100 ppm (w/w) 2,2-dibromo-3-nitrilopropionamide (DBNPA) + 100 nM (180 ppb) Peptide A (a biofilm dispersal agent) led to extra SRB (sulfate reducing bacteria), APB (acid producing bacteria) and GHB (general heterotrophic bacteria) sessile cell count reductions of 0.9-log, 0.8-log and 0.6-log, respectively, compared with the outcome obtained by 100 ppm DBNPA only. The cell count reductions led to enhanced mitigation of carbon steel corrosion weight loss and pitting corrosion, which were corroborated by electrochemical measurements.

587799 The Effects of Sodium Dodecyl Sulfate Concentration, Salt Concentration, and of Types of Salt on Oil – in – Water Emulsion Stability

Emulsions are homogenous mixtures of two immiscible liquids where one phase is finely dispersed in the other. Emulsions are thermodynamically unstable and will ultimately separate1. Surfactants are often used to decrease interfacial tension between the two immiscible phases and extend the lifetime of the emulsion2. Emulsions are commonly found in consumer products. However, in the majority of consumer products there are a large number of additional ingredients used to give products certain properties. It is important to understand how different components such as surfactant concentration, salt concentration and ionic composition affects emulsion stability.

583786 Peptoid-Functionalized Gold Nanoparticles for Use in Electrochemical Sensors

Detection of pathogenic organisms is a key part of providing medical care to global communities. Current detection methods are slow, expensive, and sometimes unable to provide the sensitivity and specificity necessary for accurate measurements and effective treatment plans. Electrochemical sensors enhanced with peptoid-functionalized nanoparticles are an alternative to traditional techniques that offers rapid, accurate detection of pathogens. These sensors work by undergoing an electrochemical reaction when their electrodes come into contact with a target molecule; peptoid-functionalized nanoparticle electrode coatings have the potential to increase the speed and specificity of these reactions. Preparation of these modified gold nanoparticles (AuNPs) involves the exchange of stabilizing molecules added during AuNP synthesis for thiol-terminating peptoids designed to provide desired functionality. This surface modification procedure is complicated by both aggregation of AuNPs in peptoid-containing solution and adsorption of peptoid-functionalized AuNPs to polypropylene centrifuge tubes. Successful implementation of functionalized AuNPs in this project requires that no aggregation occurs to limit the functionality of the peptoid or AuNP. Multiple peptoids were observed to cause AuNP aggregation and adsorption, with aggregation rates proportional to the concentration of added peptoid; changes were monitored via visual observation of solution color and solution absorbance measurements. To attempt to delay or eliminate aggregation, solutions were reacted in a variety of different conditions, including the presence of sodium hydroxide or Tween 20, refrigeration, incubation, mixing, and absence of light. Alternatives to an aqueous solution, such as phosphate buffer solution (PBS), were also tested. The identification of a method to stabilize the AuNPs during the peptoid functionalization process is essential to future work on the binding capabilities of peptoid-functionalized AuNPs with proteins found on the surface of pathogenic viruses.

584845 Phase Transitions of Titanium Dioxide Nanocrystals

Titanium dioxide (TiO_2) is a widely-used catalyst with three distinct nanocrystal phases – anatase, rutile, and brookite. The chemical properties and catalytic activity of TiO_2 vary with its crystal phase, making morphology an important determining factor in catalytic performance. A synthesis method involving thermal hydrolysis of a titanium precursor solution in the presence of urea was implemented for the direct synthesis of anatase and brookite, while rutile was produced through the high-temperature calcination of anatase and brookite materials. The anatase-to-rutile transition (ART) was investigated as a function of both calcination temperature and time to gain an understanding of how those factors affect TiO_2 phase transition. Our results revealed that partial transitions occurred in the 450-550°C range. This temperature range also makes it possible to produce customized mixtures of the anatase, brookite, and rutile phases.

587980 Eigencages: Learning a Latent Space of Porous Cage Molecules

Porous organic cage molecules harbor nanosized cavities that can selectively adsorb gas molecules, which lends them applications in separations and sensing. The geometry of the cavity strongly influences their adsorptive selectivity. For comparing cages and predicting their adsorption properties, we embed/encode a set of 74 porous organic cage molecules into a low-dimensional, latent "cage space" on the basis of their intrinsic porosity. We first computationally scan each cage to generate a three-dimensional (3D) image of its inherent pore cavity. Leveraging the singular value decomposition, in an unsupervised manner, we then learn across all cages an approximate, lower-dimensional subspace in which the 3D porosity images cluster together. The "eigencages" are the set of orthogonal, characteristic 3D porosity images that span this lower-dimensional subspace, ordered in terms of importance. A latent representation, or encoding, of each cage follows by approximately expressing it as a combination of the eigencages. We show that the learned encoding captures salient, inherent features of the cavities of porous cages molecules and is predictive of properties of the cages that arise from cavity shape. Our methods could be applied to learn latent representations of cavities within other classes of porous materials and of shapes of molecules in general.

583892 Evaluating the Fluid and Aerodynamic Properties of Uio-66 Nanoparticles

Pulmonary diseases such as lung cancer, cystic fibrosis, asthma, and chronic obstructive pulmonary diseases (COPD) are associated with alarming annual mortality and morbidity rates [1]. Evidently, current therapies for pulmonary diseases are inadequate, and the seriousness of many pulmonary diseases motivates a need for improved pulmonary drug delivery [2]. Metal-organic frameworks (MOFs), coordination compounds of metals and ligands, offer promise as nanoparticle drug delivery vehicles due to their cargo-loading capacity, tunable pore size, and potential biocompatibility [3]. The goal of this project is to evaluate relevant fluid and aerodynamic properties of a specific zirconium-based MOF, UiO-66, for drug delivery applications. The cargo-loading, fluid and aerodynamic properties were characterized, while the biocompatibility was confirmed. UiO-66 particles were synthesized under acid-free conditions, with water content, synthesis time, and ligand:metal ratio controlled to modulate the particle size. Using a fluorescent assay, the cargo-loading was characterized for a model cargo, rhodamine B (RB) using particles with varying amounts structural defects, or defectiveness. Dynamic light scattering (DLS) was utilized to measure the hydrodynamic diameter (ZAVG) of particles. The aerodynamic diameter of the particles was characterized using a next generation impactor (NGI), after particles were aerosolized using a collision jet nebulizer. Finally, cytotoxicity was analyzed through a cell titer glow assay after dosing murine-derived alveolar macrophages (MH-S) with particles. The results of SEM yielded geometric radii between 30-120 nm. The median mass aerodynamic diameter (MMAD) decreased from 0.64 μm for unloaded particles to 0.42 μm for max-loaded particles at 7.63% defectiveness. At all other defectiveness levels, a similar decrease was observed. The ZAVG diameter consistently decreased as loading increased across all defectiveness levels. Finally, after dosing alveolar macrophages, no acute toxicity or inflammatory response was observed. The observed decrease in MMAD for loaded particles is an important result, as generational pulmonary deposition depends primarily on aerodynamic particle size. The lack of acute cell immune response after dosage also supports the viability of UiO-66 as pulmonary drug delivery vehicles.

References:

[1] The top 10 causes of death, Fact sheet N°310. WHO: World Health Organization. 2012.

[2] Jeffry G. Weers, et. al. Smith. Journal of Aerosol Medicine and Pulmonary Drug Delivery. Volume: 23 Issue S2: December 6, 2010

[3] Joseph Della Rocca, Demin Liu, and Wenbin Lin. Nanoscale Metal–Organic Frameworks for Biomedical Imaging and Drug Delivery. *Accounts of Chemical Research* 2011 *44* (10), 957-968.

583153 Development and Modeling of Cu@Ag Core-Shell Nanoparticle Catalysts with Galvanic Replacement for Electrochemical CO2 Reduction

Efficient solar-driven electrochemical CO_2 reduction (CO_2RR) would provide a carbon-neutral source of fuels and chemical precursors. Copper (Cu) is a catalyst that has been widely studied for CO_2RR due to its moderate CO binding strength and the variety of products it is able to catalyze. Recent advances in CO_2RR have made use of engineered nanoscale catalysts with various sizes and morphologies of Cu nanoparticles. One of the main goals of CO_2RR is shifting selectivity towards C_{2+} products, which have higher energy densities and are more profitable when compared to other products. While a Cu catalyst alone produces mostly hydrogen and hydrocarbon products, introducing other metals at the interface changes the interaction between reaction intermediates and surface atoms, enabling tuning of the selectivity. For example, the introduction of silver (Ag) allows for a synergistic effect during catalysis that leads to a shift in selectivity towards oxygenated products, which are not observed with either Cu or Ag separately. Additionally, core-shell bimetallic nanoparticles allow the two metals to be present in close proximity instead of a physical mixture of monometallic particles. Furthermore, the distribution of the two metals can be reliably tuned via a galvanic replacement reaction (GRR). In this work, we present a colloidal synthesis of Cu-Ag bimetallic nanoparticles, employing GRR to precisely control the shell thickness with a diffusion-based self-limiting mechanism. Using Cu-Ag as an example system, we develop a mathematical model based on transport, diffusion, and surface energies that describes the formation of the Ag shell on a Cu core and explains the self-limiting nature of the reaction. The nanoparticles are characterized with bulk composition measurements like ICP-OES and elemental mapping with STEM-EDS to analyze the shell thickness as a function of several reaction parameters. The synthesis demonstrates highly uniform particles with controlled shell thicknesses where colloidal stability is maintained, an important consideration for uniformity when covering the surface of an electrode. The results of the characterization are compared with the model and demonstrate good agreement with experimental results. Since our mathematical model depends only physical material constants, it may be applied to various bimetallic systems to engineer the next generation of nanoscale catalysts and further enhance the selectivity of the CO_2RR towards desired products.

588103 Optimizing Conditions for Thermal Exfoliation of Graphite Oxide Films

Thermal exfoliation of graphite oxide is an effective and economically feasible large-scale production method for graphene-based materials. These materials are desirable for various applications such as catalysts, selective membranes, semiconductors, batteries, and sensors. In the process of GO film synthesis, the vast graphene oxide (GO) theoretical monolayer specific surface area (2600 m^2/g) is lost due the perfect face-to-face GO monolayer alignment. In GO synthesis, modified Hummers method was applied, and the GO suspension (2 mg/mL) was used to prepare drop-cast and wrinkled GO films of varying thicknesses. These GO films were thermally exfoliated in an instrument set low (10 K/min) and high ($100<HR<1000$ K/min) heating rates. At low heating rates, GO drop-cast films below thickness 36 μm exfoliated into low surface area (<37 m^2/g) and non-porous reduced graphite oxide (rGO) films. GO drop-cast films with thickness above 36 μm subjected to low heating rates yielded rGO films with modest surface area (~171 m^2/g). Independent of the film morphology (wrinkled/drop-cast) and thickness, at high instrument heating rates the graphite oxide films explosively exfoliated into rGO powder with fairly large surface area (~700m^2/g) and total porosity (~1.8 cc/g). It seems there is a critical film thickness and heating rate at which mass (product gases) and heat generation rates surpass their transfer rates out of the rGO system and GO exfoliates explosively. Wrinkled GO films did not demonstrate significant differences in thermal exfoliation compared to the drop-cast GO films. These insights are valuable for researchers trying to optimize GO thermal exfoliation process to produce graphene-based materials in large scale for variety of application venues.

585755 Optimizing the Stability and Functionalization of Imine-Linked 3D Covalent Organic Frameworks

Covalent Organic Frameworks (COFs) are a class of porous multi-dimensional polymers with well-defined molecular structures derived from their rigid starting materials. These properties give the material immense potential in applications ranging from gas storage to optoelectronics, catalysis, and water treatment. For 3D imine-linked COFs to become viable in these applications two major issues must be overcome. The first is that frameworks like COF-300 can collapse as a result of water molecules filling the pores, decreasing the overall surface area. Second, a method for simple orthogonal functionalization of the framework must also be established. Throughout our studies, we have identified two-stage synthetic conditions that preferentially generate the porous form via an amorphous intermediate state. Here we describe our work generalizing our two-stage synthetic approach and methods towards developing functionalized highly-stable 3D COFs.

587523 Confined Diffusion through Macroporous Structures

Currently, confined mass transport through subsurface environments (e.g., soil, sediment, rock) is primarily investigated using computer simulations or run-off studies of core samples. Computer simulations lack empirical grounding while core studies provide little insight into the solutes' interactions with the internal structure. We created an empirical model of these complex subsurface systems that observed individual particle trajectories. Macroporous structures (possessing a pore size greater than 50 nm) were created from silica (< 10 μm) packed on Piranha cleaned glass wafers and inside cured polydimethylsiloxane (PDMS) flow channels. We employed single-particle tracking methods to visualize the path of 40 nm carboxylate-modified fluorospheres through macroporous structures. A Nikon A1R microscope capable of Laser Scanning Confocal and Total Internal Reflectance Fluorescence (TIRF) with an attached camera was used. Two kinds of tracking experiments were conducted: static diffusion and flow. During experimental runs, the capture images were acquired at a rate of 50 ms and combined in a single movie (2400 frames per movie). The movies were analyzed using in-house single-particle tracking code. Trajectory maps, tortuosity averages and diffusion coefficients were calculated for each sample. Analyzing the static diffusion experiments, tracer particles in 4.3 μm structure experienced more tortuous paths compared to the 4.3 μm + 9.2 μm and 9.2 μm structures. All packing structures yielded lower diffusion coefficients compared to unconfined diffusion. The flow experiments posed a challenge in creating a close packed porous structure that remained immobile but still allowed fluid flow. Future work will observe flow behavior through packed structures and derive mathematical relationships. The overall goal of this project is to inform mass transport through low porosity environments such as soil systems and porous rock.

585095 Processing of Nanostructured Block Copolymer Micelle Crystals

Block copolymers in solvents that are selective to one block form intricate nanostructures that can give rise to useful material properties. The mechanical properties of the crystalline phases seen in concentrated block copolymer solutions depend on nanoscale structure and sample history. In this work, the phase behavior of a triblock polymer, Pluronic P123 (PEO_{20}-PPO_{70}-PEO_{20}), and water system was studied. These liquid crystals undergo a thermoreversible transition from a liquid-like disordered state at low temperatures ($T < 15$ °C) into a gel-like soft solid at higher temperatures ($T > 20$ °C). The goal of this work is to characterize the thermotropic and concentration-dependent mechanical behavior of this system in order to elucidate the nanoscale structures in the P123 and water system and inform future structural studies. We are interested in determining the impact of processing on the nanoscale structure and resulting mechanical properties of these materials.

To construct a phase diagram for the P123 and water system, inversion tests were first conducted to distinguish the disordered and ordered states. Samples with visible changes in appearance and viscosity with temperature were identified as target concentrations for rheological study. Subsequent rheometric temperature sweeps were conducted on these target concentrations to locate changes in modulus that may signify structural changes with temperature. Currently, we are working to develop a method to quantify birefringence under shear to distinguish isotropic phases, which exhibit no birefringence, from anisotropic lamellar and hexagonal phases, which are birefringent to different extents.

A motivation for this research is the potential application of P123 in drug-delivery systems. Due to its non-toxicity and low-temperature disordering, this polymer could potentially encase therapeutic enzymes throughout the water-filled interstitial spaces. This could facilitate storage of these therapeutic proteins without refrigeration. It is important to understand the impact of processing on these materials so that the desired structures and material properties can be achieved.

587309 Evaluating and Modeling the Strength of 3D Printed Samples Using a Design of Experiments Framework

Advanced technologies that precisely and rapidly convert designs into real-world products have the potential to transform industries ranging from healthcare to aerospace. Additive manufacturing offers advantages over subtractive manufacturing methods, which include such advantages as high accuracy prototyping, complex internal geometries, and less material waste. In particular, stereolithography (SLA) printing offers a reliable platform for generating mechanically isotropic parts with high feature resolution. In SLA printing, photo-sensitive liquid resin is selectively cured using UV light into solid parts in a layer-by-layer process based on specifications from a computer-aided design stereolithography file (STL). Potential challenges associated with large-scale use of SLA printing are the need for significant quantities of specialized liquid resins and long printing times. In this work, we designed and tested part geometries with varied void patterns and fractions to investigate the trade-off between reduced material use and decreased mechanical strength. To determine which patterns are most effective at maintaining the material strength per weight, we investigated three different void patterns with varying void fractions at two layer heights using a statistical design of experiments (DOE) methodology. The parts were designed and tested in accordance with ASTM standards and post-cured in a custom-built curing chamber. Tensile tests were used to measure the Young's modulus and ultimate tensile strength for each part using an Instron testing system. Overall, circular void patterns resulted in the highest mechanical strength for a given void fraction. Moreover, finite element analysis was utilized to inform and corroborate the experimental results. This work can inform the design of parts with decreased material use, making the process more economical.

588062 Manganese Dioxide (MnO_2) Nanowire Phase Behavior in Various Polar Solvents

The focus of this work is to investigate the phase behavior of manganese dioxide (MnO_2) nanowires dispersed in various polar solvents. MnO_2 nanowires were synthesized using a facile double-solvent synthesis method. The resulting materials were characterized using scanning electron microscopy (SEM) and spectroscopy techniques. A range of MnO_2 concentrations was then dispersed in ethylene glycol, dimethyl sulfoxide, ethanol, and methanol. UV-vis spectroscopy was performed to determine the concentration of nanowires that could be dispersed in a supernatant and quantify dispersion stability in each solvent. Since the nanowires are anisotropic, they have the potential to form lyotropic liquid crystal phases with increasing concentration; such phases facilitate processing into highly aligned solid materials. However, poor nanowire-solvent interactions can result in the formation of isotropic flocs and aggregates which are detrimental to the manufacture of uniform materials. Therefore, optical microscopy, with and without cross-polarizers, was used to characterize the microstructure in each sample for alignment. These findings were compared to predictions based on Hansen Solubility Parameter calculations. These results will contribute to the overall field of nanowire liquid crystal research and will enable understanding of how the dispersion microstructure can be linked to potential applications in aligned films, sensors and energy storage devices.

586334 Elastomer Nanofibers Towards Wearable Electronics

Elastomers are polymers with viscoelastic properties and weak intermolecular forces that allow them to recover their original shape after being stretched. The elastomer complex that is the focus of this study is poly(2-acrylamido-2-methyl-1-propanesulfonic acid)/polyaniline/phytic acid (PAAMPSA/PANI/PA). This complex has elastic and conductive properties due to the inherent properties and electrostatic interactions of the individual components. The method used to create nanofibers is electrospinning, a production method that uses a strong electric field to draw fibers from a liquid. This provides control over the morphology of the fibers due to the parameters available for optimization (e.g., concentration of polymer in solution, applied bias, needle-to-collector distance, flow rates). An area of research that has not yet been explored is creating useful elastomer nanofibers. Blends allow the PAAMPSA/PANI/PA complex to be electrospun, made with the addition of a water-soluble polymer, poly(vinyl alcohol) (PVA). Elastomer nanofibers can be used to develop new smart fabrics, biomedical sensors, electronic skin (e-skins), more efficient drug delivery, and flexible electronics.

584846 Creation of 3D Ordered Superlattices through Self-Assembly of Polymer-Functionalized Nanoparticles

Functional nanomaterial assemblies are an exciting area of study with various applications across catalysis, photonics, electronics, plasmonics, and biomedical applications. We focused on the study of metamaterials that can show electromagnetic properties not observed in nature with specific interest in their unique optical properties. Using gold nanoparticles (AuNPs) as a model for other metallic nanoshapes, we demonstrated that the respective functionalization of these nanoparticles with poly(ethylene glycol) methyl ether thiol (PEG-SH) and poly(acrylic acid)-thiol (PAA-SH) to form AuNP-PEG and AuNP-PAA core-shell nanoparticles allows for a bridge to be created between neighboring nanoparticles. These bridges were formed through the addition of free polymers in solutions of AuNP-PEG nanoparticles and inter-chain hydrogen bonding between protonated poly(acrylic acid) chains in the shells of AuNP-PAA nanoparticles. We demonstrate that these functionalized nanoparticles self-assemble into 3D ordered structures in bulk solution similar to programmable assembly using complementary DNA strands through the use of synchrotron small angle X-ray scattering (SAXS) and various optical characterization techniques. The effect of pH, salt concentration, PEG and PAA chains on the inter-particle distances in the assemblies will be discussed. Facile implementation of block copolymer based self-assembly techniques is enabled by substituting simpler synthetic polymers instead of DNA and the results showed that these superlattices are stabilized by vanderWaals forces, and that allows the structures to be stable even when removed from the aqueous solutions.

588231 Novel Strategy for Functional Oligopeptide Conjugation inside the Mesopores of Silica Nanoparticles

The functionalization of mesoporous silica nanoparticles (MSNPs) with biomolecule-selective oligopeptides is an important method for materials design in bio-separations, drug delivery, and other biotechnology applications. MSNPs are desirable for these applications because their flexible and well-controlled synthesis process provides well-defined pore spaces with high surface areas and pore volumes. MSNPs provide a unique protective microenvironment for binding between target biomolecules and conjugated polypeptides. However, procedures specifically designed for oligopeptide attachment within the pores of MSNPs still need to be developed to avoid uncontrolled reactions at the external particle surface.

Here, a functionalization process is described involving a diazirine-based hetero-bifunctional linker, sulfo-succinimidyl 6-(4,4' azipentanamido) hexanoate (sulfo-NHS-LC-diazirine). This linker allows for controlled attachment of oligopeptides to the surface of the MSNPs (average pore diameter of 8 nm and surface area of 730 m^2/g) that have been functionalized with amine ($-NH_2$) groups by condensation of aminopropyltriethoxysilane. The choice of hetero-bifunctional linker allows for specific control over whether sulfo-NHS reacts first with the peptide N-terminus or the amine groups on the surface. It is found that peptide-first attachment (reacting sulfo-NHS first with the peptide N-terminus) leads to more efficient conjugation because this approach allows for diffusion of the linker-peptide conjugate to the interior of the particle pores before attachment by UV activation of the diazirine group. This is quantified using the fluorescein isothiocyanate tagged tetrapeptide arginine-serine-serine-valine (RSSV) to give 0.5 mol peptide/mol amine. To demonstrate the versatility of the approach, three distinct oligopeptides (RSSV, Glycine-Glycine-Glycine-Glycine (4G), and a 16-mer composed of four repeated RSSV units) are conjugated by peptide-first attachment to the surface of the MSNPs. Thermogravimetric analysis (TGA) shows similar degrees of conjugation for all three peptides, suggesting versatility and generalizability of the procedure developed. Finally, these peptide-conjugated nanoparticles are used as a high-capacity adsorbent for ß-estradiol (representative of the broad estrogen family) to show specific binding activity above that of MSNPs without conjugated peptides. This study has implications for utilizing oligopeptide-functionalized MSNPs for bioactive molecule separation from living cells (nanoharvesting) or adsorption for environmental remediation (removal of endocrine disruptors from wastewater).

588237 Production of Monodisperse Yolk-Shell Titanium Dioxide Microspheres with Tunable Characteristics

Titanium dioxide is a photocatalytic material that has proven useful in the degradation of organic pollutants. In turn, this photocatalyst may prove useful in efforts to handle common greenhouse gases more sustainably. One possible application for the use of TiO2 in photocatalytic degradation of organic pollutants could possibly be a bi-reforming reaction of carbon dioxide and methane into synthesis gas. By degrading these greenhouse gases into synthesis gas, the product gas mixture of carbon monoxide and hydrogen can be used in the synthesis of hydrocarbon fuels through Fischer-Tropsch reactions. Through this reaction, greenhouse gases may be recycled into useful liquid fuels. Without the use of a photocatalyst, the high reaction enthalpy associated with the aforementioned reforming reaction (247 kJ/mol) renders the reaction non-feasible. However, the implementation of TiO2 as a photocatalyst may aid significantly in lowering heat energy requirement of the reaction.

Unfortunately, the commercially available TiO2 photocatalyst, Degussa P25, is a mixed phase polydisperse nanopowder with an anatase/rutile phase ratio of 85:15. The high degree of polydispersity coupled with the very small average size of these nanoparticles would result in undesirably dense packing and a high pressure drop across a packed bed reforming reactor which creates obstacles to achieving desirable throughputs at atmospheric pressure. In order to solve this issue, TiOz microparticles with a high degree of monodispersity needed to be manufactured while maintaining or exceeding the specific surface areas of Degussa P25.

This work explores the design and production of highly monodisperse TiO2 microparticles with tunable size, surface morphology, surface areas, and phase ratios. The particles were manufactured in a flow-focusing microreactor composed of three different capillaries that takes advantage of the immiscibility of two feed solutions to create monodisperse droplets prior to the hydrolysis reaction. The continuous phase is a hydrophilic solution containing 1 wt% Pluronic F108 (a surfactant) in formamide, while the dispersed phase is a hydrophobic solution composed of toluene, a photocurable polymer, titanium n-butoxide, and a photoinitiator. The droplets formed by two-phase flow through the outlet capillary are collected in water containing surfactant and subsequently photocured, washed, dried, and calcined. This work has shown that the aforementioned characteristics can be adjusted by varying hydrolysis conditions, precursor composition, and calcination temperatures. In this way, it has been possible to design microparticles suitable for use in photocatalysis of greenhouse gas reforming reactions.

587308 Elucidating Structure-Property Relationships of Atomically-Precise Gold Nanoclusters from Density Functional Theory

In recent years, ligand-protected Au nanoclusters (NCs) have retained tremendous interest due to their structural determination with atomic precision, unique optical and electronic properties, and great potential in diverse nanotechnological applications like catalysis, biomedicine and imaging. Since their discovery, Density Functional Theory (DFT) has been instrumental in calculating the electronic properties of NCs. Although accurate, DFT is computationally expensive, especially for NC systems consisting of a couple of hundred atoms. Thus, rapid methods that retain the accuracy of DFT are highly sought after to screen NCs exhibiting desirable electronic properties. Towards this goal, we developed a computational framework to discover structure-property relationships on ligand-protected Au NCs. Our approach connects NC descriptors based solely on experimental structural data to electronic properties calculated with DFT. Applying our framework, we developed a universal NC electronic gap model based on a sample of 22 NCs ranging from 18 to 102 Au atoms with 8 different ligand types. The model was in turn used to elucidate the structural features that dictate electronic gaps of NCs. Overall, this research accelerates materials discovery at the nanoscale by developing structure activity relationships on ligand protected NCs. These relationships elucidate how NC size (number of Au atoms) and presence of ligands (number of thiol ligands) affect the NC electronic properties and are able to identify specific NC morphologies that exhibit desired electronic properties.

585774 Analysis of Parameter Interactions with a Box-Behnken Design on the Production of Cellulose Nanocrystals through Sulfuric Acid Hydrolysis

Cellulose is a natural, abundant, and inexpensive, high-molecular-weight polymer that has attracted attention from various industrial sectors. Cellulose Nanocrystals (CNCs) are most commonly derived from acid-catalyzed hydrolysis of cellulose and are used in a variety of applications due to their remarkable mechanical, chemical, rheological, and optical properties. One barrier to mass implementation of CNCs is the high cost associated with the manufacturing process, primarily from low overall yields. Optimization of the manufacturing process offers a potential solution to increase yield, reduce cost, and enhance potential for mass production and total market approval.

The goal of this study is to explore the interaction of sulfuric acid concentration (x1), hydrolysis temperature (x2), and hydrolysis time (x3) through a Box-Behnken Design (BBD) on the overall yield of CNCs through the sulfuric acid hydrolysis process. Our prior research has explored the outer limits of the practical design space and highlighted key processing parameters through a rotatable central composite design. However, our initial experimental design consisted of conditions at axial locations that were not achievable, either because of extreme conditions that completely degraded the cellulose, or at times that are unachievable (e.g. negative time). This BBD study encompasses a down-selected design space to further analyze targeted conditions for optimization of the hydrolysis process and subsequent yield. Conductometric titration was performed to analyze sulfate group density. Further, X-ray photoelectron spectroscopy was carried out to investigate the extent of CNC degradation during synthesis. The parameter ranges selected for the study were 56-63 wt% (x1), 50-80 °C (x2), and 30-180 min (x3), with a maximum CNC yield of 73%.

585323 Poly(l-lactic acid) Nanoparticles Produced By Impingement Jet Mixing Solvent Displacement: Optimization of Process Parameters for Particle Size and Polydispersity

Many researchers are concentrating on developing polymer nanoparticles for controlled drug delivery because they are small enough to diffuse through tissues, target to specific tissues and in some cases, penetrate through cell membranes to deliver the drug more locally, mitigating side effects seen with intravenous injection. There are still difficulties in producing polymer nanoparticles at scale and in a size range (60 to 80 nm in diameter) that are large enough to avoid many of the biological barriers such as macrophages and other immune cells, and yet small enough to have reasonable diffusion rates through tissue. Another outstanding issue is to find out how the polymer nanoparticles can be purified from excess surfactant used when making the particles.

In this work, we looked at the effect of process parameters (amount of surfactant, location of surfactant, quench volume, and flow rate) on the polymer nanoparticle diameter produced with impingement jets mixing in a batch-fed process through solvent displacement. Solvent displacement uses a solvent for the polymer that is miscible with water. When the polymer solution mixes with water, the polymer precipitates and the solvent diffuses into the water. Excess surfactant is added to stabilize the polymer nanoparticles in order to prevent aggregation. We find that the quench volume and the surfactant concentration have the biggest effect on polymer nanoparticle diameter and the other parameters have little effect.

587864 Fabrication of PbSe Quantum Dot Thin Films with an Inkjet Printer

The fundamental processes associated with the fabrication of epitaxially-connected quantum dot solids is neither well understood nor controlled. As a result, many of their anticipated acclaimed properties have remained elusive. Inkjet printing offers rigorous process control, enhances manufacturing scalability, enables film integration into a wide range of device platforms, and allows for patterned designs without further processing.

PbSe quantum dot thin films are created at the interface of two immiscible solvents. The PbSe quantum dots are in the upper solvent phase, or superphase, and become epitaxially connected when a chemical trigger binds to the ligands on the facets of the quantum dots and removes them. Then, the chemical trigger-ligand complex enters the lower solvent phase, or subphase, promoting uniform thin film formation. For such a system to work in an inkjet printer, the subphase can be initially printed onto a wafer. Then, the superphase can be printed on top of the subphase. Finally, a chemical trigger can be applied and the film formation would occur.

We optimized the jetting of the superphase onto a silicon wafer substrate. We varied the jetting conditions, such as the waveform, and the composition of the superphase, such as changing the concentration of PbSe or the solvent. We balanced the constraints due to the fluid mechanics and thermodynamics involved with droplet formation and uniform thin film formation. Our optimal superphase solution was a PbSe ortho-dichlorobenzene solution. In future studies, we will optimize jetting this superphase onto subphases and then initiate film formation.

586771 Optical Characterization and Purification of DNA-Wrapped Single-Wall Carbon Nanotubes at Controlled pH

Single-wall carbon nanotubes (SWCNTs) have unique optical, electronic, and chemical properties which can be utilized for an extensive range of potential applications, including biosensing and bioimaging. The DNA wrapping of SWCNTs has been proven an efficient method for dispersing nanotubes in an aqueous medium and their subsequent purification of chirality-defined SWCNTs. By inducing DNA conformational changes on the sidewall of SWCNTs through precise control of solvent pH, stable hybrids of DNA-wrapped SWCNTs (DNA-SWCNTs) with increased photoluminescence can be obtained. In this work, we monitored the effect of pH on optical properties of DNA-SWCNTs using single-stranded DNA of $(GT)_n$ sequences with n = 6, 10, 15, 20 at a broad range of pH 1-13. It was hypothesized that the site-specific deprotonation of nucleotide bases in basic regime at pH ~ 12 promotes hydrogen bonding between the bases in $(GT)_n$ DNA; therefore, causing conformational changes in DNA coating structures. This conformational change led to increased fluorescence emission of nanotubes up to 7 times by effectively shielding quenching species in solvent from the surface of SWCNTs. In addition, the effect of nanotube concentration and excess unbound DNA on optical properties of DNA-SWCNTs were explored. The phenomenon of a pH-induced conformational change of DNA on nanotubes was further explored for SWCNT purification using a polymer aqueous two-phase separation method. Our preliminary work showed the potential of separating single-chirality SWCNTs by precise control of pH using specific DNA sequences. Future work will focus on developing a scalable method for separating chirality-defined SWCNTs, that can be utilized for biological sensing and imaging applications.

583149 Development of a Flow-Free Gradient Generator Using a Thiol-Acrylate Microfluidic Resin/Hydrogel (TAMR/H) Hybrid System

Microfluidic gradient generators have been used to study the migration, growth, and drug response of numerous biological systems including cancer cells and algae. Traditional approaches using serpentine channels produce highly stable gradients; however, these devices require direct flow over the cells, which limits their compatibility with suspension cultures. Flow-free gradient generators have been utilized, but they are often hindered by the need for negative flow or external clamps to maintain fluid-tight seals between the polydimethylsiloxane (PDMS) and hydrogel. The goal of this work was to develop new polymeric materials using thiol-acrylate chemistry to produce a two-layer, flow-free microfluidic gradient generator capable of rapid and facile bonding to maintain fluid-tight channels. The geometry of the device consisted of three parallel channels where a chemoattractant or chemorepellent was flowed through the outermost channel to create a chemical gradient within the center 'flow-free' channel. The fluidic channels were imprinted into a rigid top layer, fabricated using thiol-acrylate microfluidic resin (TAMR). The TAMR was bound to a thiol-acrylate hydrogel (H), which served as the bottom layer of the device, and allowed for the flow-free diffusion of chemicals between three parallel channels. The TAMR/H hybrid device consisted of activated acrylate resin made by covalently bonding an amine-curing agent into the acrylate resin coupled with a thiol-containing reagent that is thoroughly mixed into the resin. The hydrogel consists of thiol and acrylate groups within an aqueous buffer. The diffusion of biomolecules through the TAMR/H system and development of a stable chemical gradient was confirmed both experimentally using fluorescent microscopy and computationally using COMSOL. The stability, stiffness, and adhesive strength of the TAMR/H device was thoroughly characterized to confirm proper device operation. The performance of the TAMR/H device was compared to a control PDMS/agarose device with a similar geometry. As proof of concept, the TAMR/H microfluidic device was used to study the chemotactic response of a highly motile strain of GFP-expressing *Escherichia coli*. The migratory response of both wild type and mutant *E. coli* was observed in both the TAMR/H and PDMS/agarose microfluidic devices. Population-based analysis confirmed a similarity in performance between the two materials verifying that TAMR/H was a viable alternative to PDMS-based microfluidic devices. These findings highlight the utility of the TAMR/H hybrid system, which has the potential to perform a suite of experiments spanning 3D cell migration and dynamic co-culture between different cell types commonly found in heterogeneous environments like biofilms and cancer.

583656 Synthesizing and Modeling the Optical Properties of Silica Aerogel

The purpose of this project is to model the optical properties of silica aerogel by solving the 1-D radiative transfer equation (RTE). Silica aerogel is an optically transparent and thermally insulative material with a very low density. Due to its microstructure, which consists of a thing solid backbone of silica particle aggregates enveloping thousands of nanoscopic pores, the material suppresses the three modes of heat transfer. The unique combination of physical and chemical properties allows silica aerogel to be an excellent candidate for solar thermal energy applications. The forward solution to the 1-D RTE will predict the transmittance and reflectance of the aerogel as functions of both scattering albedo and extinction, which are inherent properties of the material. By compiling the numerical results of the forward solution, separate 2-D contour plots can be constructed wherein scattering albedo is plotted versus optical thickness, and the contour gradient represents the theoretical transmittance and reflectance. To determine the intrinsic properties of a specific sample, the transmittance and reflectance of that particular sample must first be determined experimentally using a UV-Vis spectrophotometer equipped with an integrating sphere to account for losses. Once values for transmittance and reflectance are available, the values may be plotted as single contour lines on the previously constructed contour plots which constitute the backward solution to the RTE. When these two plots are superimposed upon one another, it is found that the curves intersect at exactly one data point. This solution point provides the exact scattering albedo and extinction coefficient for a specific sample of silica aerogel. Repeating this process for several samples of aerogel produced by various synthesis processes and with different thicknesses will allow the team to predict the transmittance of a sample synthesized by a specific procedure and of a particular thickness. Furthermore, obtaining the intrinsic properties of a sample will allow for the calculation of the scattering coefficient of that sample, which can be used to determine the approximate pore diameter of the sample. The calculated pore diameter will be verified using Small-Angle X-Ray Scattering. Since the pore size of the resulting aerogel directly affects the degree to which light is scattered throughout the medium, it would be useful to predict the transmittance of a sample characterized by a certain pore size and vice versa; this will indeed be possible using the model formulated in this project.

586026 Hydrogel Network Characterization Using Low Field Nuclear Magnetic Resonance

Hydrogels, defined as linked polymer chains with high water content, present a growing interest as a drug delivery system. In order to optimize their use in the field, important characteristics such as mesh size and structure need to be understood. However, challenges arise when relying on the use of traditional methods such as Small Angle X-Ray Scattering (SAXS) or BET gas adsorption, which might be either inaccessible or require time-consuming sample preparation. As a result, the use of low field nuclear magnetic resonance (LF-NMR) has become an attractive alternative for hydrogel characterization. With this method, a magnetic field induces the alignment of the samples' NMR-active nuclei, such as ^{1}H, followed by a radio frequency pulse which disrupts their alignment. As the magnetic field is held constant, the nuclei tend to "relax" back while the equipment collects the output signal. The relaxation time that characterizes the nuclei dephasing in the x-y axis (T_2) can then be used to understand the surface-solvent interactions and provide insightful information for the mesh size calculation of hydrogel samples. The present study focused on the use of LF-NMR to characterize samples made with poly(ethylene glycol) diacrylate (PEGDA) at high (8000 Da) and low (700 Da) molecular weights using Carr-Purcell-Meiboom-Gill (CPMG) pulse sequencing and CONTIN algorithm. Accuracy of LF-NMR was assessed by comparing the obtained results to SAXS results from literature for PEGDA with the same molecular weight. The same trends were observed in the mesh size calculations for both LF-NMR and SAXS, with very similar results between the two methods for similar hydrogel compositions and differences of less than three nanometers in mesh size calculations. Our results indicate that the interaction between hydrogels and water can be captured by NMR, with noticeable differences among various molecular weights and polymer concentrations. Therefore, NMR appears to be a valuable characterization method for hydrogels and similar wetted systems.

585875 Release of N-Acetyl Cysteine from Chemically Modified Polydimethylsiloxane (PDMS) Hydrocephalus Catheter over Time

One of the biggest causes for hydrocephalus shunt failure is cell adhesion and eventual blockage of catheter holes. With an estimated 50% of shunts failing in the first two years of insertion in pediatric patients, decreasing cell attachment and catheter hole blockage can prolong the lifetime and effectiveness of shunts in hydrocephalic patients. From previous studies, it has been shown that treatment of the polydimethylsiloxane (PDMS) surface of a standard catheter with an N-Acetyl-Cysteine layer (NAC/EDC/NHS) increases the wettability of the surface and decreases cell adhesion. Other studies show that NAC's antioxidant behavior induces glutathione and in turn modulate cell inflammatory pathways. The current study explores the release of this NAC coating from the surface of the catheter over time. Using contact angle, nanodrop, a pressure assay, and SEM imaging, this is quantified for shunt samples incubated for 0, 10, 30, 60, and 90 days. Preliminary contact angle results show a decrease in wettability of the surface as incubation time is increased, suggesting the NAC coated surface releases the coating over time. SEM imaging has revealed a uniform coating on the surface of the catheters overtime indicating a uniform release. Nanodrop testing further verifies the discharge of the NAC coating from the catheter surfaces. This release is further confirmed with the other methods mentioned above, with in vivo experimentation ongoing. Confirming this release occurs, this creates a need for future studies on the effect of release on cell activity surrounding the shunt. Future work will consist of exploring cell activity through a pro-inflammatory cytokine panel. The effects of NAC on cell activity may transform the way hydrocephalus is treated in the future.

585262 Gelation of Hyaluronic Acid for Drug Delivery and Tissue Engineering Applications

Hyaluronic acid (HA) is a naturally occurring biopolymer found in most vertebrates that regulates cell migration, differentiation, inflammation responses, and angiogenesis. Crosslinking of HA improves mechanical properties and can increase the stability in water, making HA an ideal candidate for drug delivery and tissue engineering applications. Here, methacrylated hyaluronic acid (MeHA) was synthesized to add functional groups for crosslinking purposes and dithiothreitol (DTT) was used to crosslink the HA into a hydrogel. Control of crosslinking of HA in solution yields nanogels suitable for drug delivery, or macroscopic gels useful as tissue scaffolds. A fluorescent tag was added to enable quantification of HA nanogel concentration. Fluorescence spectrometry of samples with varying ratios of fluorescent methacrylated hyaluronic acid (Fl-MeHA) to MeHA revealed a linear correlation between fluorescence intensity and the concentration of Fl-MeHA. In addition, oscillatory rheology was used to determine the stiffness of 3 wt% MeHA gels using various concentrations of DTT crosslinker, and these experiments revealed that gels exhibited varying stiffness over time. These preliminary results show that hyaluronic acid gels have the potential to be used in a variety of contexts including drug delivery and in vitro model systems.

585748 Development of a Robust Procedure to Make Non-Swelling, Fast Curing Degradable, Injectable Hydrogels from Thiocure 1300 and Peg 575 Dimethacrylate

Hydrogels are an important class of biomaterials that consist of a water-swollen polymer. They can be either physically crosslinked or chemically crosslinked. Many chemically crosslinked hydrogels require UV or thermal methods to undergo crosslinking. Injectable hydrogels are hydrogels that can pass through a syringe needle into the body and therefore must either be shear thinning or undergo a reaction in a short period of time to crosslink and form a solid hydrogel once in the body. In 2009, Langer and coworkers developed a degradable injectable hydrogel from poly(ethylene glycol) diacrylate (PEG-575-DA) and ethoxylated trimethylolpropane tri-3-mercaptopropionate (ETTMP, Thiocure 1300) dissolved in phosphate-buffered saline at pH 7.5. The two components were separately passed through basic alumina before separately being cooled for 24 h in a refrigerator. Next the ETTMP and PEGDA were mixed together at 4-°C. Curing occurred within 5 minutes. We tried to replicate this procedure but found no success in making hydrogels. In this work, we developed a robust method to cure ETTMP and PEGDA into hydrogels at pH 7.4 within 2 min.

The synthesis of these hydrogel is based on using a calculated volume of precursors that results in an equimolar stoichiometric ratio of thiol and acrylate groups. We varied the height of the alumina column to determine the effect on gelation time. A range of 0.5 in. to 2.5 in. were tested in increments of 0.5 in. Gelation time decreased as the column height increased. A 2.5 in. column resulted in a 1. 33 min cure time and a 1 in. column resulted in a time of 2.3 min. We then varied the pH of phosphate-buffered saline from 7.2 to 7.6. Gelation time decreased with higher pH levels. A pH 7.2 solution cured in 3.32 min. and a pH 7.6 solution cured in 1.38 min. Based on our findings we determined the optimal parameters were a 1.5 in. column and pH 7.4 buffer. This method allows hydrogels to consistently cure in 1.5 min. We then performed a number of tests to analyze the mechanical properties of the hydrogel. Dynamic rheology was used to determine the gelation time, which was confirmed by a frequency sweep to ensure the data was within the linear viscoelastic range. Tensile testing was conducted to verify the elastic modulus of the hydrogels.

588378 Incorporation of Allograft into Chitosan Hydrogels for Treatment of Vertebral Compression Fractures

In the United States, 1.5 million osteoporotic fractures occur annually, almost half of which are vertebral compression fractures (VCFs). Currently, the most common non-invasive treatment options for addressing VCFs are kyphoplasty and vertebroplasty. These both include the injection of poly(methyl)methacrylate (PMMA), which is non-biodegradable and has much higher compressive strength than native bone, resulting in an increased risk of adjacent future fractures.

We have been exploring chitosan hydrogels as a potential replacement for these established treatments. Chitosan hydrogels have become widely used for tissue engineering applications due to their injectability, biodegradability, and bioactivity; however, poor mechanical properties are an apparent disadvantage of these biomaterials. Recently, a combination of physical and chemical crosslinking (with carbonate and genipin, respectively) as well as physical incorporation of cellulose nanocrystals (CNCs) have been shown to produce hydrogels with desirable degradability and low toxicity. Furthermore, incorporation of calcium phosphates into these biomaterials facilitates the release of calcium ions (Ca^{2+}) and phosphate ions (P_i) within their osteoinductive therapeutic window, conveying desirable bone-developing bioactivity. Calcium phosphate incorporation was also found to increase hydrogel compressive strength. Though promising, these hydrogels are still significantly weaker than native bone tissue, leaving them unsuitable for load bearing applications. Recently, we have incorporated crushed cancellous allograft (MTF Biologics) to address this issue. Our results show that these novel complex hydrogels possess physical properties similar to those of healthy bone tissue.

In vitro studies will be performed on murine mesenchymal stem cells (MSCs) to test the biocompatibility and osteoinductivity of the allograft/hydrogel composite. Additionally, mass loss, swelling, and compressive strength studies will be conducted to ensure that the hydrogels have desirable ion release profiles and mechanical properties for VCF repair applications. It is expected that these complex hydrogels will have the same osteoinductive properties as the original hydrogels while also having the improved strength necessary to support load bearing bone regeneration.

587563 Fabrication of Microporous Hyaluronic Acid Hydrogels through Salt Leaching

Water-absorbing polymeric constructs, such as hydrogels, have been increasingly utilized for tissue engineering purposes, as well as *in vitro* models to study disease. Specifically, hyaluronic acid (HA)-based hydrogels have been extensively utilized, as HA is a major component of the extracellular matrix. Herein, we report a simple procedure on the incorporation of a microporous structure within an HA hydrogel. By utilizing methacrylated hyaluronic acid, dithiothreitol (DTT) as a crosslinker, and implementing salt particles of various diameters to act as porogens, a microporous structure was obtained through a salt leaching technique. Scanning electron microscopy, rheological testing, and swelling studies were carried out to analyze the microporous HA hydrogels. It was found that an increase in salt particle diameter resulted in microporous hydrogels with increased overall pore areas that were all significantly different from control hydrogels. The microporous hydrogels were also found to be able to retain significantly larger quantities of water than control hydrogels; however, the microporous hydrogels exhibited significantly lower storage moduli than control hydrogels. Additional studies to determine cell viability and degradation behavior using the hyaluronidase enzyme are currently being carried out to further characterize the hydrogels. Such a system could provide a useful tool for studying cell-matrix interactions *in* vitro in a physiologically relevant setting.

585133 Formulation of a Fiber Reinforced Silicon Carbide Colloidal Gel Ink for Direct Ink Writing and Unassisted Sintering

Silicon carbide (SiC) is an advanced ceramic material made desirable by its low cost, high hardness, and excellent durability. For these reasons, it is used in a wide variety of applications including ceramic plates in ballistic vests, brake surfaces for cars, and abrasive tools for grinding. However, SiC is susceptible to cracking and ultimately failure due to a lack of plastic deformation, leading to low toughness. Here, a SiC fiber reinforced composite (FRC) material was made into a colloidal gel ink for direct ink writing (DIW) and sintering. DIW enables the fabrication of parts with complex geometries that are difficult or impossible to make with conventional pressing methods, as well as the ability to do rapid prototyping. In this work, 30 vol% SiC whiskers are added to SiC particles to toughen the resulting SiC matrix, while yttrium oxide (Y_2O_3) and aluminum oxide (Al_2O_3) were added in small quantities as sintering aids. The solid ceramic particles were suspended in water using a polyelectrolyte dispersant, and an oppositely charged polyelectrolyte was used as a flocculant. A viscosifier was used to increase ink viscosity and prevent ink phase separation. An ink formulation (>45% by volume solids) has been developed that exhibits shear thinning behavior and has high solids loading. Push-rod dilatometry was used to characterize the densification behavior of the printed part during unassisted sintering. Scanning electron microscopy (SEM) was used to characterize the microstructure after sintering, and to determine if whisker fracture occurred during ink processing or printing.

587120 Nanocomposite Based Tough Hydrogels for Fabrication of ECG Monitoring Device

Hydrogel is a three-dimensional (3D) network of hydrophilic polymers that can hold a lot of water while keeping up the structure because of concoction or physical cross-connecting of individual polymer chains. By definition, water must establish at rate 10% of the all-over weight (or volume) for a material to be a hydrogel. It likewise has a level of adaptability fundamentally the same as regular tissue because of their huge water content. They are delicate material that carries on like skin. Intense hydrogels are delicate however have high perseverance to strain with satisfactory versatility. Hydrogels with directing properties and sturdiness are broadly utilized in numerous sensor gadgets that can be utilized in biotechnology and biomedical applications. In our work we have synthesized a polymer-Ag nanocomposite based ionic and conductive hydrogel to utilize it as a sensor material. These hydrogels are intense enough to oppose cut by a knife with 900% stretch capacity. The conductive hydrogels were utilized to create ECG sensor. Arduino based code was additionally used to build up an easy to use application for continuous ECG monitoring and information securing. Along these lines, the hydrogel sensors are promising competitor in numerous medical gadgets and other biomedical application.

584463 Predicting Polyelectrolyte-Micelle Phase Transitions: A Study in Charge Densities

In solution, electrostatic complexation may drive oppositely-charged macro-ions to undergo associative phase separation into a complex coacervate phase – a dense, polyelectrolyte-rich liquid with widespread applications across a range of industries — and a dilute phase, the supernatant. The widespread applications of coacervates for molecular separation and controlled delivery have resulted in extensive analysis of the phase behavior and properties of polyelectrolyte-micelle systems. However, much of the work to date has focused primarily on the effects of micellar surface charge density with model polymer-surfactant complexes. Thus, our objective is to explore the effects of stoichiometry and micellar chemistry over a larger phase space, with the goal of constructing a predictive framework for polyelectrolyte-micelle phase separation based on species' charge densities. Here, we utilize a combination of turbidimetry coupled with optical microscopy to characterize the phase behavior of complexes between the cationic polymer poly(diallyldimethylammonium chloride) (PDADMAC) with surfactant micelles containing a mixture of the anionic surfactant sodium dodecyl sulfate (SDS) and the neutral surfactant Triton X-100 (TX100) as a function of composition and salt concentration. Morphology mapping of coacervate formation will help inform future experiments focusing on understanding the effects of both micelle chemistry and polymer charge density. The results from this study will broaden the array of polyelectrolyte-micelle systems for applications such as biomolecular encapsulation and drug delivery.

587961 Asymmetric Hybrid Solid Electrolyte with High Conductivity and Low Interfacial Resistance for Lithium Batteries

Asymmetric Hybrid Solid Electrolyte with High Conductivity and Low Interfacial Resistance for Lithium Batteries

Zachary D Althouse†, Yubin He†, Haochen Yang† and Nian Liu*†

† School of Chemical and Biomolecular Engineering, Georgia Institute of Technology, Atlanta, GA 30332-0100, USA

Abstract

Lithium-ion battery technology is already nearly ubiquitous in portable electronics and is increasingly being put forward as an option in electric vehicle applications. However, current Lithium-ion batteries (LIBs) possess intrinsic safety hazards because of their use of flammable organic liquid as the electrolyte. This makes solid state electrolytes (SSE) a very promising area of research as they lack the volatility concerns of their liquid counterparts. There are two main categories of solid electrolytes: polymer and ceramic. Polymer electrolytes have better interfacial contact and electrochemical stability, while being less mechanically strong. Ceramic electrolytes have good conductivity and mechanical stability but have poor interfacial contact. To mitigate the inherent weaknesses of each material, combining the two technologies into a hybrid electrolyte has recently been investigated. In this work, preliminary investigation into an asymmetric solid "sandwich" electrolyte has occurred. This design uses crosslinked polyethylene oxide (CPEG) and polycarbonate (CPC) SSE on either side of a ceramic electrolyte backbone (LLZO).

Preliminary results have shown that the CPEG-SSE and UVPC-SSE exhibit high conductivity (0.36 mS/cm and 0.18 mS/cm, respectively) at room temperature. A high conductivity of 0.30 mS/cm was also observed for the asymmetric SSE (CPEG/LLZO/UVPC). The Li/CPEG/Li symmetric cell shows excellent cycling stability at 0.2 mA/cm^2 for 500 hours, demonstrating that CPEG SSE possesses good stability against Li metal anode. However, CV results suggest that CPEG-SSE has low anti-oxidation ability at higher voltage (>4 V vs Li^+/Li). On the other hand, UVPC SSE showed an extended electrochemical window up to 4.6 V vs Li^+/Li, but poor cycling stability in a Li/UVPC/Li symmetric cell. The Li-$LiFePO_4$ batteries with a symmetric SSE (CPEG/LLZO/CPEG) have demonstrated low interfacial resistance (<160 ohm) as well as excellent rate performance (155 mAh/g at 0.2 C, 93 mAh/g at 1 C, and 57 mAh/g at 1.5C). Further studies will be conducted to determine the electro-chemical stability and battery performance of the hybrid membrane. This design, if effective, could lead to increased development of a modular electrolyte system, in which specific polymers would be chosen based on the cell requirements and identity of the cathode and anode materials.

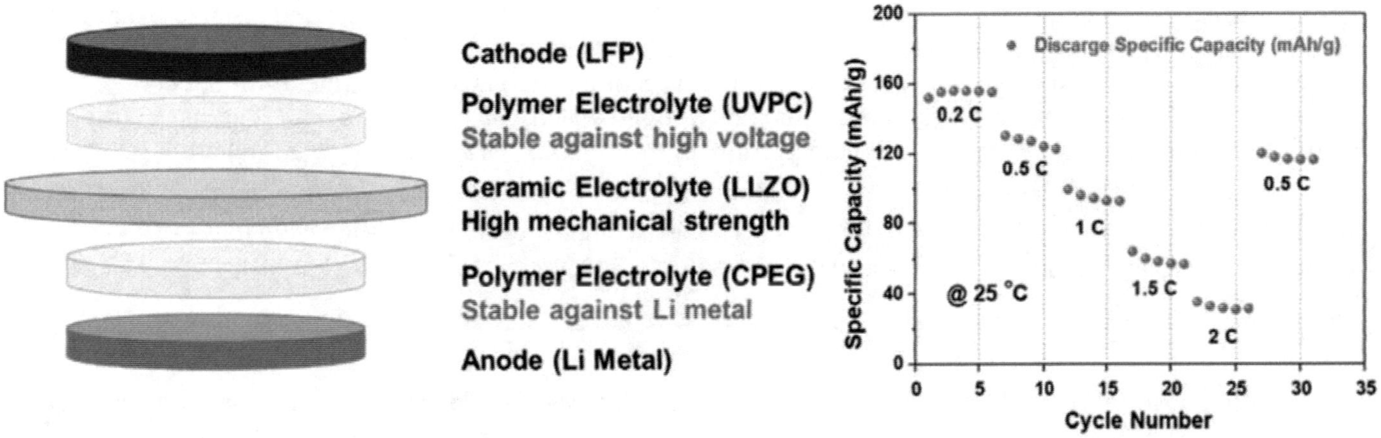
Cathode (LFP)
Polymer Electrolyte (UVPC)
Stable against high voltage
Ceramic Electrolyte (LLZO)
High mechanical strength
Polymer Electrolyte (CPEG)
Stable against Li metal
Anode (Li Metal)
Discarge Specific Capacity (mAh/g)
Specific Capacity (mAh/g)
Cycle Number
0.2 C
0.5 C
1 C
1.5 C
2 C
0.5 C
@ 25 °C

583820 Effect of Functional Groups on the Adsorption of Lighthydrocarbons in *Fmj*-Type Metal–Organic Frameworks

Adsorption is considered to be an effective method of light hydrocarbon storage which has advantages in safety and efficiency. Metal–organic frameworks (MOF) , a group of crystalline porous materials with periodic network structure assembled from coordination bond by metal ions/clusters and organic ligands, are considered as promising candidates for light hydrocarbons adsorption due to their highly regulated channel and modified skeleton.The functionalized modification of targeted porous metal–organic frameworks (MOFs) is important to improve the adsorption capacity of light hydrocarbon gases. In this study, five isomorphic MOFs with different functional groups (F,Cl, NH_2, CH_3, and OCH_3) are successfully synthesized by systematic functional modification of UMCM-151. By studying the adsorption properties of C2 and C3 hydrocarbons (C_2H_2, C_2H_4, C_2H_6, C_3H_6, and C_3H_8), it is found that the electron-withdrawing group functional UMCM-151-F and UMCM-151-Cl exhibit strong affinity for C_2H_2 (98.71 cm^3g^{-1} for UMCM-151-F and 90.29 cm^3 g^{-1} for UMCM-151-Cl), while the electron-donating group functional UMCM-151-NH_2, UMCM-151-CH_3, and UMCM-151-OCH_3 have strong affinity for C_2H_4 (97.89 cm^3 g^{-1} for UMCM-151-NH_2, 90.22 cm^3 g^{-1} for UMCM-151-CH_3 and 94.13 cm^3 g^{-1} for UMCM-151-OCH_3). The differences in affinity of electron effects for light hydrocarbon provide an experimental basis for porous MOFs to improve their light hydrocarbon storage capacity.

588057 Development of an Environmentally Friendly, Biodegradable Temperature Sensor

Temperature sensors are a useful tool to monitor the temperature of an object. There are a few temperature sensors on the market that can be 3D-printed, but they are not biodegradable. There are also temperature sensors that are biodegradable but not used in additive manufacturing. The temperature sensor that will be developed in this experiment will possess both qualities. Biodegradable sensors are important for waste recyclability—being able to dispose of the sensor without harming the environment. The device is composed of D-Sorbitol because it is biodegradable due to its presence in plants during photosynthesis, which means it can be consumed naturally. The other major material used in the sensor that is environmentally friendly and biodegradable is cellulose nanofibrils (CNF). Electrodes for the sensor can be biodegradable, as well. 3D-printing—additive manufacturing—is useful when in mass production. By producing many devices in a batch or continuous process saves time, which in turn saves money for the manufacturing company. It makes this quality very valuable to potential in the market; therefore, the device could be integrated into a marketable product that can be used as a wireless, wearable temperature sensor. In order to achieve this goal, the voltage and temperature changes must be analyzed thoroughly, as well as an analysis of the materials being used. Finally, the materials must be able to be prove that additive manufacturing is a viable option for production. The purpose of this study is to determine how the voltage is produced when a temperature difference is applied and be able to predict the temperature of the heat source given the produced voltage.

588092 Stability of Solid Oxide Fuel Cell Cathode Material Under Reducing Conditions

The performance and durability of solid oxide fuel cells (SOFC) are currently being held back by cathode materials. The cathode materials that this poster will take into question are LSCF, LSM, and different variations of PNO. LSM and LSCF are materials that are already being used as cathode materials and are being tested alongside the PNO to observe how they compare to one another. In terms of diffusion coefficients, lowest activation energy for oxygen reduction, surface exchange, and electrode polarization, PNO shows the highest potential for new SOFC. However, the instability of these materials can affect their long-term use and degradation over time. This poster will discuss the quantitative data of the materials' phase transformation and degradation over time when exposed to extended periods of heating under reducing conditions. Cathode materials are being tested by being heated for upwards of 500 hours at 750°C, while x-ray diffraction analysis is completed at various intervals in order to gather phase data. This process is being completed under different reducing conditions, meaning the cathode samples will be heated in the presence of different gases: Nitrogen, Air, and Hydrogen. While under these conditions normal PNO can experience degradation, instability, and phase transformation. In order to stabilize the phase without changing the material, other compositions of PNO are being tested alongside the original.

588100 An Open-Source Program to Simulate Time-Resolved Photoluminescence and Electron Transport in Semiconductors

Time-resolved photoluminescence (TRPL) is a powerful tool to extract the material and optoelectronic properties of semiconductors. However, extracting relevant information from TRPL data can be challenging as many complex electron dynamics can exist in these materials. Therefore, simulating TRPL data as a function of various material properties is of interest to better understand these electron dynamics and correlate with measured data. Dedicated commercial software for simulating transient electron dynamics is, however, limited and cost-prohibitive. Here we introduce an open-source program that we have developed to simulate transient electron dynamics in semiconductors. This program uses numerical methods to solve non-linear coupled differential equations as these equations governing the transient diffusion, drift, and the decay of electrons – whose recombination is a precursor to photon emission – can generally not be solved analytically. Our Python application for generating, visualizing, and exporting electron dynamics and TRPL data will be shown. In addition, example TRPL simulations in comparison to experimental data will be shown for thin-film photovoltaic absorbers and semiconducting nanowires.

583016 An Ultrafast Rechargeable Aluminum Battery

Demand for electric vehicles is rising due to environmental concerns, government subsidies, and improved battery performance. Lithium ion batteries may not be able to scale with the increasing demand due to their limited amount of stored charge, long re-charging times, high cost and risk of fire. Aluminum batteries are a promising replacement for lithium-ion batteries not only because aluminum is much more abundant and less expensive than lithium, but because aluminum batteries are non-flammable, have 4x the volumetric charge capacity of lithium ion batteries, and have ultrafast recharging capabilities. The focus of this research was creating an ultrafast, safe, cheap and long-cycling-life rechargeable aluminum battery for electric vehicles. The prototype batteries are composed of an aluminum-indium anode, a selenium cathode, and a molten salt ($NaCl$-KCl-$AlCl_3$) electrolyte. While lithium ion batteries, have a charging rate of about 1.0 C, our aluminum batteries reach a rate of 200 C (18 seconds) due to the molten salt promoting a high power. Our aluminum batteries demonstrated excellent rate performance, high coulombic efficiency (~90%) and good stability during early testing. Our prototypes are a promising energy storage alternative that could help to propel the electric vehicle industry forward with safe, affordable, high capacity and fast re-charging batteries.

588196 Artificial Photosynthesis: Photoelectrochemical Water Splitting Using Cu_2o

Photosynthesis is a biological process that converts energy from sunlight into chemical energy as an organic fuel. Cuprous oxide (Cu_2O), a p-type semiconductor, was used to mimic this process as a cathode material for a photoelectrochemical (PEC) water splitting reaction. Cu_2O can absorb energy from sunlight and convert it into a chemical form as hydrogen. The produced hydrogen can then be used as a clean fuel in devices like hydrogen fuel cells or further converted to ammonia (NH_3) for transport. In this study, thin layers of Cu_2O crystals were deposited on a planar Cu substrate using chronoamperometry (CA) and characterized using scanning transmission electron microscopy (STEM), X-ray diffraction (XRD), and Mott-Schottky analysis. The peak current density for the water splitting reaction was measured in a PEC cell using linear sweep voltammetry (LSV). The effect of deposition time was investigated as the Cu_2O layer thickness has shown to affect the activity. The Cu_2O prepared using a 5-minute deposition time showed the best performance relative to deposition times of 10, 20, 30, 40, 50, and 60 minutes. The 5-minute deposition sample produced the highest peak current density of approximately -1.70 mA/cm^2. Longer deposition times showed decreased performance.

584831 Conductivity, Cushioning, and Improving Lithium Ion Battery Performance

Batteries are currently the most efficient way to store high quality electrical energy. The capacity of renewable energy storage, transportation, and usage depends heavily on the performance of batteries, which must be improved to meet consumer demands. This can be done by altering the chemistry of the three major battery components: anode, cathode, and electrolyte. Graphite has been the primary material for anodes and can be improved by mixing silicon with it, but silicon has problems with extreme volume fluctuation therefore different compositions and porosities of the anode must be studied. Electrochemical impedance spectroscopy was utilized to evaluate how porosity affects varying characteristics contributing to the observed impedance.

585763 Investigating the Redox Activity and Lithiation of Boron Icosahedrons in the Solid State

In 2009, the Department of Energy established the Energy Frontier Research Center (EFRC) program to advance the scientific understanding of transformational energy storage technologies. With the support of the EFRC, the redox activity and lithiation of boron icosahedrons in the solid state was examined. These compounds consist of a 12 coordinated boron cluster decorated with a series of organic ligands. Borane clusters offer an intriguing opportunity for battery research, as organic compounds are not traditionally considered for redox activity. Borane clusters, however, exhibit three-dimensional aromaticity, resulting in delocalized electron density that allows for accessible redox activity, which is unexpected of an organic compound. With that in mind, these borane clusters can be thought of as an organic analog to the transition metals that are traditionally used in electrodes. In addition to being a unique cathode material, the organic nature of borane-based cathodes offers a high degree of redox tunability through variance of the attached functional groups during the scalable organic synthesis.

The electrochemical behavior and structural dynamics of these materials was investigated through a comprehensive series of studies. Fundamental electrochemical studies under experimental conditions used in traditional battery cell design with solid film electrodes, including lithium counter electrodes and lithium hexafluorophosphate electrolyte, were conducted. Through cyclic voltammetry cycling, the properties of borane clusters that allow them to compete with traditional energy storage cells are further examined. Preliminary electrochemical cycling reveals that a derivative borane cluster decorated with lithium carboxylate exhibits redox activity at 3.4 V with a high degree of reversibility while cycling lithium. We have demonstrated the reversible (de)intercalation of Li^{+}in borane clusters cycled in the solid state, offering a promising alternative to transition metal-based electrode materials.

588105 Improving Electrical Conductivity of Carbon Fiber By Metal Electrodeposition Method

Carbon fiber is a carbon-based material with the advantages of high surface area, high tensile strength, high chemical resistance, low density, and high stiffness along the fiber. These advantages build carbon fiber a very promising role as a substrate (e.g., an electron collector) in energy storage devices which can highly increase the gravimetric energy density. Therefor, carbon fiber could be the key component assembling a flexible energy storage device for wearable electronics. However, carbon fiber is not a very good electrical conductor and will increase the internal resistance and lower the output voltage compared with metal substrate and aluminum or copper electron collector. This work focuses on improving the electrical conductivity of Carbon Fiber Mat (CFM, 50mm × 20mm) by depositing a thin layer of pure metal on the surface. Several Metal/CFM composites were synthesized for this purpose. The best one shows an electrical conductivity one hundred times higher than that of the same size carbon fiber mat.

584857 Waste Biomass-Derived Anode for Application in Superior Lithium Ion Batteries

Given the simultaneous demand for lithium-ion batteries and environmentally-friendly practices, there is a distinct opportunity for batteries derived from waste biomass. Carbonaceous battery anodes synthesized from avocado seeds upcycles a completely valueless waste material into a highly demanded product. The novel electrode demonstrates promising performance, outperforming graphite at every applied current density from 25 mAg^{-1} to 400 mAg^{-1}. The novel electrode performs comparably to graphene electrodes in terms of cycling (315 $mAhg^{-1}$ at 100 mAg^{-1}) but does not share graphene's instability at high (400 mAg^{-1}) applied current densities. By characterizing the material using X-ray Powder Diffraction, Raman spectroscopy, BET analysis, and extensive cyclic voltammetry, the amorphousness of the biomass-derived anode was found to contribute 54% of lithium storage to capacitance-driven mechanisms, compared to 35% in graphite and graphene. However, the high amorphousness and high surface area facilitates the formation of a large solid-electrolyte interphase, causing a significant first-cycle irreversible capacity loss. Despite this, its combined superior performance and environmentally-friendly nature causes the presented anode to be a practical upgrade from the current commercial graphite anode.

588112 Synthesizing Silver Selenide Thin Films Via the Cation Exchange Process for Thermoelectric Applications.

Thermoelectric (TE) devices are unique semiconductor devices that can directly convert heat to electricity and vice versa to serve applications such as waste heat recovery and solid state cooling. Bismuth-Tellurium and other commonly used alloys have been shown to be high performance TE materials but are composed of elements that are toxic, rare and expensive to process. To accelerate the widespread adoption of TE devices, safe and abundant materials need to be integrated into TE devices at an economical cost. Prior research has involved synthesizing Ag_2Se nanoparticles, however this is an energy-intensive and time consuming technique with poor yields. To tackle this problem, we optimize the TE performance of Ag_2Se by producing it via a low-cost, solution-processable approach relying on the cation exchange of $Cu_{2-x}Se$ to Ag_2Se. Cation exchange has recently emerged as an extremely versatile tool that allows manufacturing of a variety of materials whilst maintaining the morphology of a known anionic structure. We report a maximum power factor (PF) of ~800 $\mu Wm^{-1}K^{-2}$ achieved via a simple thin film soak. This PF is the highest value reported to date for a solution-processable Ag_2Se thin film at room temperature. This soaking technique allows for a safe, economical and scalable solution to using Ag_2Se in a variety of TE applications.

588167 Improved Lithium-Ion Battery Cathode Cycling Stability at Higher Cut-Off Potential Using a Novel and Scalable Dual Coating Method

The electrification of transportation and the need for grid energy storage to accompany renewable energy require secondary batteries with higher energy density, longer useful lifespans and lower cost. In order for lithium-ion batteries, the most developed battery technology suitable for these uses, to achieve greater energy density at the cell level, the operating voltage window tolerated over the course of long-term cycling must be increased. In this work we have demonstrated a promising dual coating system of Li2SrSiO4 and Al2O3 achieved via an easily scalable sol gel technique and confirmed via XRD, used on commercial LiCoO2 cathode powder. We demonstrate our finding that the optimal thickness of each coating can lead to increased specific capacity (~185 mAh/g at 0.5C) at a cut-off potential of 4.5 V and greater stability through multiple cycles at very high C rates (up to 10 C). Cyclic voltammetry was used to demonstrate reduced polarization, and therefore increased stability, after cycling in the dual coated cathodes relative to bare or Al2O3 coated material. Fluorescence mapping of elemental distribution provides evidence of mechanical property advantages of the dual coating system, and the ability of the dual coating to preserve the cathode particles despite inhomogeneous thickness. The mechanism of the superior performance relative to bare and Al2O3 coated cathodes was studied using scanning and transmission electron microscopy, electrochemical impedance spectroscopy and synchrotron X-ray spectroscopy in pristine and cycled cathodes. These findings have implications for the search for still better materials for which our simple coating method could be utilized. Additionally, we aim to gain insights into the transferable nature of the mechanisms by which this coating stabilizes cathodes at high voltage between the LiCoO2 utilized in this work, and experimental, lower cobalt cathode materials such as NMC811 or lithium-excess materials.

584065 Run-to-Run Reproducibility in Electrochemical Impedance Spectroscopy

Electrochemical impedance spectroscopy (EIS) is a powerful method that can be used to detect analyte interactions on the surface of the working electrode. However, reproducibility issues associated with a repeated impedance measurements of similar conditions using EIS is known.[1] We have identified methodology that can be used to improve the reproducibility of repeated EIS impedance measurements. Multiple factors were taken into consideration for run to run measurement variation such as buffer solution concentration (*e.g.* 20mM vs 1 mM ferri/ferrocyanide as a redox couple), buffer solution composition (*e.g.* phosphate buffer solution vs. deionized water), and EIS setup options (*e.g.* DC offset conditions and N_2 purging). Additionally, techniques such as removing electrodes and rinsing them were tested (*i.e.* the working, auxiliary, and reference electrodes were removed and rinsed in between EIS measurements). Some techniques were found to hinder or not affect EIS measurement reproducibility, while others supported reproducible measurements. Taking these methods into account, the most reproducible model of repeated EIS impedance measurements will be discussed.

(1) Bogomolova, A.; Komarova, E.; Reber, K.; Gerasimov, T.; Yavuz, O.; Bhatt, S.; Aldissi, M. Article ASAP •. *Anal. Chem* **2009**. https://doi.org/10.1021/ac9002358.

585514 Fabrication of Porous Metal Via Free Corrosion of Hypoeutectic Compositions of Al-Cu Alloy

Nanoporous metals have recently been studied for their diverse applications in catalysis and gas separation. Often, they are made of costly metals such as gold. The present project sought to obtain a porous metal via selective dissolution of hypoeutectic aluminum-copper alloys, as the parent material. An Al-50 wt. % Cu master alloy melt was diluted with pure aluminum to create target composition ingots. These were cut into 10 x 10 x 2 mm samples, which were subsequently weighed, annealed, and fine-polished for corrosion. The corrosion in an 8% v/v NaOH solution was traced at different time intervals. Afterwards, optical microscopy allowed capturing micrographs that were studied using image analysis to measure the porosity. The corroded samples were analyzed using x-ray diffraction to verify that the selective dissolution was effective.

588307 Development of Accurate Eam Potentials of Gold with Bayesian Uncertainty Quantification

One of the most critical limitations to performing successful molecular dynamics (MD) simulations of metal structures in many fields of study is the availability of suitable interatomic potentials. Embedded atom potentials (EAM) in particular provide a unique solution due to their low computational cost and ability to accurately model bulk properties of pure metals and their alloys. Many EAM potential parameters for gold have been developed to reproduce experimental and/or quantum calculations data. However, when these parameters are used to simulate systems beyond their calibrated environments, the data presented can be invalid or open to interpretation. We present accurate EAM interatomic potentials for gold that was developed to reproduce its experimental properties by integrating particle swarm optimization (PSO) with molecular dynamics (MD) simulations. The resulting EAM parameters of gold were validated by performing Bayesian uncertainty quantification. Overall, these new parameters of gold are robust and transferable across a range of different conditions. The interactions between these new gold parameters and water were also developed to reproduce its interfacial tension at 300 K.

585900 Highly Flexible and Transparent Conductor for Solar Cell

Flexible and wearable electronic technology has recently gained a significant amount of attention because of its potential applications for life improvement. Application of many existing and nouveau devices, such as electronic skin, human-machine interfaces, and health monitoring sensors, can increase comfort for better user experience. Because the most commonly used transparent conductor in photo-voltaic devices, indium tin oxide (ITO), is rigid and brittle, it cannot be used in flexible devices. Given that, as the wearable electronic and optoelectrical technology grows, the demand for a highly flexible electrode to accommodate human body movement like folding limbs or stretching muscles for such devices emerges. In this paper, a flexible electrode comprising of fused silver nanowires (AgNWs) and a conductive polymer blend having proficient conductivity and high optical transparency has been developed. The solution-processed electrode can be made at low temperature and cost, making it efficient for mass production. After multiple bending and twisting cycles, the flexible electrode film still has a sheet resistance of 24 ohm/sq. and average optical transmittance of 85%. The developed transparent electrode shows proficient flexibility and competent optoelectrical performance compared to commercial ITO, making it a promising candidate for flexible, transparent electrodes in solar cells.

585994 3D Layer-By-Layer Electrostatic Assembly of Graphene-Noble Metal Thin Films for Electrocatalysis in PEM Fuel Cells

3D Layer-by-Layer Electrostatic Assembly of Graphene-Noble Metal Thin Films for Electrocatalysis in PEM Fuel Cells

Mark H. Jaskot, Joshua M. Musiol, Kamil Woronowicz, Pamela L. Sheehan[1], Preston C. Haney[1], Harry L. Moore[1], H. Daniel Bahaghighat, Enoch A. Nagelli*

Department of Chemistry & Life Science, Chemical Engineering Program

United States Military Academy, West Point, NY 10996

[1]Explosive Ordnance Disposal, Demilitarization & Experimental Directorate, Army Futures

Command, CCDC Armaments Center, Picatinny Arsenal, NJ

Abstract

Graphene-metal composites offer great opportunity for cheaper and more efficient conductive materials to be applied in fuel cells. Using layer-by-layer self-assembly, graphene oxide-noble metal thin films were synthesized using electrostatic coordination between the carboxylic acid edge functional groups and platinum (Pt^{2+}) or palladium (Pd^{2+}) ions. Reduction of the films created free-standing sheets of interconnected graphene and platinum or palladium nanoparticles, held together using a poly(methyl methacrylate) (PMMA) polymer. The films were created using platinum and palladium ions to test differences in behavior between metal ions but yielded similar results in both cases. Using UV-Vis spectroscopy, Raman spectroscopy, and SEM imaging, the films showed distinct presence of both graphene and the different noble metals while also displaying alignment and assembly of metal ions at the edges of each 2D graphene sheets. This 3D assembly between the noble metal nanoparticles and the graphene sheets creates the potential for these films to be used as novel nanocomposite electrodes for electrocatalysis in proton exchange membrane (PEM) fuel cells.

*Contact: Dr. Enoch Nagelli, Email: enoch.nagelli@westpoint.edu

584587 pH-Dependent Degradation of Poly(ethylene oxide)-*Block*-Polycaprolactone Films

Currently 90% of pharmaceuticals in the discovery pipeline are hydrophobic, yet such hydrophobic compounds account for only 40% of drugs that are FDA-approved and marketable. Consequently, researchers recognize that there is an urgent need for better delivery mechanisms for hydrophobic drugs to increase efficacy and reduce negative side-effects. Poly(ethylene oxide)-block-polycaprolactone (PEO-*b*-PCL) is an FDA-approved, biocompatible, amphiphilic block copolymer, making it an intriguing candidate for hydrophobic drug encapsulation and controlled release into the body. In this work, a variety of characterization techniques, including DSC, NMR, and SEM, are employed to investigate the degradation of PEO-*b*-PCL films within aquatic environments. Previous work shows that the degradation of these films is accelerated in both basic and acidic environments, however there are opportunities to better understand the degradation rate as it relates to the physical structure of the films. This information suggests that pH could be used as an external stimulus to guide the release of hydrophobic compounds encapsulated within the PCL. Through examining the effect of block fraction and crystallinity on the degradation rate and pathway at various pH, the potential of PEO-*b*-PCL as a delivery agent can be better understood.

587448 Cyclic Voltammetric and Spectroelectrochemical Characterization of Polymer Thin Films

Electrochemistry is a well established field in analytical chemistry. One property of great interest in electrochemistry is the apparent charge transfer diffusion coefficient of polymer thin films. Combining electrochemical techniques with spectroscopy, new insights can be gained on the characteristics of the polymer film. Cyclic voltammetry was used to prepare polymer films of tris(5-amino-1,10-phenanthroline)ruthenium(II) and iron(II) on Indium Tin Oxide (ITO) surfaces. When the film was electrochemically reduced, a visible color change was measured spectroscopically and related to the charge transfer rate. A correlation was studied between the spectrochemical, electrochemical and impedance measurements. While under diffusion controlled conditions, this correlation was used to find the apparent charge transfer diffusion coefficient. This coefficient is useful for applications in other research and provides a better understanding of polymer thin films.

584804 Powering Clean Energy with Seawater: Understanding the Selectivity of Oxygen/Chlorine Evolution for Metal Catalysts

Energy is a globally valued resource. In order to meet energy demands in the face of climate change, finding new methods to store renewable energies is essential. A clean method to store energy is through the electrolysis of water. Such a reaction can be driven by renewable energy sources like solar energy. In this project, an abundant resource is explored as an electrolyte to store energy—seawater. Seawater contains many different salts, often salts containing chlorides. The electrolysis of water evolves hydrogen and oxygen gas, which can act as fuels when recombined in a fuel cell to generate electricity, heat, and desalinated water; however, the addition of chloride containing salts in the electrolyte results in the evolution of chlorine gas, a toxic byproduct, competing with oxygen evolution reaction. The goal in electrolysis of artificial seawater is to understand and control an electrode's selectivity towards oxygen evolution by analyzing an electrode's performance via cyclic voltammetry and chronoamperometry. By comparing the performance of various electrodes, earth-abundant metals with high catalytic activity towards oxygen evolution over chlorine evolution can be confirmed, and a greater understanding of what materials are efficient and selective towards oxygen evolution can be reached. With this understanding comes better electrode designs in the future to improve selectivity towards oxygen evolution. In this project, the tested electrodes have been developed by Pacific Northwest National Laboratory, the Epitaxial Complex Oxide Lab from Sungkyunkwan University, and in Dr. Kelsey Stoerzinger's lab from drop casting and electrodeposition.

587689 Developing Small Organic Molecule Sensors from Single Stranded DNA Aptamers

By combining a biophysical model with a genetic algorithm, we computationally designed single stranded DNA (ssDNA) sensors for detecting small organic molecules of interest. These sensors have notable advantages over current methods of small molecule detection, being significantly less expensive than liquid or gas chromatography with mass spectrometry and have much higher throughput than enzyme-linked immunosorbent assays. When bound by their ligand, the sensors undergo structural conformational changes that lead to cleavage activities that are readily detected by inline nanopore-based sequencing, such as the mobile Minion sequencing platform. These ssDNA sensors utilize aptamers, which are structured nucleic acids that bind to ligands of interest, for molecular recognition with high specificity. As a key novelty, multiple ssDNA sensors can be combined and multiplexed with each output distinguishable by unique barcoded nucleic acid sequences. We have successfully designed sensor sequences for tobramycin, 17-B-estradiol, and ATP to produce two-state behavior and ligand-activated sensors.

587697 Confining Electrodeposition of Metals in Structured Electrolytes

The 2019 Chemistry Nobel Prize for Li-ion batteries substantiates the global importance of these devices in contemporary electronics. Even so, developing electrochemical systems with higher energy density than Li-ion batteries is critical for decarbonizing our society's energy systems. One pathway towards improving Li-ion battery involves replacing the graphitic anode with Li metal. This change can increase the anodic energy capacity 10-fold and enables the use of high capacity, Li-free cathodes. However, Li metal batteries are inherently unstable because of the high reactivity of Li metal with organic electrolytes that results in the formation of a solid electrolyte interphase (SEI), an insulating layer on the Li anode surface. The SEI's inherent inhomogeneity causes some regions on the anode surface to be more conductive than others. Thus, Li ions deposit unevenly on the metal anode, leading to the formation of sharp nucleates. Continual deposition due to the high local electric field at the tip of the nucleates forms 'needle-like' dendrites capable of piercing through the battery separator, causing short circuits and initiating thermal runaway reactions in organic electrolytes. These catastrophic processes can result in lethal battery fires. There have been several studies dedicated to the prevention of dendrite growth and thermal runaway reactions by means of a high modulus physical barrier. One notable study concluded that a modulus at least twice that of Li (~10 GPa) is required to achieve stable, smooth electrodeposition. However, these high strength "solid-state electrolytes" (SSEs) have low ionic conductivity, which reduces the power these batteries can produce, and therefore limits their practical use. Thus, engineering techniques and materials to break this tradeoff are critical concerns.

Recent theoretical studies have reported that rough dendrite morphology can be mediated using battery separators having pore-diameter lower than the critical (smallest) size of the dendritic nucleate. To evaluate this, we synthesized cross-linked polymer composite electrolytes with tunable pore size, which host organic liquid in these pores. We systemically varied pore size within our electrolyte systems and found that the materials are resilient towards dendritic growth below a critical pore size, as predicted by the theoretical studies. Further, by the housed organic liquid solvent within the polymer network, our electrolyte system could maintain ionic conductivity at room temperature. These results show that the tendency for battery failure by dendrite-induced short-circuits can be reduced in rationally designed electrolytes, opening the gateway for future work nanoporous electrolytes.

587705 Evaluating the Sensitivity of the Momentum^{TM} Magnetic Particle Imager for Ferucarbotran Iron Oxide Nanoparticles

Magnetic particle imaging (MPI) is a promising imaging technology that enables non-invasive quantitative visualization of the distribution of biocompatible superparamagnetic iron oxide (SPIO) tracers. MPI signal is proportional to SPIO mass, does not suffer from artifacts arising from tissue background signal, has no limitations in tissue depth penetration, and does not use ionizing radiation. These features make MPI ideal for characterizing *in vivo* nanoparticle biodistribution quantitatively. The objective of this study was to characterize the sensitivity of the new Momentum^{TM} MPI imager (Magnetic Insight, Inc.) using the commercially available SPIO nanoparticle ferucarbotran.

A serial dilution of ferucarbotran iron oxide particles was prepared with the iron mass determined using the 1,10-Phenanthroline colorimetric assay for iron quantification. All samples contained 1uL of solution and were measured in triplicate, in two different positions of the field of view. MPI scans were made using four imaging modes available in the Momentum: default, isotropic, high resolution and high sensitivity. Images were analyzed using VivoQuant to evaluate signal intensity in regions of interest containing the tracer samples. The limit of detection was calculated on the basis of analyzing the spontaneous background signal of empty scans and the maximum intensity signal of samples at low concentrations.

It was found that *'high sensitivity'* scan mode (3 T/m) had a limit of detection ~38 ng Fe, whereas *'high resolution'* scan mode (6 T/m) had a limit of detection of ~200 ng Fe. We also observed differences in signal shape and background across all different imaging modalities.

Determining the MPI limit of detection and quantification of different SPIO tracers is important for the development of a variety of applications. Understanding the optimal magnetic characteristics of SPIO tracers and their behavior on the Momentum^{TM} MPI will provide a better understanding of the capabilities of this instrument for particle characterization and biomedical application. This will ensure that reliable data can be obtained and understood. Future studies will include determining the limit of detection for in-house synthesized particles to understand the behavior and properties of particles in the MPI, facilitating optimization of tracers for MPI imaging.

587714 Thiophene-Ethynylene-Based Viologen Derivatives: A Novel Platform of Photosensitizer for Photodynamic Antimicrobial Chemotherapy

Photodynamic antimicrobial chemotherapy (PACT) has an excellent efficacy on infections caused by bacteria, fungi, and especially resistant bacteria. Photosensitizers are an important impact factor in PACT. To address the lack of broad-spectrum property and relatively few alternatives of traditional photosensitizers, a novel thiophene-ethynylene-based viologen derivatives platform represented by 4,4'-(thiophene-2,5-diylbis(ethyne-2,1-diyl))bis(1-hexylpyridin-1-ium) bromide (TEHPB) is presented. TEHPB was synthesized via the reaction of 2,5-bis(pyridin-4-ylethynyl)thiophene and 1-bromohexane. As a quaternary ammonium compound, the cationic pyridine group enhances the hydrophilia of TEHPB. This study presents a new design to incorporate thiophene and ethynylene structure into viologens while preserving its conjugated system. Our study also provides a deeper understanding of structure-property relationships of viologen derivatives. We use the colony counting method to determine the phototoxicity and dark toxicity of TEHPB against the gram-negative bacteria *Escherichia coli* and the gram-positive bacteria *Staphylococcus aureus* by graphing the sterilization rate vs. concentration of TEHPB. The dark toxicity testing indicated that the minimum bactericidal concentration for *E. coli* was 0.98 μM, and the minimum bactericidal concentration for *S. aureus* was 1.95 μM. In the case that the light intensity of white light is 20 mW/cm^2 and the illumination time is 1 hour, the sterilization testing results show that with increasing concentration, TEHPB has a significant bactericidal effect. Herein, our experiments indicated that the minimum bactericidal concentration for *E. coli* was 0.25 μM, and the minimum bactericidal concentration for *S. aureus* was 0.125 μM. Because of its good antibacterial property against gram-positive and gram-negative bacteria, TEHPB has a great potential to be used as a novel broad-spectrum antimicrobial photosensitizer, and the platform of thiophene-ethynylene-based viologen derivatives may open a new avenue to develop and synthesize novel photosensitizer system.

587776 Robust, Crosslinked Anion Exchange Membranes for Fuel Cell Applications

Fuel cell technology is a growing field of research and use due to potential applications in military, transportation, and space application. While internal combustion engines are less than 25% efficient at converting chemical energy into actual power, fuel cell technology can reach efficiencies up to 70%. In the last decade, research related to anion exchange membrane fuel cells has increased since these fuel cells allow non-noble metal catalysts, such as nickel, instead of expensive catalysts like platinum. Here, the goal is to develop stable membranes with adequate ion exchange capacity and conductivity as well as good mechanical strength and durability. A topdown approach using a commercially available cationic polymer creates an anion exchange membrane through crosslinking and casting. Further study on the anion exchange membranes includes measuring ion exchange capacity, water uptake, and rheological properties. By measuring the properties of membranes with different compositions and thicknesses, this research will advance the understanding of anion exchange membranes.

587791 3-D Printing of Energetic Materials

3-D Printing of Energetic Materials

Over the last decade 3-D printing has been at the forefront of materials processing. However, for energetic materials 3-D printing is in its infancy and many critical material parameters are unknown. In this work, we explore ceramic fuel based pyrotechnics, also known as pyrolants. For this study, a ceramic fuel (MgB2) and a fluoropolymer (THV) were formulated in five different ratios of 30/70, 40/60, 50/50, 60/40, and 70/30. The combustability, combustion velocities, density, thermal characteristics, and aging were all studied in the course of this project. All five ratios were printed as inks and 30/70, 50/50, and 70/30 were printed as filament for comparison. As the amount of MgB2 was increased, the combustion velocity increased significantly with the 30/70 ratio at 4.016 mm/s while the 70/30 ratio combusted at 35.197 mm/s. In addition to the combustion velocities, the density was also assessed. The printed inks had a density of 0.4-0.7 g/cc lower than the filament most likely due to the porosity from air pockets that occur during printing. This also explains why the ratio of 60/40 was lower than the 50/50 ratio, when it was expected to be higher. Using a Brookfield Rheometer, it was determined that the 60/40 ratio fell out of the expected range with an apparent viscosity of 2,000,000 cP higher than the other four samples. Over the course of eight weeks in an aging oven, the filament aged extremely well. There was no significant change in color or in burning rate leading to the conclusion that the printed pyrolant is reliable over time. Future studies should consider optimizing printing parameters to reduce porosity.

586845 Vapor Induced Phase Transition of Cesium Lead Halide Perovskite Thin Films for Novel Use in Solar Cells

In recent years, perovskite solar cells have emerged as a promising new alternative to the current silicon-based panels which have dominated the market. Research was done to characterize the black to clear phase transition undergone by thin films of cesium lead halide perovskite when exposed to various solvent vapors. The thin films of a cesium lead halide perovskite were exposed to a constant partial pressure of different solvent vapors, and the kinetics of the transition from the perovskite phase to the non-perovskite d-phase were tracked by absorption spectroscopy. We find that other common solvents, in addition to the previously known water and alcohols (MeOH and EtOH), will catalyze this phase transition. Interestingly, we find sigmoidal kinetics for this process, indicating a cooperative nature whereby the transition of one domain further catalyzes the subsequent transition of other domains. Furthermore, we find that the rate of the phase change appears to be affected but not wholly governed by molecular dipole strength. This foundational work yields deepened understanding of the mechanisms of the clear to black phase transition in halide perovskites. In future, perovskite solar cells could be designed to resist undergoing this phase transition, so that they remain black and photovoltaically active, or they could potentially be designed to exploit this transition in a product such as a dynamically switching photovoltaic window.

583987 Formation of Poly(ethylene glycol) Coatings on Silicon Surfaces Via Nucleophilic Substitution Reactions

Attachment of 4-vinylbenzyl chloride (VBC) monolayers to hydrogen-terminated silicon through UV-induced hydrosilylation generates a surface coating (g-VBC) that can be further modified via substitution reactions at the benzyl chloride moiety. To apply g-VBC in synthesizing functional coatings, we demonstrated the attachment of poly(ethylene glycol) methyl ether 2000 (PEG2000) via nucleophilic substitution. Two reaction systems were used to produce PEG-modified silicon substrates. In the first approach, the deprotonated hydroxyl end of PEG2000 was reacted with surface bound benzyl chloride sites. A variety of reaction conditions were compared to determine those most conducive to high conversion at the surface. In a second approach, VBC was reacted with PEG2000 in solution, and the resulting compound was then grafted to hydrogen-terminated silicon via UV-induced hydrosilylation. For both of these approaches, characterization methods included contact angle measurements to assess wetting and ellipsometry to determine surface thickness. The fouling resistance of the PEG-modified substrates were compared using a FITC-albumin assay to test for non-specific protein adsorption. The developed reaction schemes are of particular interest as they yield Si – C bound films expected to exhibit greater hydrolytic stability than alkoxysilane- and chlorosilane-based methods reliant on siloxane bonds to attach coatings on silicon.

588174 The Role of Cryogenic Treatment on the Properties of Various Metals and Polymers

Cryogenic treatment of materials is used to improve and strengthen the structure, grain, and elasticity of materials including ceramics, polymers, and metals. Process and material type used can influence the improvement of material properties that undergo cryogenic treatments, and can be performed in adjunction to heat treatment, quenching, and cold work [1]. The material properties may be improved including but not limited to hardness, wear resistance, material fatigue, and conductivity including superconductivity. Improvement in machining tools to increase the lifetime of work, stainless steel was one of the first materials to undergo cryogenic treatment [1]. Other metals and materials including brass, aluminum, titanium, nickel alloys, nylon, and plastics have been tested and treated to understand how cryogenic treatment changes the material structure and conductivity. Cryogenic treatment is used in a variety of treatments for metals and plastics that are used in different operations. Recent research has made discoveries in tool machining and material integrity as well as the behavior of materials strength and conductivity. Four research papers were reviewed to understand how cryogenic metal and polymer treatments are being used to strengthen and change material structure for prevention of fatigue. Metal machining is conducted at high temperatures, which decreases tool life. The cryogenic treatment of titanium alloys was investigated through thermal analysis of cutting temperature, tool wear, cutting operations, and dry machining [3,4]. Surface integrity of the titanium and machining performance was also used to determine that cryogenic treatment improved machining performance and tool wear [3]. Polymers and nylon were determined to have changes in mechanical properties to increase tensile strength and material fatigue [5].

[1]Cryogenic Society of America, Inc. (October 2019). *Cryogenic Treatment* . Retrieved from Cryogenic Society of America, Inc. : https://cryogenicsociety.org/resources/defining_cryogenics/cryogenic_treatment/

[2] Fetecau, C., Timotin, P., Rosculet, R. T., Stanciu, N. V., & Stan, F. (2018). Mechanical Behavior of LDPE/MWCNT Composites After Fatigue and Cryogenic Treatment. *Proceedings of the ASME MSEC.* College Station, TX.

[3] Kaynak, Y., & Gharibi, A. (April 2019). Cyrogenic Machining of Titanium Ti-5553 Alloy. *Journal of Manufacturing Science and Engineering, 141.*

[4] Shi, B., Elsayed, A., Damir, A., Attia, H., & M'Saoubi, R. (February 2019). A Hybrid Modeling Approach for Characterization and Simulation of Cryogenic Machining of Ti-6Al-4V Alloy. *Journal of Manufacturing Science and Engineering, 141.*

[5] Shaffer, D., Reinstadtler, C., Roth, J. T., & Ragai, I. (2017). Analysis of Cryogenically Treated Sheet Nylon 6/6. *Proceedings of the ASME MSEC.* Los Angeles, CA.

585266 Recyclable Shape-Memory Elastomers

Shape-memory (SM) polymers are an important class of materials with a wide range of application areas, including biotechnology, textiles, and aerospace engineering. These materials undergo triggered actuation from kinetically metastable deformed states to their original, or permanent, shapes. Semicrystalline SM materials with complete shape-recovery conventionally require covalent crosslinks, thereby precluding melt-processing and recycling. Typical SM polymers are thus constrained to one shape-specific application per product and inherently limited in the sustainability of their life cycles. The objective of this project is to replace covalent crosslinks with strong and reversible hydrogen bonding interactions to enable melt-processable SM polymers without sacrificing performance. Two high molecular weight melt-processable systems have been successfully synthesized, with excellent SM fixities and strain recovery before and after processing. Dynamic mechanical analysis reveals a stiffness plateau that extends over a large temperature range (60-170°C) that is well above the shape-triggering temperature (~50°C). Stress relaxation studies suggest that chain (dis)entanglement kinetics strongly influence the materials' viscoelastic behavior, including the conditions of temperature and stress under which viscous flow is achieved. Dynamic, time-resolved, isothermal experiments building on these results will better illuminate the relationship between environmental conditioning and stress decay timescales and how it can be used to optimize processing conditions. The role of interchain H-bond strength in determining thermomechanical behavior at high temperatures is also currently being investigated through the use of linkage groups with weaker regimes of interchain association strength.

588247 Additive Manufacturing with Semi-Crystalline Thermoplastics: Properties for Printability

Temperature control is an integral part of success in 3D printing. Slight changes in printing conditions can drastically affect the printed part. This project is aimed to investigate the thermal properties of common semi-crystalline 3D printing thermoplastics in fused filament fabrication (FFF) with thermoplastic polyurethane (TPU) and polylactic acid (PLA). Melt flow behavior of the filaments was studied by printing lines on the printer bed to determine the mass flow rate over a range of nozzle temperatures from a high temperature of 255ºC to the values where no filament is extruded by 2ºC increments. The time remained constant throughout the experiments for each filament. However, the mass flow rate drastically decreased as the nozzle temperature dropped below 209ºC and 183ºC for TPU and PLA, respectively. Filament behavior changed with nozzle temperature, so while the filaments could be extruded at lower than manufacturer suggested temperatures, the overall printability is drastically reduced below a certain threshold of nozzle temperature. This behavior is attributed to change in melting range of the polymers with the heating rate. Therefore, the data was compared to DSC measurements at different heating rates to better understand the range of melting temperatures of the semi-crystalline polymers. This data can be used to achieve more successful 3D prints and an overall better understanding of printing behaviors at various temperatures.

585992 Controlled Drug Release from Polymeric Coatings Fabricated Using an Electrospraying Technique

Electrospraying is an effective technique for producing polymeric coatings with smooth topographies. This technique is unique based on its ability to create reproducible coatings in ambient conditions from a plume of charged droplets. These polymeric coatings can be homogenously fabricated to incorporate potential drug molecules for delivery. The main objective of the study is to create drug loaded films that exhibit controlled released into a media, Phosphate Buffered Saline (PBS). *In vivo* conditions were simulated by utilizing a thermal shaker and replenishing the PBS after each model drug concentration measurement. Tailored drug release is desirable for many medical treatments, including cancer treatment. The total cumulative percentage of the model drug Rhodamine B (RhB) released over time was determined. Apart from *in vivo* conditions, controlled release was observed at different temperatures, by exploiting the change in the glass transition temperature (of the polymer, Polyvinyl Acetate (PVAc). By altering the temperature, the rate at which the dropped seemed to increase. We infer that the thickness of the coating also has a role in determining the change in the . In the future we will also explore the effect of varying the coating thickness as well as the polymer on the release of RhB. We will also fabricate functionally graded coatings to produce tunable and unique release curves.

587330 Optimizing the Alignment of Electrospun Pnipam:PLGA Blendednanofibers for Cell Sheet Engineering

Cell sheet engineering has emerged as a new tissue engineering strategy to harvest cells with their deposited extracellular matrix (ECM) as intact sheets. To prevent disrupting their ECM, thermoresponsive culture dishes are utilized for harvesting cells instead of using enzymatic treatment. One such thermoresponsive polymer that has been used for the culture of cell sheets is poly(N-isopropylacrylamide) (PNIPAM), which exhibits a reversible phase change at its lower critical solution temperature (LCST) of 32°C. At an incubation temperature of 37°C, PNIPAM exhibits a hydrophobic state, allowing for cells to adhere and proliferate. Various tissues in the human body, such as vascular smooth muscle cells (SMCs), tendons, bones and skeletal muscles, rely on cellular alignment for proper function; therefore, it is favorable to engineer aligned cell sheets for successful transplantation. In order to culture aligned cell sheets, nanofibers can be used as a 3D scaffold for cell growth and proliferation. Electrospinning is a promising technique for fabricating scaffolds due to its ability to easily and inexpensively manipulate nanofiber morphology. In this study, we optimized the electrospinning process for PNIPAM, PLGA, and PNIPAM:PLGA blended nanofibers using a statistical design of experiment of key process parameters. Through extensive experimental testing and statistical DOE, we were able to achieve an alignment coherency of over 80% for PLGA, PNIPAM, and blended fibers. Results showed that reducing polymer solution concentration resulted in fibers with smaller diameters and increasing the speed of the mandrel increased nanofiber alignment. Future work will assess material thermal responsiveness by measuring cellular attachment and release.

588323 Packing Density and Thermoresponsive Morphology of PNIPAM-Co-PAA Microgel Coatings

Microgels that alter their size or shape in response to external stimuli such as temperature, pH, and light are gaining considerable attention for their application potential in “smart” membrane technologies. By analyzing microgel depositions on solid substrates, one could better understand how these microgels deposit and function which is critical to advancing their implementation. This study investigates the effects that temperature and pH have on the deposition density of poly[(N-isopropylacrylamide)-co-(acrylic acid)] hydrogels and seeks to find a relationship between density and functional surface roughness change. The microgels were deposited upon silicon wafers primed with poly(octadecene-alt-maleic anhydride) and polyethyleneimine which created monolayer systems. The POMA layer provided a strong base layer due the 18-carbon tail hydrophobic interaction with the wafer. The branched PEI then reacted with POMA’s anhydrides which provided a top layer that was abundant with hydrogen bonding sites for the microgels. Microgels were deposited via spin-coating and incubation at temperatures both below and above the lower critical solution temperature of pNIPAm. Optical imaging and Matlab software were used to determine the densities of the monolayer systems, while a laser scanning microscope and MultiFile Analyzer software determined the microgel surface roughness upon temperature change. As could be expected, temperature and pH played a key role in how the microgel adherence to the solid substrate: increasing temperature and pH showed a positive correlation with density measurements. Following, a relationship was also observed between the density and the temperature associated surface roughness change. Overall, deposition conditions and methods have profound effects on surface morphology that can be qualitatively and quantitively described and related to functional surface roughness change.

584473 Thermal Stabilization of Invasion Plasmid Antigen D (IpaD) Using Silica Gels

The structure of vaccine proteins is vital to maintaining vaccine effectivity; however, exposure to heat can disrupt intermolecular interactions maintaining the secondary and tertiary structures of proteins, thereby making the vaccine ineffective. The cold chain is the system of maintaining vaccine potency by transporting them at constant 2-8°C environment. Unfortunately, it is estimated that 50% of vaccines are wasted before use due to breaks in the cold chain, according to the World Health Organization. Many preservation methods have been applied including entrapment (via sol-gel matrix), adsorption, and covalent attachment. Of these, adsorption has the simplest chemical procedure and does not require harsh chemicals. Silica is often used as an adsorbent due to its porosity, hydrophilic properties, and high thermal stability. Silica is also used in nature to protect organisms from heat denaturation. For example, small organisms such as diatoms and radiolaria are known to wrap themselves in a silica coating for heat protection. Due to these properties, silica has high potential to biomimetically protect and preserve vaccine proteins during transport. Here, we examine the adsorption of Invasion Plasmid Antigen D (IpaD) onto various mesoporous silica gels and study the protein's increased thermal stability on these solid supports for creating a cold chain independent vaccine. The solution depletion method using UV-Vis is utilized to study the adsorption of IpaD onto silica gels. Then the silica-IpaD complex is heated above the denaturing temperature of the protein to determine if the silica gel maintains the secondary structure of IpaD during heating. These proteins are then removed using N,N-Dimethyldodecylamine N-oxide (LDAO) detergent, and their secondary structure is tested using circular dichroism. The results demonstrate that silica gel maintains a high percentage of the native unheated alpha-helical structure of IpaD, thereby maintaining vaccine potency.

588349 Development of a Rechargeable Antimicrobial Textile

Functionalizing cotton with antimicrobial properties shows promise for applications in hospitals, for the military, and the general public. There has been a great deal of attention paid to methods of attaching n-halamine compounds to textiles for this purpose. Over time, the halogens present on these compounds dissociate into the surrounding environment and function as biocides. After dissociation, the n-halamines can be recharged by exposure to a solution containing halogen ions. One such n-halamine is 5,5-dimethylhydantoin (DMH), which is commonly brominated to a biocidal form for water purification. Previously in this project, a procedure for the synthesizing and purifying a DMH derivative, allyl-5,5-dimethylhadantion (ADMH), had been optimized. This compound retains the antimicrobial functionality of DMH, yet also can be chemically bonded to cotton. Once bonded, the strength of the antimicrobial properties was evaluated by exposing the fabric to the bacteria Staphylococcus aureus and Serratia marcescens and observing the growth inhibition in comparison to a positive and a negative control. Currently, tests have shown bacterial growth inhibition on only the fabric surface, suggesting that no chlorine is leaching into the environment surrounding the sample. This is desirable since skin irritation is noted as a possible drawback of using chlorinated n-halamines. Previous work on this project followed a "bottom-up" approach to the fabric modification; each necessary molecule is added to the fabric individually concluding with ADMH and its chlorination. This process does not allow for quantitative assessment each reaction's efficiency. It is desired to produce the final biocide in solution so that it may be attached in a single step which is ideal for synthesis on an industrial scale. Thus far, the molecule that links ADMH to the fabric has been synthesized and precipitated out via acidification. No IR spectroscopy data exists for the linker currently, and therefore this route of analysis has been performed by matching peaks with bonds known to exist on the linker. The final goal of this project is to develop a water-soluble molecule for which the process of synthesis and fabric attachment can be easily scaled up. This allows the treated fabric to be available to a wide range of consumers and may be implemented in a variety of fields where textiles are in continual contact with skin.

585756 Monitoring Temperature Induced Poly(N-isopropylacrylamide) Concertation with an Applied Electric Field

Poly(N-isopropylacrylamide), or pNIPAM, is a free-radical polymerized polymer, that is commonly studied for uses in surface coatings, tissue engineering, regenerative medicine, biosensing, and more, due to its temperature responsiveness. Dynamic Light Scattering (DLS), used to measure temperature induced coacervation in determining a lower critical solution temperature (LCST), which is typically 30-35°C for pNIPAM. We initially confirmed the LCST to be thermodynamically reversible and reproducible for both pNIPAM and an electrochemically active poly(*N*-isopropylacrylamide)-*block*-poly(ferrocenylmethyl methacrylate) block copolymer, or p(NIPAM-*b*-FMMA).

A voltage was applied through the solution using an electrode lined cell, a technique with DLS that has not been previously reported. We created the cell by inserting copper tape to two sides of a DLS compatible cell, and the leads were connected to a voltage supply to generate an applied electric field across the light path. During the applied voltage, an increase in temperature ("ramp up") and decreasing in temperature ("ramp down") was used to look at hysteretic effects on the thermodynamic coacervation.

Our results were similar to published values for the LCST of pNIPAM for both increasing and decreasing temperature ramps in the absence of an Applied Electric Field (AEF). However, we observed a decrease in LCST during a temperature increase in the presence in of an AEF (1 V/cm). Additionally, during the temperature decrease, the polymers remained coacervated and were no longer thermosensitive. Upon further experimentation, the same phenomenon occurred from at least 0.4 V/cm and upwards in stepwise 0.1 V/cm increments.

586889 PET Can be a Sustainable Option for Replacing Polyols in the Production Process of Polyurethane Foams

Nowadays there is a constant increase in the consumption of plastic worldwide, its growth rate is about 6% every year, which is causing plastic residues end up into water sources and ecosystems. The main polymer found as residue is PET since it is widely used as a packing material in food industry.

One way to mitigate this waste is through recycling, some companies try to develop new uses for this polymer. One of the industries interested in its use are companies manufacturing rigid polyurethane foams. A polyurethane foam is a material with a high thermic performance, useful for maintaining a constant temperature inside industrial equipment and reducing energy consumption; it is also impermeable and due to its capacity of creating a sealing layer it can avoid air and water leaks. These foams are produced from a reaction between polyol and an isocyanate.

585698 Hydrophobic Non-Fluorinated Polymeric Coatings on Cellulose Paper Via Initiated Chemical Vapor Deposition (iCVD)

Functional polymeric coatings present on materials serve to strengthen surface properties and enhance the versatility of said materials for applications such as packaging, flexible electronics, and microfluidics. While fluorinated polymeric coatings are commonly used for increased hydrophobicity, they can be detrimental to the environment and public health. In this work, we attempt to combat this safety issue by exploring alternative polymeric coatings that boast similar functional benefits to the surface properties of fluoropolymers using a solvent-free vacuum deposition process. Siloxane polymers were explored as an alternative to fluoropolymers, as they are common in many commercial and industrial applications and boast hydrophobicity and flexibility in coatings. Whatman cellulose chromatography papers and silicon wafers were coated with polydivinylbenzene (PDVB), poly(1,1,3,3,5,5-hexamethyl-1,5-divinyltrisiloxane) (PS5) and P(DVB-*co*-S5), a cross-linked polymer containing PDVB and PS5, via initiated chemical vapor deposition (iCVD). PS5 alone failed to deposit while PDVB and P(DVB-*co*-S5) resulted in successful coating depositions. The coatings were characterized using FTIR spectroscopy, and the hydrophobicity of the coatings were measured using contact angle goniometry. In addition, the thermal and mechanical stabilities of the coatings were demonstrated by annealing and folding samples. Consistent hydrophobicity of the coatings before and after stability tests was confirmed. Hydrophobicities used to demonstrate thermal and mechanical stabilities of PDVB and P(DVB-*co*-S5) were compared, with P(DVB-*co*-S5) coatings exhibiting the larger hydrophobicities on both cellulose papers and silicon wafers.

587455 Measuring the Surface Tension of Different Surfactants to Make Better Polyhipes

Foams and emulsions are created by the combination of either two immiscible liquids or one liquid and a gas with the addition of a surfactant. Water and a hydrophobic phase, such as a monomer, are two liquids that are commonly used. As the water and monomer do not mix, a surfactant must be used. A surfactant is a molecule with hydrophilic and hydrophobic components, which allows the two immiscible materials to mix. The first step in creating a polymer foam is creating an emulsion that consists of the monomer (continuous phase), water (dispersed phase), and a surfactant. Then the emulsion is polymerized resulting in a polymer foam filled with water. These voids are often interconnected in the polymer, so liquid is able to flow in and out. The water is then dried out of the foam to result in a polymer foam with voids. These polymer foams have been proven useful in many engineering applications such as tissue engineering, cosmetics, and food science. The goal when making these emulsions is to increase the surface area of the voids in the polymer foam to enable more interaction between the surfactant and the fluid. This is a result of many smaller voids being distributed throughout the foam rather than larger voids existing in a smaller quantity. The way void size is controlled is through the manipulation of surface tension. Surface tension is a resistance force that exists on the surface of a liquid. It is important for the creation of emulsions because the lower the surface tension, the higher the emulsion surface area. After testing the surface tensions of different surfactants with other monomers, a relationship between surface tension and concentration can be quantified. As more surfactant is added, the surface tension will decrease with the goal of determining the lowest possible surface tension that can be achieved. The concentration at which this occurs is called the critical micelle concentration. As a result of how low the CMC found is, it can be determined which surfactants work best with different mixtures and better emulsions can be achieved. Using a tensiometer, concentrations of the surfactants Triton X-100, C-TAB, Kaolin, Span 80, SDS, and Hypromellose in the monomers Styrene and 2-Ethylhexyl acrylate were evaluated.

585758 Determination of Reverse Micelle Diffusion Coefficients in Polymeric Organogels

The goal of this project is to determine diffusion coefficients for sodium bis(2-ethylhexyl) sulfosuccinate (AOT) reverse micelles through styrene-ethylene-butylene-styrene (SEBS) copolymer organogels at varying concentrations of SEBS. Characterizing the diffusivity of AOT through these gels will benefit future applications like transdermal drug delivery through informed structure-property (*i.e.*, nanostructure-diffusion) relationships. Previous work in our group has shown that gel nanostructure is tuned by varying the amount of SEBS copolymer. For example, increasing the SEBS concentration increases the size of physical crosslinks and the concentration of chains occupying space in the gel, while decreasing the space between physical crosslinks. The current work uses Fourier-transform infrared spectroscopy (FTIR) to track changes in gel AOT concentration over time for gels ranging from 10 wt% to 40 wt% SEBS. The acquired data is modeled using Fick's laws to yield a diffusion coefficient for each gel formulation. The resultant diffusion coefficients decrease with increasing SEBS concentration. We hypothesize that the combination of aforementioned nanostructure trends are the culprit for our observed decrease in diffusion coefficient with increasing SEBS concentration. Understanding the relationship between the gel nanostructure of polymeric organogels and diffusion coefficients of these organogels will provide key insight for biomedical applications.

587492 Synthesis and Characterization of Multifunctional Bio-Based Polybenzoxazines and Their Blends with Epoxy Resins

Synthesis and Characterization of Multifunctional Bio-Based Polybenzoxazines and their Blends with Epoxy Resins

Sarah Salazar, Amanda McCahill, Alexandra Chong, Joseph F. Stanzione III

Department of Chemical Engineering, Rowan University, Glassboro, NJ 08028

Abstract

The uncertainty of the petroleum industry and increase in environmental concerns has caused a growing interest in bio-based polymers. Polybenzoxazines are polymers that possess high glass transition temperatures, high char yields, good mechanical properties, and low shrinkage upon curing. In addition, the structures of benzoxazines allow for remarkable molecular design flexibility and can be used in numerous applications, such as lightweight aviation composite parts. Bio-based chemicals have the potential to replace petroleum-based benzoxazine platform molecules such as BPA, which has displayed toxic effects on human health and the environment. Additionally, benzoxazines are known to produce brittle polymers with low crosslink densities, so blending with epoxy resins have shown to improve crosslink densities resulting in better mechanical properties of the polymers.

Higher degrees of oxazine functionality have shown to display better mechanical and thermal properties of polybenzoxazines. Bio-based mono-, di-, and tri-functional benzoxazines were synthesized and characterized based on varying degrees of functionality. The benzoxazines with each targeted oxazine functionality were synthesized via Mannich condensations involving guaiacol and vanillin combined with furfurylamine and paraformaldehyde. Monofunctional benzoxazines were blended with different weight percents of diglycidyl ether of bisphenol A (DGEBA) and subsequently cured. These uncured and cured resin mixtures were characterized using DSC, TGA and DMA to understand their structure-property relationships. Overall, bio-based benzoxazines with increased oxazine functionality and their epoxy blends were investigated to produce environmentally sustainable polymers with enhanced thermal and mechanical properties.

584851 Melting Temperature Effects on the Crystallization Order of PEO-b-PCL Copolymers

Various samples of poly(ethylene oxide)-b-poly(e-caprolactone) (PEO-b-PCL) of different molecular weights were dissolved either in toluene or chloroform, melted at different temperatures, and cooled to investigate which block would crystallize first. Depending on which block crystallizes first, it may crystallize more thereby influencing the macroscopic properties of the copolymer film. Analysis was conducted using Differential Scanning Calorimetry (DSC) and FTIR. Previous observations have indicated that during evaporation crystallization, the polymer film preferentially crystallizes PCL first but under certain conditions, PEO crystallizes first. These conditions included but were not limited to the solvent used to solvate the film and the molecular weight of each block. So, the goal of this project was to investigate whether the melting temperature influences the order of crystallization of the film. In doing so, the researcher is ultimately trying to find out if it is possible to control the properties of the polymer film.

585199 Residence Time Distribution and Specific Mechanical Energy in Solid-State Shear Pulverization: Processing-Structure Relationships in a Chilled Twin-Screw Extruder

Solid-state shear pulverization (SSSP) is a continuous polymer processing methodology based on a modified twin screw extruder. The unique application of low barrel temperatures and mechanochemistry have contributed to the development of an extensive range of polymer-based materials, from environmentally responsible polymer blends to nanocomposites, for more than 30 years. The complex processing-structure-processing relationships in SSSP can be elucidated by evaluating the interplay between numerous processing parameters using integrated, measured covariants. Residence time distribution and specific mechanical energy are evaluated in a base case polypropylene study under a full factorial experiment involving screw design, screw speed, and throughput parameters. These factors are in turn correlated to dispersion morphology and thermal property results from a parallel study based on a model polypropylene/carbon black composite. This investigation highlights the tunability of SSSP processing parameters for tailored output with desired purposes and applications. In particular, enhanced residence distribution can be achieved with low screw speed and high throughput settings, leading to high levels of material mixing and shear.

585202 The Study of Segregation of Dyes in PEO-*B*-PCL

As an amphiphilic biodegradable copolymer, PEO-*b*-PCL is useful for many biomedical applications. The segregation of medicine in the PEO-*b*-PCL has implication on the release rate of medicine inside a biological system. The objective of the project is to study the segregation of medicine in PEO(5000)-*b*-PCL(16300) diblock copolymer using azobenzene dyes and bromophenol blue to mimic the presence of the medication. PEO(5000)-*b*-PCL(16300) films were integrated with methyl yellow, methyl red and bromophenol blue dyes with different mass ratio to determine the effect of dyes concentration on the physical and chemical properties of the copolymer including melting point, crystallization temperature and percent crystallinity of each block. As shown by DSC analysis, the melting point and the crystallization point of PEO-*b*-PCL both decreased as the concentration of bromophenol blue increased while they both increased as the concentration of methyl red increased. Also, both DSC and ATR-FTIR data showed that the percent crystallinity of PEO block was influenced by the presence of methyl red. However, methyl yellow had no significant influence on the melting point and crystallization point of the PEO-*b*-PCL. The preliminary results for the dyes segregation in PEO-*b*-PCL films are: methyl red mostly in PEO block; bromophenol blue distributed among PEO and PCL blocks and methyl yellow stay on the surface of the films.

587835 Coal-Plastic Composites for Use in Construction Application

Wood-Plastic Composite (WPC) decking offers significant improvements to traditional treated lumber decking material such as increased weather resistance, durability, and easier maintenance. Despite superior quality, sales of WPC decking have been held back by its significantly higher price; a board of WPC costs more than three times the price of a traditional treated lumber board. Cost could be reduced by use of less expensive filler, by increasing the filler content, or by lowering the energy consumed to produce a board. In this study, coal is evaluated as an alternative to saw dust as a filler, due to its lower cost and potential to be used at higher filler contents. Novel Coal-Plastic Composites (CPCs) with coal contents up to 60wt% are characterized by testing if they meet certain building codes for decking applications at various filler contents and by comparing their traits to those of current commercially available WPCs. Some of the building requirements tested include mechanical strength, flammability, and service life. Research so far has yielded positive results: CPCs with 60wt% filler, the same filler content as most commercial WPCs, have been found to meet the building requirements. Further, some advantages over WPCs such as superior flammability properties and increased strength have been discovered, with comparable service life.

586932 Multimaterial Aerosol Jet Printing of Functionally Graded Nanocomposites

Advanced Materials Laboratory, Sandia National Laboratories, Albuquerque, NM, USA

Aerosol jet printing offers a digital, non-contact, high-resolution direct-write capability with broad materials compatibility. In this work, a multimaterial aerosol jet printhead is designed for in-line mixing of multiple functional inks. Drawing on fluid dynamics modeling and 3D printing for custom printhead design, this enables direct patterning of compositionally-graded films. In this case, COMSOL modeling supports static mixer development tuned for relevant aerosol parameters, with geometry optimization based on mixing efficacy, mixer volume, and particle transfer efficiency. A proof-of-concept experiment using a fluorescent tracer in epoxy and acrylate matrixes validates this approach and allows straightforward visualization and characterization. In this case, composition information is obtained from fluorescence microscopy and profilometry measurements to calibrate deposition parameters and assess print quality. This materials system is then extended to incorporate ZrO_2 nanoparticles to tailor optical and dielectric properties. Overall, this multimaterial aerosol jet printing method provides a rapid, digital, and low-waste method to control print composition on the fly, establishing this nascent technology as a compelling platform for diverse applications including, but not limited to, functionally-graded electronic, optical, and magnetic materials.

Sandia National Laboratories is a multimission laboratory managed and operated by National Technology & Engineering Solutions of Sandia, LLC, a wholly owned subsidiary of Honeywell International Inc., for the U.S. Department of Energy's National Nuclear Security Administration under contract DE-NA0003525.

[Abstract for submission for a poster at the AIChE 2019 Student Conference in Orlando, FL on November 11, 2019]

587004 Design and Application of High-Performance Ionene Composites

Media for state-of-the-art gas separation membranes draws from material classes including ionic liquids (ILs), poly(ILs), and supported liquid membranes (SLMs) as well as more robust materials including polymers with varied functionality. Ionenes are polymers which contain ionic moieties along the main chain, rather than as pendants, which can be designed to harness the advantageous properties and structural elements typically associated with high-performance (HP) engineering polymers (i.e. aromatic character, imide or amide linkages). We have recently investigated the design, synthesis, and application of a new class of HP ionenes, which incorporate imidazolium cations (paired with Tf_2N anions) along the main chain, alternating with diverse functional groups. The modular design of functional bis(imidazoles) and bulky dihalide monomers linked via the Menshutkin reaction, we have synthesized numerous derivatives of these ionenes which can be processed and tuned for a variety of applications. These materials exhibit intriguing self-assembly and nanostructuring, which is dictated by the complex combination of ionic and functional intermolecular interactions, leading to sophisticated materials with good thermal and mechanical stability. This structuring can be further tailored via the addition of "free" imidazolium-based ILs or functional ionic fillers, which serve as a non-covalent crosslink between polymer chains, stabilizing the polymer matrix and increasing the free volume between chains. These composites are processed into thin films (solvent-cast or melt-pressed) and the performance of these materials as membranes for light gas separations is analyzed. Due to the diverse potential structural combinations which lead to unique properties and behaviors, certain derivatives have also been investigated as functional coatings, elastomeric shape memory materials, additives, and fibers for additive manufacturing. Here, we present the development of these ionene composites resulting from complex monomer units, and the impact of design and processing on structure-property relationships and performance in the aforementioned applications.

585733 Solid-Phase Extractive Polymerization As a New Method for Polyanhydride Synthesis to Limit Oxidation during Polymerization and Improve Purity for Controlled Drug Delivery

Polyanhydrides are an attractive class of erodible polymer for controlled drug delivery because they can exhibit bulk or surface erosion and they have been shown to stabilize several classes of drugs including macromolecular therapies.

The traditional method of polymerizing polyanhydrides is melt polymerization where an activated acetylated diacid prepolymer is heated at 180 °C under vacuum for 1 h to 3 h. The vacuum helps to remove the condensation by-product, acetic acid, generated during polymerization. While this method is effective at producing polymers with molecular weights greater than 15 kg/mol, the intense heating can lead to thermal degradation, oxidation, and discoloration of the polymers. Another issue with this method is the quick increase in the viscosity of the melted prepolymer and growing polymer during polymerization leads to poor stirring and removal of acetic acid.

To overcome the drawback of melt polycondensation we have developed a less thermally aggressive method of polymerization that we call solid-phase extractive polymerization. In this method, we use a solvent to facilitate the polycondensation while also extracting the acetic acid by-product with basic aluminum oxide.

Using sebacic acid as a model diacid, we studied the effect of a solvent boiling point on the polymer molecular weight. Interestingly, we find that high boiling point solvents allow for temperatures close to that of melt polycondensation but they are not efficient at removing the acetic anhydride because the number of solvent cycles is low. Low boiling point solvents cycle many times but they do not heat the prepolymer enough to get appreciable polymerization. Methylcyclohexane and acetonitrile provide a balance in temperature and number of cycles through the basic aluminum oxide extraction media to produce polymers in the 15 kg/mol to 30 kg/mol range in 1 h. The polymers made with this method also do not exhibit significant signs of oxidation or discoloration.

We believe this method may be applicable to many polymerization methods that generate a by-product.

587118 Efficient Removal of Microcystin-LR Cyanobacterial Toxins from Contaminated Freshwater By Biocompatible Polymer Coacervates

Microcystin-LR (MC-LR) is one of many toxins released upon the lysis of cyanobacteria, blue-green algae, and some cells in Great Lakes and other natural freshwater sources, which imposes potential health risk to humans and animals. The safe level of MC-LR in drinking water according to the World and Health Organization (WHO) is 1μg/L, yet its concentration in natural freshwater could reach to 0.2-1800μg/L when cyanobacteria bloom in the summer, making perilous for consumption and daily uses. In this research, we explore a new polymer flocculate based on the coacervate complexation of biocompatible polyethylene oxide (PEO) and polyacrylic acid (PAA) via hydrogen bonding to demonstrate efficient coacervate removal of MC-LR from aqueous solutions. Gel-like PEO-PAA complexes are added to model contaminated water of varied concentration from 1-5000 μg/L; the supernatant aqueous phase is separated from the polymer coacervate complex after mixing for 5-30 min and the residual MC-LR concentration in aqueous solution is quantified by UV-vis spectroscopy and mass spectroscopy. Preliminary results show that our coacervate flocculates could remove up to 95% of MC-LR with an initial concentration of 1000μg/L. The efficiency of coacervated complexes at varied MC-LR concentration and PEO-to-PAA molar ratios are under current investigation, seeking an optimal PEO-to-PAA composition ratio for maximal efficiency of MC-LR removal.

587153 Development and Characterization of Bio-Based Thermosets from Birch Bark

Pollution from plastic waste is a growing global issue. Traditional petroleum-derived plastics can take hundreds of years to degrade naturally and petroleum stores are quickly depleting. As such, bio-based plastics are increasing in popularity in both scientific and consumer spheres as they offer a sustainable alternative to petroleum-based plastics. In this work, we investigate the use of betulin, a naturally-occurring diol that can be extracted from birch bark in high yields, as a building block for the formation of bio-based polyester thermosets. We systematically assess the choice of co-monomers on the thermomechanical and degradative properties of the resulting polymers. Our results show that the room temperature moduli range from 2.0 to 790 MPa and the glass transition temperatures are tunable over a 100°C range depending upon the co-monomer selected. Moreover, all samples exhibit thermal stability above 300°C in both nitrogen and air. This work is part of a larger study focused on the valorization of betulin for commodity plastics. The results presented here demonstrate that betulin is a promising building block for the design of bio-based thermosets.

587159 Synthesis of Polymeric Membranes with Carboxylic Acid and Thiol Groups for Ion Adsorption Applications

The removal of heavy-metal ions from industrial wastewater is highly important due to their health and environmental effects. Additionally, the recent increase in the usage of electronic devices has led to an increased usage of lanthanides, which should be recycled due to their high value. Adsorption and ion exchange technologies have been used for these applications and further development of sustainable technology in these areas is crucial. This work involves the characterization, fabrication, and performance testing of polymeric membranes containing carboxylic acid and thiol groups, which are suitable for these applications. Membrane fabrication began with the polymerization of polyacrylic acid (PAA) in the pores of commercial polyvinylidene fluoride (PVDF) membranes to add carboxylic acid groups that can form complexes with cations, such as La^{3+}. Furthermore, membranes containing thiol groups, which have a strong affinity for heavy metal ions, were fabricated by covalently attaching cysteamine to PAA. The incorporation of PAA into a PVDF membrane was verified by a change in surface charge of the modified membrane near the pK_a of PAA (~4-5). 3.7 g La^{3+}/m^2 membrane was captured by a PVDF-PAA membrane via convective-flow ion exchange. The binding of cysteamine to PVDF-PAA membranes was confirmed by energy-dispersive X-ray spectroscopy, which indicated the presence of adsorbed silver in the membrane pores. When treating a feed containing 40 - 50 ppb of Hg^{2+} and 70 ppm of Ca^{2+}, a thiol-functionalized membrane captured 82% of the Hg^{2+} in one pass. The capture of La^{3+} by a PVDF-PAA membrane and Hg^{2+} by a cysteamine-functionalized PVDF-PAA membrane demonstrated the versatility of the PVDF-PAA membrane as a base for multiple applications involving the capture of metal ions.

Dr. Evan Hatakeyama of Chevron Energy Technology Company provided guidance on this research. This research was supported by Chevron Energy Technology Company, NSF EPSCoR, and NIEHS-SRP.

586615 Synthesis of Temperature Responsive Polymeric Flocculants with Cationic Moieties for Environmental Remediation

Perfluoroalkyl substances (PFAS) are a prolific and persistant contaminant in the environment, and our group is developing a novel copolymer sorbent with 2-phase LCST behavior to aid in PFAS removal. The hypothesis in question was that the incorporation of cationic comonomers will aid capture of anionic contaminants such as types of PFAS in temperature-driven polymeric flocculation. A model anionic organic dye, indigo carmine, was used as the preliminary contaminant in the pursuit of a conclusive visual remediation effect. The copolymers are primarily n-isopropylacrylamide (NIPAAm) with low levels of cationic comonomers like dimethylaminoethyl acrylate (DMAEA) or dimethylaminopropyl acrylamide (DMAPAA). The copolymers were polymerized via a free radical polymerization, and the structure and properties were characterized by NMR, GPC/SEC, FTIR, and UV-Vis spectroscopy.

Comonomer was successfully incorporated as shown qualitatively in NMR spectra and an LCST shift upwards. The molecular weights were between 2 and 4 million g/mol. FTIR data was noisy and analysis did not clearly support comonomer incorporation. LCST was obtained via UV-Vis spectroscopy. The LCST of the new copolymers ranged from 33.75 to 40.2 °C, and was shown to be dependent on wavelength, unexpectedly. This result could challenge the industry norm of reporting LCST at a single wavelength of 500 nm. LCST of the poly(NIPAAm) (PNIPAAm) synthesized were between 30.5 and 33.75 °C, depending on initiator concentration. The copolymers were brought to flocculate in the presence of indigo carmine to determine if removal could visually or quantitatively be observed, but the efforts thus far have not been successful. The hypothesis cannot yet be proved or disproved as no contaminant removal has been demonstrated, but the comonomers were successfully incorporated into the copolymers and further optimizations to the organic anionic dye buffered flocculation study have been recommended.

586731 Synthesis and Characterization of Pnipaam-Au@Ionp Thermo-Responsive Nanocomposites

Gold-shelled iron oxide nanoparticles (Au@IONPs) were synthesized and characterized for use in environmental remediation efforts using a variety of techniques to optimize the uniformity and tunability of the gold shell coating for the subsequent attachment of a polymeric coating using poly(*N*-isopropylacrylamide) (PNIPAAm). A gold shell on iron oxide nanoparticles has been shown to prevent further oxidation and aggregation of the magnetic nanoparticles, while also having negligible effects on the inherent superparamagnetic properties of the IONPs.

Comparisons were made between aqueous and organic media direct-coating procedures for the synthesis of the Au@IONPs. Aqueous methods employed a hot, citrate reduction of gold chloride onto the surfaces of coprecipitated IONPs. Organic methods used oleylamine for the reduction of gold chloride onto the surfaces of IONPs, synthesized by thermal decomposition, and provided added stability as a surface capping agent. DLS sizing data showed greater control over IONP aggregate sizes when using thermal decomposition methods as opposed to coprecipitation methods, and SEM/TEM images were collected of the nanoparticles. The reduction of gold chloride onto the thermally decomposed nanoparticles in organic media allowed, too, for the fine-tuning of the gold-shell thickness that was not achieved using only aqueous reduction methods. Surface-initiated, atom transfer radical polymerization (SI-ATRP) attachment of NIPAAm was then performed on each set of nanoparticles, and FTIR spectroscopy was employed to verify the successful attachment.

587796 Diffusion and Mechanics of Styrenic Block Copolymer Organogel Formulations: An Overview

Block copolymer gels have gained significant interest as a useful material platform for numerous applications including drug delivery vehicles. Non-aqueous block copolymer gels, *i.e.*, organogels, offer advantages over their widely popular hydrogel counterparts including elimination of drying out due to solvents with negligible vapor pressure as well as the ability to create a hydrophobic environment for drug delivery. Here, we characterize the diffusion of model nanocarriers, swelling behavior, and mechanical profiles of various block copolymer organogel formulations. These gels are comprised of styrenic block copolymers, aliphatic oil, and a nanocarrier-forming surfactant, and are uniquely formulated to provide insight into the role different components play in gel properties. While mechanical properties are probed using standard quasi-static uniaxial testing, diffusion is quantified using a novel method that directly measures the retained nanocarrier mass in gels through FTIR spectroscopy. Absorbance measurements are taken at specific time intervals and a plot of retained mass versus time is generated. Utilization of this method allows the retained nanocarrier mass to be tracked even with changing pathlength and volumetric expansion, both of which result due to gel swelling. The current overview study indicates that the mechanical profiles and internal diffusion of block copolymer organogels are tunable through simple changes in formulation. In fact, the data suggests that the two gel properties can be altered independently.

586767 Photonic Crystals Sensors Enabled By Peroxide-Responsive Shape Memory Polymers

Here we report novel chromogenic sensors that exhibit a change of color when exposed to an external stimulus, such as heat, pressure, shear stress, acoustic waves, and exposure to various liquids or gases. By fabricating shape memory polymers (SMPs) into the structure known as a photonics crystal, we have developed SMP sensors with dual-shape capabilities. Such sensors work by changing its shape from a predetermined shape to its permanent photonic crystal shape when exposed to the appropriate stimulus. Having potential applications ranging from user-friendly environmental monitoring to specifically detecting chemicals, these sensors introduce a more efficient alternative because they are easily perceived, inexpensive, reusable, have a small footprint, and are time and energy efficient. The active components of the SMP sensors are a few μm thick macroporous photonic crystal layers, which are fabricated using self-assembled, 3-D highly ordered colloidal crystals as structural templates. This microscopic thin-film configuration renders orders of magnitude faster response speed than bulky SMP samples in traditional applications. Moreover, by leveraging easily perceived color changes, which is associated with the unconventional all-room-temperature shape memory cycles enabled by the recent discovery of a new series of multi-stimuli-responsive SMPs, detection of certain analyte in complex solutions, such as hydrogen peroxide in commercial products, has been demonstrated with a detection limit as low as 1%.

587935 Graphene Nanoscrolls Induced Unique Crystallization of Poly Lactic Acid

Poly (lactic acid) (PLA) has gained attention for its use in biomedical applications due to its biocompatibility and optical, mechanical, and barrier properties. However, it is usually found in an amorphous or semi-crystalline state, which causes it to be brittle and have poor toughness. In order to improve the mechanical properties of this polymer, extensive research has been put into effective ways to crystallize PLA. This is difficult as PLA crystallization is a slow process that does not comply with large scale crystallization measures. A way to counteract this problem is by incorporating a nucleating agent into the polymer to lower the activation energy needed to start crystallization and affect the shape of the crystals, thereby affecting the mechanical properties of the polymer.

Graphene nanoscrolls (GNS) are rolled up sheets of carbon, created by overcoming an energy barrier of about 100s to 1000s J/mol applied to graphene nanoplatelets. Once this energy is added, the flat sheets roll up and do not unscroll due to p bond interactions between the carbon molecules in different layers keeping the structure strong. Graphene itself is known for its chemical functionalization capabilities and physical properties such as high mechanical strength and thermal conductivity. GNS contains these properties from its graphene source, but also differs from other carbon-based nanomaterials such as carbon nanotubes, as its unfused edged creates an open morphology. This unique shape allows for a lattice strain which can provide more mechanical strength and create new crystallization structures.

In this study, graphene nanoscrolls were fabricated from graphene nanoplatelets and used in PLA crystallization to improve the mechanical and physical properties of PLA. This was done by sonicating a mixture of GNP and toluene for scrolling and then using liquid nitrogen for structural stability to create GNS. The GNS then dried at room temperature before being incorporated into a melt-mix crystallization with PLA. Different weight percentages of GNS to PLA were used to see the affect on crystallization. Samples were then characterized by SEM, TEM, FTIR, XRD, and underwent a nanoindentation test to determine how the crystal structure of PLA changed, if any chemical interactions occurred between the PLA and GNS, the crystal structure and changes in mechanical properties. Through these analyses it was determined that the addition of GNS to PLA increases the crystallinity, crystal structure, modulus and hardness without any chemical changes to the polymer.

583901 Tuning the Adsorption of Elastin-like Polypeptides (ELPs) on Ots and Its Surfaces By Manipulating Environmental Conditions and Surface Hydrophobicity

Elastin-like polypeptides (ELPs) are thermoresponsive randomly coiled polypeptides that permit the binding of leucine zipper fusion proteins and immobilize onto silane self-assembled monolayers (SAMs) via bioconjugation. The functionalization of ELP can be terminated by hydroxyl (OH) and carboxylic (COOH) groups thus inducing micelle conformations on the surfaces of trichloro(octadecyl)silanes (OTS) and isooctyl trimethoxysilanes (ITS). The present study synthesized SAMs via chemical vapor deposition as a basis for ELP thin film micelle and unimer formation. Toluene and chloroform organic solvents restricted ITS and OTS silanization on Si-wafers to enable ELP unimer-to-micelle transitions while contact angle goniometry (CAG) and quartz-crystal microbalance with dissipation monitoring (QCM-D) characterized the surfaces. Using a UV-Vis spectrometer, 350-410 nm adsorption analysis was conducted for 0.1, 0.5, 1.0, and 2.0 mg/mL ZR-ELP in 1xPBS. Prior to adjusting PBS NaCl concentrations, dynamic light scattering profiled various ZR-ELP concentrations. The following NaCl concentrations were studied: 0.154, 0.300, 0.450, 0.900, 1.500, 2.000 M. QCM-D Voight-based viscoelastic modeling determined the thickness of ZR-ELP on OTS and ITS doped Si-wafers. Deposition frequencies (ΔF), dissipation energies (DE), and viscoelastic behaviors of ZR-ELP were monitored using QCM-D as well. After OTS and ITS SAMs were synthesized, contact angle goniometry suggested that OTS-doped Si-wafers were considerably hydrophobic compared to std. Si-wafers and ITS-doped Si-wafers. The following contact angle measurements were retrieved: 43.52° ± 2.98 std.-Si, 53.11° ± 2.85 ITS-Si, and 107.64° ± 2.05 OTS-Si.

Reference List

[A] Costa, R. R.; González-Pérez, M.; Herrero-Gutiérrez, M.; Pires, R. A.; Alonso, M.; Rodriguez-Cabello, J. C.;

Reis, R. L.; Pashkuleva, I. Tuning the Stiffness of Surfaces by Assembling Genetically Engineered

Polypeptides with Tailored Amino Acid Sequence. *Biomacromolecules* **2018**, *19* (8), 3401–3411.

https://doi.org/10.1021/acs.biomac.8b00723.

[B] Li, L.; Li, N. K.; Tu, Q.; Im, O.; Mo, C. K.; Han, W.; Fuss, W. H.; Carroll, N. J.; Chilkoti, A.; Yingling, Y. G.; et

Functional Modification of Silica through Enhanced Adsorption of Elastin-Like Polypeptide Block

Copolymers. *Biomacromolecules* **2018**, *19* (2), 298–306. https://doi.org/10.1021/acs.biomac.7b01307.

[C] Jang, Y.; Champion, J. A. Self-Assembled Materials Made from Functional Recombinant Proteins. *Acc. Chem.*

Res. **2016**, *49* (10), 2188–2198. https://doi.org/10.1021/acs.accounts.6b00337.

585724 Adhesion and Shear Modulus in Block Copolymer Gels

The goal of my research is to quantify the adhesive strength of polymer gels as their molecular connectivity is decreased. Additionally, I aim to identify why such a relationship exists. Polymer gels exist throughout nature; gastropods, such as snails and slugs, produce mucus that helps them adhere to surfaces during their locomotion. Snail slime is so dilute, ranging from 91%-99% water, that it is often used as a lubricant. So, it is scientifically intriguing that it can also function as an adhesive. Thus, we have been analyzing dilute polymer organogels to determine how the decrease in connectivity increases the adhesiveness of the gel. These polymer gels are made of mineral oil and a combination of di-block and tri-block copolymers. As the composition of di-block polymer, in the gels, increases and, correspondingly tri-block copolymer decreases (since their total is held fixed), the molecular connectivity of the gel decreases. Adhesive strength was tested in shear yield-point experiments. Gels composed of 20 wt% total copolymer and 100 wt% to 40 wt% triblock copolymer (0 wt% to 60 wt% diblock copolymer) were using this experimental approach. It was determined that adhesive strength increased as the di-block composition increased while shear modulus decreased. The adhesive strength trend conflicts with that expected based upon qualitative observation. In the future, tests on a series of gels with 10 wt% total polymer composition will be conducted, as well as, alternative techniques explored to validate the observed trends.

585909 Industrial-Applicable Method to Recycle High-Value Engineering Thermoplastics

High-value engineering plastics are increasing in demand in specialty industrial applications as well as everyday consumer items in today's society. These materials have a high processing cost and can retain value due to its chemical stability. Just like commodity plastics, these engineering plastics face the challenges of one-time use and negative impacts on the ecosystem. For environmental and economic reasons, a method to efficiently recycle these specialized polymers would be beneficial. This research poster will describe and discuss a new method to allow these post-consumer and post-industrial engineering thermoplastics to be recycled and reprocessed with minimal changes in properties compared to the virgin polymer. Structural characterization is conducted using scanning electron microscopy as well as Fourier-transform infrared spectroscopy. Static mechanical tests and hardness test will be used to measure the mechanical property changes. In the future, the same processing technology can be expanded to include other polymers that are difficult to recycle.

584371 The Effect of Short-Range Attractions on Sequence-Defined Polyelectrolyte

Complex coacervation is the liquid-liquid phase separation of polyelectrolytes in aqueous salt solution into a polymer-dense phase, the coacervate, and a polymer-dilute phase, the supernatant. Previous work using Monte Carlo simulations demonstrated that changing the sequence of charged and neutral monomers on polyelectrolytes while keeping the charge fraction constant alters the extent of phase separation. However, previous data does not account for the hydrophobicity of different neutral monomers. To understand the coacervation of polymers with various chemical structures, van der Waals interactions at various strengths are included in Monte Carlo simulations to show hydrophobic effects in the system. Comparisons are made to existing Monte Carlo simulations and experimental data. Understanding these patterning effects will enhance the knowledge of biomacromolecule phase separation, as well as expand the understanding of sequence-dependent polymer physics.

585393 Understanding the Properties of Imidazolium Ionenes through Systematic Variation of Monomer Structures

Ionenes are polymers that are made up of ionic groups in the backbone. The most common ionic groups found in ionenes are quaternary ammonium groups. Imidazolium ionenes have been far less studied than ammonium ionenes and as such these ionenes will be studied. In this project, the Menshutkin reaction will be used to polymerize 1,10-dibromodecane (or similar a,ω-dihaloalkanes) with 1,4-Bis((1 H-imidaxol-1-yl)methyl)benzene (and similar derivatives) via condensation (step-growth) mechanisms. The ionene product will feature imidazolium cations in the backbone with "free" bromide (Br^-) anions which can then be exchanged to bistriflimide (Tf_2N^-) anions to provide a means of making the ionene hydrophobic and more thermally stable. The influence of structural variables will be examined as relating to the polymer properties including solubility in various solvents, glass transition temperature, thermal stability, polymer nanostructure/organization and mechanical properties.

588079 Self-Assembly and Solution Structure of Supramolecular Bottlebrush Polymers

Supramolecular polymers are unique compared to traditional polymers in that they are composed of key reversible and directional secondary bonding. Secondary bonds, most notably hydrogen bonds, can be fine-tuned and can consequentially alter the polymer's material properties. Studies on self-assembly of linear chains with sidechain pendants (approximating the structure of graft polymer) typically use short small molecules for hydrogen bonding, yet it is unclear how well short polymers (<5K Daltons) will assemble. To study the impact of how sidechain length impacts supramolecular assembly, hydroxyl terminated polystyrene (PSOH) side chains were blended on to a poly-4-vinyl pyridine (P4VP) backbone. P4VP was chosen as the backbone due to the strength of the pyridine groups as an H-bond acceptor while PSOH of varying molecular weights were used as the pendant chains. Fourier transform infrared spectroscopy (FTIR) was primarily used to investigate the hydrogen bonding occurring in the blended samples. Employing FTIR methods provides fundamental results in analyzing the amount of the pyridine groups in P4VP that are participating in supramolecular self-assembly. Small angle neutron scattering (SANS) located at the NIST Center for Neutron Research (NCNR) was also used to track the mesoscopic solution structure. Final results from the FITR showed that a single hydrogen bond (OH-P4VP) appeared to be insufficient to drive self-assembly for even oligomeric polymer chains. These findings contrast from SANS results on a lauryl gallate (surfactant) polymer blends. These findings show promise in supramolecular graft polymer being based on the multiple hydrogen bonding sites.

Sustainable Light Weight Plasticized Concrete An Evolution of Sustaining Sustainability.

Jeet N. Gajera (Author)
Dwij Pandya (Co-author)
Sanskar Patel (Co-author)
Mohd. Faizan Sakariyawala (Co-author)

Sardar Vallabhbhai National Institute of Technology, Surat, Gujarat, India.

Abstract:

The most versatile man-made product used in the world is concrete. Concrete uses mainly natural resources like sand, coarse aggregate and fine aggregate, thus depleting these resources exponentially. So, to overcome these major issues we made “SLWP concrete” (The Sustainable Light Weight Plastic concrete). The natural resources like fine and coarse aggregate used in concrete were replaced by 3rd grade waste plastic like HDPE, LDPE, PP, PET in three different variations such as, first replacing coarse aggregate by waste plastic moulds of size 20mm down in two batches (i.e. 50% and 100% by volume). Then, secondly by replacing fine aggregate by waste plastic beads 4.75mm down in again two batches (i.e.50% and 100% by volume) and replacing both fine and coarse aggregate with same combinations. We also conducted Slump test, Compressive Strength test, Flexure test, Tensile Strength test on the sample. We also measured Bulk Density, Water Absorption Test, Specific Gravity, and Sieve Analysis on the plastic aggregate samples.

Keywords: SLWP Concrete, Beams, Cube, Compressive and flexure strength, Bulk density, Specific Gravity.

Need of this research project:

Recently, there was a very serious and threatening notice by the UN delegates quoting "**CRITICAL POINT OF NO RETURN"** by 2030. This meant that if sufficient efforts are not made till 2030 to reduce global warming, then all the efforts made after 2030 will be in vain. So there is an urgent need to reduce this as soon as possible.

Plastic is one of the prime reason for global warming. Plastic warms the earth twice as the aviation does. Plastic takes millions and millions of years to decompose, thus creating a huge problem for its disposal and it can't be burnt as well. Thus plastic has to be used productively somewhere such that it does not reach the earth's surface.

Concrete is the most versatile man-made material and the second most consumed material after water. Concrete uses a lot of natural resources like fine aggregates, coarse aggregates, sand, etc. Because of the excessive requirement of concrete, these resources are depleting day by day.

Project Work:

- **Methodology and Innovation:**

Firstly waste plastic waste like HDPE, LDPE, PP, PET etc. was collected and it was sorted as following:

- Plastic Bottles (PET waste)
- Crates and container (HDPE and LDPE waste)
- Crates, packaging (PP waste)

Thus, waste was sorted in this way and used in the different ways by processing.

The above mentioned waste was also containing the 3rd grade plastic waste and which could'nt be further recycled like black coloured Poly-Propylene.

This type of waste was collected in large amount in the chemical and polymers industries. Plastic Waste like PET and LDPE having low density, low compressive strength compared to that of HDPE and PP was melted in furnace and mixed with river sand to make plastic sand brick and it showed the positive results in terms of compression and fire resistance test compared to normal brick.

Image of plastic sand brick made of 3rd grade plastic waste. Also, red oxide can be added as colouring agent.

Then other types of waste like HDPE and PP having higher strength and density as compared to previous mentioned plastic they are further used in making the innovative **"SLWP-concrete"** Sustainable Light Weight Plasticized Concrete. This innovation was made to overcome the problem of excess plastic waste on the earth and to develop an era of sustainability.

The plastic waste like Black Coloured PP was taken and burned till it gets melted and the gases like VOC's, HCl, SO2 etc. were evolved. So, proper system to remove these gases was adopted. Now the melted Black PP was poured to Cuboidal Mould to get the box shape and was also poured in the cylindrical container. Now this Melted plastic was allowed to be cooled. It can be cooled in two ways:

- Forced cooling (Brittle material)
- Natural cooling (Brittle-Ductile material)

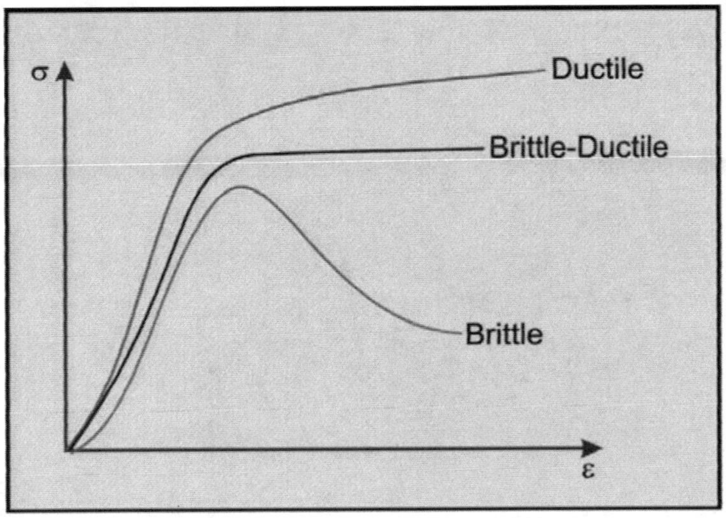

So, it was observed that natural cooling was giving good results when it is crushed. The block was allowed to cool for 1hr and then it was taken out from the mould. Mould was nearly 20mm-25mm thick to get Coarse aggregate of 20mm down. Now Plastic slab was put into the jaw-crusher to get the crushed aggregate of the desired size

Jaw Crusher

Further, crushed aggregates were sieved i.e. sieve analysis was performed and the aggregates of different sizes like >20mm, 20mm down, 10mm down, 4.75 down are separated and used for different purposes.

Now, firstly the concrete blocks of volume 15cm X 15cm X 15cm were made by replacing only natural coarse aggregate which are in high dearth by 20mm down plastic aggregates in the batches of 50% and 100% replacement. The tests showed the positive results in the compressive strength and the weight reduction of 15%-30% was observed in the 50% and 100% replacement category. Also, the beams of volume 10cm X 10cm X 50cm were made in same the batches of proportions as mentioned above. Beams were made to measure the Flexure strength.

Plastic aggregate of size 20mm down
Used to replace only coarse aggregate

Secondly, the concrete blocks of volume same as in first case was made by replacing only grits(used in concrete) which are 10mm down by the plastic beads of 10mm down in the batches of 50% and 100% replacement and this test also showed the positive results that can replace convectional concrete. Also, beams of the same volume and same batches of proportion was made to measure the flexure strength.

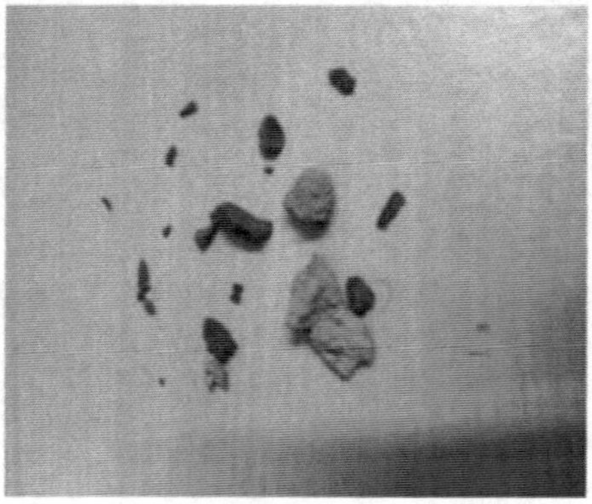

Plastic aggregate of size 10mm down used to replace only grits

Also for the both type of replacement Slump test was carried out to check out for the workability of concrete

Slump testing equipment

Results showed that the concrete was workable as convectional concrete and slump of less than 25cm was formed which showed that concrete was neither too wet nor too dry.

Following are the images of block and the beams made of **SLWP** concrete

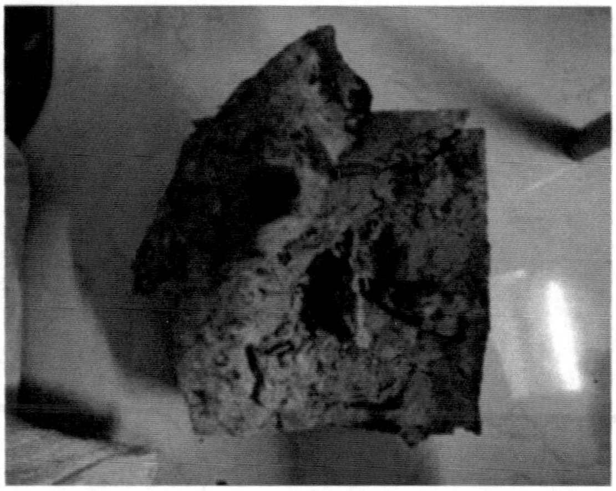

Image of SLWP concrete. Red outlined areas shows the plastic aggregate

Image of Beam made of SLWP concrete

As, we have replaced the coarse aggregates, grits and fine aggregates by the plastic aggregates of the suitable sizes it is necessary to find the properties of the plastic aggregates.

Following properties were measured:

- Bulk Density
- Specific gravity
- Water absorption test
- Fineness Modulus of aggregate

Compression Strength and flexure tests results of the SLWP Concrete will be revealed in the poster along with above mentioned aggregate properties at ASC 2019.

Also, one of the prime factor which should be taken care of is whether the **SLWP Concrete** is economical or not.
Cost of melting is added in SLWP concrete as compared to Convectional concrete. But, overall it is economical and has almost same cost as that of Convectional Concrete.

Conclusion:

Overall this project can be varied with many variations and can be used in multiple places from paving blocks to construction of building. In initial scale this project should be given place and can be adapted in small scale for making of paving blocks, Toilets at construction site and then by proper supervision and modification this can be adapted to replace convectional concrete. So, large amount of plastic waste generated on the earth can be recovered in an excellent way and natural resources can be preserved and thus we can create an era of **Sustainability**

NOTE: The results of all the experiments and tests mentioned above are certified by **National Institute of Technology Surat** Concrete lab in comparison with convectional Concrete.

584125 Layer-By-Layer Films of Photosystem I and Conducting Polymers

The integration of Photosystem I (PSI), an essential protein in photosynthesis, into photoelectrochemical cells for environmentally sustainable solar energy conversion has gained attention recently because of the protein's nearly perfect quantum yield, vast abundance, and nanoscale size. My research examines PSI's capability to form organic thin films with electrically conducting polymer interlayers. This strategy allows for parts of the polymer to extend into the PSI structure and act as an electron transfer medium between the electrode and protein or between protein layers. I form these films using layer-by-layer (LBL) Deposition, where each molecular layer on the electrode is electrostatically attached to each other by alternating charge. In this case, a layer of PSI containing mixed charge is alternated with a layer of the mixed-charged conducting polymer PEDOT:PSS for up to 10 layer pairs. Film thickness increases by an average of 1.9 nm per polymer layer and 6.2 nm per PSI layer. Infrared spectra are consistent with the presence of both the protein as well as both parts of the PEDOT:PSS polymer in the film. Contact angles were used to assess the composition of the outermost surface of the film after each deposition. Changes in contact angle are consistent with distinct layers adding to the total film for up to ten total layer pairs. Electrochemical impedance spectroscopy shows that the polymer entraps ions well to provide minimal change in impedance with increasing film thickness. Finally, photochronoamperometry shows that the photocurrent of the layered films on an electrode increases up to a critical number of layer pairs.

585225 Degradation Induced Shape Recovery of Anhydride-Based Shape Memory Polymer Composite

Shape memory polymers (SMP) are smart polymeric materials that can recover from a temporary shape to their permanent shape in response to an external stimulus. The permanent shape is achieved through physical or chemical crosslinks, while the temporary shape is a result of deformation and immobilization of the polymer chains. SMPs have become high in demand due to their many applications, ranging from medical devices and actuators to smart adhesives and sensors. We have recently created a reconfigurable shape memory elastomeric composite (Re-SMEC) featuring a water-degradable, elastomeric polyanhydride (PAH) matrix as the fixing phase, and elastic fibers as the permanent, memory retaining phase. This project explores the recovery of the original shape as a response to the degradation of the matrix. The morphology, mechanical properties, phase transitions, and degradation profile are reported.

587189 Temperature Swing Adsorption of Perfluoroorganics By Poly(N-isopropylacrylamide) Hydrogels and Membranes

This research is aimed at evaluating the effect of temperature on the removal of perfluoroorganic compounds in water filtration. Polyvinylidene difluoride (PVDF) microfiltration membranes are functionalized with poly-N-isopropylacrylamide (PNIPAAm), and its temperature responsive behavior is studied as it relates to water flux and adsorption/desorption of perfluorochemicals. PNIPAAm has a relatively low Lower Critical Solution Temperature (LCST) of around 32°C, which makes it appealing for industrial use. PNIPAAm has a sharp transition from hydrophilic behavior below its LCST to hydrophobic behavior above its LCST, thus, when the PNIPAAm is exhibiting hydrophobic behavior, it retracts/collapses upon itself in an aqueous environment. In doing so, the isopropyl groups dehydrate first, causing the polymer chain to collapse, thus causing hydrophobic contaminants to partition into that domain. Reversely, bringing the temperature down past the LCST will rehydrate the isopropyl groups, which reduces adsorption and allows desorption to occur. The stability and reusability of these membranes are demonstrated as well. Specifically, this study's goal is to evaluate the feasibility of temperature swing adsorption/desorption of Perfluorooctanoic acid (PFOA) in water filtration. This research is supported by the NIEHSSRP grant P42ES007380, and by the NSF KY EPSCOR program. Full-scale PVDF membranes were developed through collaborative work with Nanostone-Sepro (Oceanside, CA, USA).

588140 Fabrication of a Novel Reactive Membrane for Degradation of Polyfluorinated Compounds

Nanomaterials are typically used in the field of membrane technology to impart a desired characteristic on the membrane so as to enhance selectivity of the membrane and increase the life span of the membrane. Photocatalytic membrane processes generate oxygen-reactive radicals under irradiation by light that could degrade pollutants in feed solutions. Phosphorene is a recently discovered two-dimensional material that shows great promise as a photocatalyst because it can generate reactive oxygen species (singlet oxygen, hydroxyl radicals and superoxide radicals). The main objective of this collaborative research is to develop and validate environmentally safe nanomaterial-based approaches for treatment of drinking water including degradation of per- and polyfluorinated chemicals (PFAS). PFAS are surfactant chemicals commonly found in household and commercial products, but they are also very persistent in the environment and are known human carcinogens. They are very resistant to traditional drinking water treatment technologies and hence a reactive membrane that can degrade and remove these harmful chemicals during the treatment process will be worthwhile. Specifically, we aim to study the in vivo toxicity of phosphorene in its free form and in filtrates after nanoparticles are embedded into the membrane on *C. elegans* mortality, reproduction, behavior and antioxidant responses. Among PFAS that will be used, are representatives of long-chain PFAS including perfluorononanoic acid (PFNA), perfluorooctanoic acid (PFOA), perfluoroheptanoic acid (PFHpA), and perfluorooctane sulfonate (PFOS) as well as newer short-chain PFASs, such as GenX. In this research, phosphorene will be fabricated by the liquid exfoliation of black phosphorus. Membranes will be made via the non-solvent induced phase separation technique. Phosphorene will be physically incorporated into the membrane. The membranes will be characterized using several microscopic and analytical techniques and finally tested for the degradation of PFAS in solution.

587706 Direct CO_2 Capture from Air Using Poly(ethylenimine)-Loaded Polymer/Silica Fiber Sorbents

Direct CO_2 capture from atmospheric air is gaining increased attention as one of the most scalable negative

carbon approaches available to tackle climate change if coupled with the sequestration of CO_2 geologically. Furthermore, it can

also provide CO_2 for further utilization from a globally uniform source, which is especially advantageous for economies without

natural sources of carbon-based feedstocks. Solid-supported amine-based materials are effective for direct air capture (DAC)

due to their high CO_2 uptakes and acceptable sorption kinetics at ambient temperature. In this work, we describe the

application of polymer/silica fiber sorbents functionalized with a primary amine-rich polymer, poly(ethylenimine) (PEI), for

DAC. Monolithic fiber sorbents composed of cellulose acetate and SiO_2 are synthesized via the dry-jet, wet quench spinning

technique. These fibers are then functionalized with PEI (Mw 800 Da) in a simple and scalable postspinning infusion step and

tested for CO_2 capture under pseudoequilibrium conditions as well as under breakthrough conditions. An investigation to study

the effect of feed flow rate, adsorption temperature, and presence of moisture in the feed on the CO_2 breakthrough performance

of a densely packed fiber sorbent module is conducted to highlight the potential application of this class of structured contactors

in direct air capture. The pressure drop of these contactors at high gas velocities is also evaluated. Finally, a vacuum-assisted

desorption step is demonstrated for production of high-purity CO_2 from both dry and humid ambient air mixtures.

588257 The Modeling and Measurement of Solute Solubility and Partitioning for a Hybrid Supercritical Fluid/Ionic Liquid Extraction Process

Chemical separation is an important aspect of chemical engineering, and many techniques for these processes are large energy sinks and produce a lot of waste. Ionic liquid enhanced supercritical fluid extraction is one way to reduce both the energy required and waste produced. Ionic liquids (ILs) are organic salts that have a melting point of 100°C or below and have practically no vapor pressure. Supercritical fluids (SCF) are compounds that are above both their critical temperature and pressure. These compounds have gas-like transport properties and liquid-like solvation properties. Both ILs and SCFs can be tuned to perform specific tasks. IL's ions can be modified by changing the functional groups to alter certain properties like solvation affinity for desired compounds, thermal properties, and physical properties such as density and viscosity. SCFs can be modified by changing temperature or pressure to attain transport or solvation properties that are more like a gas or a liquid, respectively. IL enhanced SCF extraction utilizes these properties to create a separation technique that may be able to selectively separate a mixture into its base compounds.

The current project is to demonstrate this separation technique by utilizing the Peng-Robinson Equation of State with Van der Waals mixing rules to model the solubility of different dyes in the $SCCO_2$ to determine the pressure needed to ensure the maximum amount of solubility within reason. Preliminary experiments have been conducted to transfer dye from a solid mixture to an IL and from a dye-saturated IL to a pure IL. All experiments conducted have supported the hypothesis that SCFs can be used to as a pathway to transport compounds from solid/liquid to another IL. A computer program, COSMO-RS, has been utilized as a screening method to analyze ILs and various dyes to calculate the activity coefficients of the ionic liquid. This allows a partition coefficient to be calculated and a combination of dye/IL to be selected to use in upcoming experiments. The current goal of the work is to selectively separate two different dyes into two different ILs, using disperse dyes as model compounds to demonstrate the utility of non-contact separation. The eventual goal, however, is to move from dyes to more practical implications such as extracting pharmaceutical compounds using this method, improving separation techniques in the chemical engineering industry.

587717 *Concentration of Sugarcane Juice Using Membrane Distillation.*

The Membrane Distillation (MD) is a thermally driven separation process in which only vapors are allowed to pass through a porous hydrophobic membrane. It has many applications like desalination, process water treatment, one of them is a concentration of Sugarcane juice. MD operates at low temperature and pressure conditions as compared to other conventional techniques like Multistage Flash (MSF), Multi Effect Evaporator (MEE), etc.. The MD system consumes low-grade heat ,which make it more attractive for economic separation by integrating solar collectors. Many developing countries from South America along with India and Thailand are the largest producers of sugar all over the world .The major problem faced by them is Carbon emission . Results show that sugar production has a carbon footprint value of 0.43-0.63kg CO2 e/kg sugar out of which 65% is contributed by agriculture. The rest 35%, which is 0.229 kg CO2 e/kg sugar is contributed by industry.The higher carbon footprint of production raises concern of global warming .Thus the conventional techniques need to be revolutionized with new modern techniques .The poster highlights the experimental data acquired from such modern technology and reports for concentration of sugarcane juice .The integration of solar energy and MD working in synergy will pave the way for sustainable production.

585481 Aminopropylsiloxane Networks for Regenerative Carbon Capture Sorbents

With over half of the world's energy generated from the combustion of fossil fuels, the demand for developments in carbon dioxide (CO_2) capture technology has surged over the last decade. The end goal for captured CO_2 is to utilize it by chemically converting it into value-added products. Currently, amine sorbents are the most prominent and effective mechanism for sequestering CO_2 at both high and low concentrations. Historically, liquid amines have been used for adsorption of CO_2 from flue gas on the commercial scale. However, aqueous sorbents capture CO_2 by the physical formation of bonds (physisorption), therefore requiring more energy to regenerate the sorbent compared to a solid sorbent that adsorbs CO_2 by chemisorption. Currently, research efforts are focused on increasing the adsorption capacity of solid materials in order to be cost-efficient and comparable to that of aqueous capacities. Here, a silica-supported, solid sorbent was functionalized with propan-1-amine groups using aminopropylsiloxane networks. Nitrogen loading on the sorbent was measured using a LECO elemental analyzer. Absolute CO_2 adsorption capacity was obtained by thermogravimetric analysis (TGA). Monolayer uptake measurements, corresponding to active surface area of the sample, were taken by a Quantachrome Gas Sorption Analyzer. Monolayer uptake and nitrogen content were compared to gain an understanding of the aminopropylsiloxane assembly and active site availability to incoming gas. Thermal cycling stability of the sorbents was also characterized using TGA to test the sorbent's ability to regenerate active surface area upon CO_2 desorption. These solid sorbents are considered "low temperature" with adsorption and desorption steps both occurring at temperatures lower than 200°C. Sorbents with different mass loadings of nitrogen and varying functional group length were analyzed and compared.

585623 Effect of Incomplete Purging on Carbon Capture By Pressure Swing Adsorption with High-Pressure Rinsing

Pressure swing adsorption (PSA) has the potential to be used for post-combustion CO_2 capture, as well as CO_2 capture from other gas sources. However, carbon capture based on adsorbent equilibrium selectivity is not viable with basic PSA cycle scheduling since recovery of reasonably pure CO_2 requires pressure ratios which are unrealistically large and expensive. High-pressure rinsing is required to remove residual air from the bed prior to blowdown, ensuring the recovery of pure CO_2 at more moderate pressure ratios. The process with rinsing might still be improved by leveraging an incomplete purge, using a volume of sweep gas less than the minimum needed to ensure a clean bed, which has been known to improve recovery in O_2 and N_2 production. [1] A PSA unit recovering post-combustion CO_2 from a 2.6 MMSCFD flue gas source with zeolite 13X was modeled based on the equilibrium theory of Kayser and Knaebel. [1, 2] The sorption/desorption pressures, system temperature, and extent of purging was optimized by minimizing OPEX and the effects of varying the cost of electricity and CO_2 offset were investigated. The energy costs associated with capture were then compared to standards for traditional gas absorption using amine solvents.

References:

1) D.M. Ruthven, S. Farooq, and K.S. Knaebel, Pressure Swing Adsorption, John Wiley & Sons, New Jersey, 1994.

2) J.C. Kayser and K.S. Knaebel. Chem. Eng. Sci., 44, 1989, 1 – 8.

585775 Engineering Membrane Adsorbers for Pfas Mitigation from Water

Per-and polyfluoroalkyl substances (PFAS) and related compounds such as perfluorooctanoic acid (PFOA) and perfluorooctane sulfonate (PFOS) are labeled as persistent organic pollutants by the US Environmental Protection Agency (EPA). PFAS were used in firefighting foam, non-stick cookware, waterproof outdoor gear, food packaging,and more. PFAS have contaminated the drinking water supply of many countries around the world, including the US.According to the EPA, PFAS accumulate and stay in the human body for long periods of time and are harmful to the body's reproductive and immunological systems. In 2019, the EPA recommends PFAS levels of no more than 70ppt in drinking water and millions of people in the US are regularly drinking water contaminated with more than this level. There is a need to effectively and efficiently remove PFAS compounds from drinking water. Currently granular activated carbon and ion-exchange/non-ion-exchange polymers in the form of resin beads are used to adsorb PFAS from water in water treatment plants. These methods utilize diffusive flow to adsorb PFAS which is a slow process. Nanofiltration and reverse osmosis can also be used to remove PFAS from water but are expensive to perform due to pressure requirements. If membrane adsorbers utilizing convective flow were able to effectively capture PFAS from water they would enable a new treatment technique for PFAS mitigation. This would allow the adsorption of PFAS to be regulated by slightly pressurizing the contaminated water to pass it through the membrane, resulting in a more efficient method of purification.

Polyethersulfone membranes were utilized to remove sodium dodecyl sulfate (SDS), a surrogate of PFAS,from water. UV photopolymerization was utilized to graft a polymer layer from the membrane surface.2-[(methacryloyloxy)ethyl]trimethylammonium chloride (MOETMA), 2-(dimethylamino)ethyl methacrylate (DMAEMA) , 4-vinylpyridine, and butyl acrylate (spacer molecule) were the polymers experimented with. The membranes weretested with and without the addition of a cross-linker, methylenebis(acrylamide) (MBAA),which acts to stabilize the adsorbing polymer chains on the surface of the membrane. Bromoethane was used to quaternize poly(4-vinylpyridine). The positively charged polymer chains allowed the membrane to adsorb the negatively charged SDS from solution. Each membrane was flux tested to determine if modification was successful. ATR-FTIR spectroscopy was used to determine the extent of modification. Contact angle goniometry was used to determine the effect of modification on the hydrophilicity of the membrane. Static adsorption testing was done to determine the membrane adsorption capacity for SDS.

585985 Graphene Oxide Membranes: Synthesis Methods, Unique Properties, and Potential for Perfluorooctanoic Acid (PFOA) Removal

Graphene oxide (GO) creates a thin membrane with tight interlayer spacing that allows it to maintain high permeability along with high particle selectivity. A GO membrane was created using a vacuum filtration cell to cast a dilute, aqueous solution of graphene oxide onto a polyvinylidene fluoride (PVDF200) substrate. A GO membrane with a 33% decrease in graphene oxide loading showed decreased membrane thickness, increased permeability, and decreased rejection of dextrose molecules suggesting that the incidence of defects increased when GO loading was reduced. The graphene oxide adopted a negative charge at any pH above ~3.5 and exhibited a maximum negative zeta potential of -22.5 mV at pH=8.2. Graphene oxide's negative surface charge in neutral water, improves its rejection of negatively charged particles such as perfluorooctanoic acid (PFOA) by charge repulsion. PFOA is a harmful, biologically persistent compound that contaminates water supplies, particularly near chemical plants where it is manufactured or disposed of. PFOA rejection was found to be highest in a high pH solution with a low PFOA concentration. The GO membrane rejected 58% of a 20-ppm PFOA solution in distilled water (pH=7) but rejection increased to 75% when a 1-ppm PFOA solution was used in the same conditions. It is predicted that lower concentrations of PFOA, like those present in drinking water, could achieve near complete rejection. When a lower pH solution (pH 3) was used, the retention appeared to be nearly as high, but the "rejection" proved to be entirely due to adsorption on the membrane surface based on a large disparity in the mass balance. PFOA rejection dropped to 15% when tested in a 90/10 isopropanol solvent with 1 ppm PFOA concentration, suggesting that charge plays a major role in rejection.

586295 Self-Regulating Forward Osmosis-Reverse Osmosis Hybrid System

Self-Regulating Forward Osmosis-Reverse Osmosis Hybrid System

Colin Fitzsimonds, Noah Ferguson, Dr. Jeffrey McCutcheon

Department of Chemical and Biomolecular Engineering, University of Connecticut, Storrs, CT

Although reverse osmosis (RO) is a commonly used desalination technology, the susceptibility of these membranes to foul makes them challenging to operate in wastewater reuse applications. Pretreatment of these feed solutions is one method of reducing the fouling of RO membranes. Such systems often significantly increase the footprint of the desalination plant and the CAPEX for construction. Using forward osmosis (FO) as a pretreatment method in a hybridized system has been considered as an alternative method to dealing with difficult to treat feed waters. The potential complexities to balancing the fluxes between the FO and RO systems and overall uncertainty about the operation of these systems act as a deterrent to their use in a desalination plant. In this work, we demonstrate self-regulating behavior that could allay fears of needing complex process control regimes in these types of hybridized systems. We demonstrate this behavior using a module-scale testing system that mirrors the behavior of a larger scale system. The system shows flux convergence between the FO and RO modules when disrupted by a sudden change in the RO module pressure and therefore does not require complex controls to ensure flux convergence between the two processes.

583614 A Bead-Based Microfluidic Testbed for an Enhanced Lab-on-a-Chip Device for Shear Enhanced Purification of Biomolecules

- Download/View Extended Abstract

Protein purification is cumbersome, expensive and requires a number of steps. Post purification, the process equipment often requires extensive, costly maintenance that needs harsh chemicals. Here, we present a new methodology of using shear forces that offers an alternative approach to protein purification that reduces the number of steps, making it easier, greener, scalable and cheaper. Analogous to affinity chromatography, a stationary phase – Carbon Nanotubes (CNTs) - will be used to purify proteins. Protein A will be attached to CNTs using EDC-NHS chemistry. Post attachment, CNT's will be tightly packed in a microfluidic channel (akin to a packed bed reactor) and are immobilized (against the fluidic drag force) by a transverse electric field from non-planar interdigitated electrodes. At low flow rates, the target proteins (monoclonal antibodies) attach to Protein A while the rest of the solution flows by (filtration). The flow rate is then suitably modified to increase the shear force enough to release the biomolecules from the CNT complex into an elution buffer while maintaining their bio-functionality (purification). Hence, the shear force approach drastically reduces the number of steps required for the desired purification. Post elution buffer, prolonged washing with a solution allows us to regenerate the CNT's for the next round of purification. To develop optimal conjugation protocols for the process, the whole attachment of protein A to separation has to be done in situ. We have created a testbed with 1 μm carboxylated silica particles to mimic the CNT. Unlike CNT (quencher), silica particles allow us optical (fluorescence) visualization. We have studied in detail the effect of fluid flow, pH, and temperature on the purification efficacy of our platform using fluorescent antibodies. Further the in-situ protocols for EDC-NHS functionalization, antibody attachment and stripping has also been studied using different fluorescent antibodies. The optimization on the silica beads allowed us to translate to a CNT based device which offers tremendous enhancement in shear force among others. We believe that this technology not only benefits the protein manufacturing scene in general (i.e., production of mainstream proteins such as insulin), but also other contemporary, specific applications such as using monoclonal antibodies from cancer treatment to sensors.

588320 Towards Fractionation of Soybean Meals in a Water- and Chemical-Free Environment

Traditional wet fractionation techniques have many drawbacks due to the utilization of acid, alkali and chemicals which cause denaturation and, consequently, loss of protein solubility. In addition, plenty amount of resources such as water and energy are required to produce protein isolates using wet methods. In contrast, dry separation processes such as air classification, sieving and tribo-electrostatic separation preserve native functionality of proteins. The aim of this research is to implement the novel tribo-electrostatic separation technique for the production of protein-rich and fiber-rich fractions from defatted soybean (*glycine max*) flour followed by optimizing the separation operating parameters.

The tribo-electrostatic separation is based on the phenomena that proteins have better chargeability properties than carbohydrates. This can be explained by the presence of many ionizable groups in proteins such as amine groups, carboxyl groups and side chains. Our custom-built separation unit including a tribo-charging tube, separation chamber with two oppositely charged electrodes, and fluidized bed was implemented in the Bioprocess Engineering Research Laboratory at Howard University. In this study, soybean particles were first charged in the charging tube and then separated in the fractionation chamber based on the acquired charges under the influence of external electric field. The optimization of process parameters was performed. We tested two different air flow rates (7LPM and 15LPM) and three plate voltages (±1kV, ±3.5kV, and ±6.5kV).

At the most optimal processes conditions, the initial protein and fiber contents of soybean flour were increased from 55.3% and 15.3% to 58.4% and 19.6%, respectively. These accounted for 66.5% of the total protein and 60.8% of the total fiber. This novel technology is water-free, chemical-free and eco-friendly alternative to the traditional wet protein separation methods.

585330 Comparing the Efficacy of Enveloped Virus and Non-Enveloped Virus Purification in Aqueous Two-Phase System

The existing need of viral vaccines in the less developed regions and growing demand for gene therapy vectors and oncolytic virus therapy has demanded the biopharma companies to upgrade the current production strategies. Downstream processing to purify viral particles from the host cell contaminants is one of the major contributors to not only limiting the throughput but also cost of vaccine production. Current methods based on chromatography and filtration are facing a major challenge to meet the required high yield production. Aqueous two-phase system (ATPS) has been identified as one of the potential candidates to address these issues in the vaccine manufacturing. ATPS formed by two partially miscible aqueous solutions can be composed of polymer-polymer or polymer-salt. However, the discrepancy in understanding the biomolecule partitioning mechanism has stalled the industrial implementation. Our previous study with porcine parvovirus, a non-enveloped virus, has inferred that two major forces, hydrophobic and electrostatic interactions, drive viral particles and contaminant proteins to the opposite phases in a polyethylene glycol (PEG, MW 12000 Da) and citrate salt system. This work explores the purification of structurally different category of viruses than the previous study to generalize the notion of above-mentioned driving forces and reduce the experimental space to optimize the purification. This study used model enveloped virus (herpes simplex virus, HSV-1), which has an extra bilipid membrane around the protein capsid enclosing the genetic material. The recovery of the virus in the PEG-rich phase was studied with an increasing PEG and citrate concentrations at various tie line lengths to study the partitioning behavior of viruses. Further, water structure amending compounds, osmolytes (glycine, betaine, and TMAO), were included in the system to compare the enhanced driving forces to boost the purification at lower compositions. The results from our experiment can be used to provide higher virus recovery and help lower overall vaccine production cost.

588335 Ionic Liquid Enhanced Supercritical Fluid Extraction

Ionic liquid enhanced supercritical fluid extraction (ILESCFE) is a novel, environmentally benign extraction process that carries the potential to be a widely applicable, tunable alternative to traditional extractions. Traditional extractions rely heavily on volatile organic solvents which are often toxic and can easily escape into the environment as fugitive emissions. ILESCFE eliminates the issue of fugitive emissions due to the fact that the supercritical fluid (SCF) is environmentally benign and ionic liquids (ILs) effectively lack a vapor pressure. The properties of the SCF can be tuned by slight adjustments to temperature and pressure and ILs can be tailored to specific tasks through the addition of different functional groups and by using different cation-anion pairs. ILESCFE operates under the 0th Law of Thermodynamics to accomplish no-contact *in situ* concentration of the solute into the desired ionic liquid. Systems of multiple ionic liquids can be designed to achieve high selectivity of the desired solute in an extraction. This research focuses on modelling with COSMO-RS which is a quantum chemistry-based method that can predict the behavior (activity coefficients) of species in a mixture. COSMO-RS can be used to quickly screen countless ionic liquids to determine which ILs exhibit the proper selectivity for a given solute. Preliminary screening has indicated efficacy of COSMO-RS for the scope of this project. Qualitative proof-of-concept experiments have been performed in a Parr bomb reactor at 40 °C and 1450 psi with various hold times. Solid Disperse Orange 1 has been successfully concentrated *in situ* to a generic IL (1-hexyl-3-methylimidazolium bis(triflouromethylsulfonyl)imide). The transfer of Disperse Orange 1 saturated in a generic ionic liquid to the same pure ionic liquid has also been successfully demonstrated in a proof of concept experiment. These experiments demonstrated transfer of a solute from both its pure solid form and saturated in a liquid. Selectivity has been demonstrated as well through the *in situ* concentration of a solid mixture of Disperse Orange 1 and Methylene Blue to a generic IL with the end result showing a higher affinity for Disperse Orange 1. Dyes have been used as model compounds due to the ease in which quantitative measurements can be taken through the use of UV-Vis spectroscopy.

588366 Residence Time Distribution and Holdup in Centrifugal Extractors

Annular centrifugal extractors (ACE) are liquid-liquid extractors which achieve high extractive efficiencies with a fast residence time. However, due to a fundamental lack of understanding of the driving force and the kinetics of ACE units, scale-up is, currently, not feasible. This study seeks to quantitatively understand the hydrodynamics of the ACE units through phase holdups and residence time distributions (RTD). Ultimately, data from this study will help build mathematical models to improve the physical description of the system. To experimentally determine RTD and holdup, organic (30% tributyl phosphate in isopar) and aqueous (0.0071% nitric acid solution) streams were brought to steady-state through a 4-pack extractor unit. Then, Sudan I dye was added to the inlet organic stream, and concentration data was collected over a 10-minute period from the organic outlet stream. Subsequently, RTD and holdup were quantitatively calculated with the concentration data.

583800 Liquid-Liquid Equilibria Studies of Potential Organic Solvents As Ethanol Extractants from Aqueous Solutions

Recent years have brought concern for the environment to the forefront, specifically with negative impacts of fossil fuels. Jet fuels are estimated to contribute to 2% of CO_2 emissions and are predicted to rise. Many alternatives to petroleum based fuels are being explored, including biofuels which have the potential to reduce CO_2 emissions by 85%. However, biofuels do not come without challenges. Inevitably, these fuels will have high water content, due to water formed during biological reactions. Unfortunately, water can be damaging to engines, requiring removal for optimum performance rates. Hydrocarbons, such as ethanol, have unideal interactions with water, forming an azeotrope, and cannot be fully separated via simple chemical engineering separations such as distillation, making the process highly energy-intensive and expensive. Liquid-liquid extraction could be a more cost-effective solution, with a proper solvent, because it can be implemented with smaller equipment and lower energy consumption. This project analyzes the effectiveness of using three green, organic solvents: castor oil, dodecane, and 2-methylcyclohexanol, to extract ethanol from water. Liquid-liquid equilibrium was determined using the cloud point method and data for the three ternary systems was collected using headspace gas chromatography at 25°C. A ternary diagram for each organic solvent, water, and ethanol was obtained. Preliminary liquid-liquid extractions using a laboratory scale mixer-settler were used to compare the effectiveness of each solvent in separation ethanol from the aqueous stream. Parameters gleaned from the ternary systems and the preliminary liquid-liquid extractions studies were used to analyze the potency of each solvent in the separation of ethanol. Ultimately, this work provides new insights into the use of green solvents as possible ethanol extraction agents in liquid-liquid extractions processes.

586241 The Effect of Calcination Extent on Sorbent Utilization in Calcium Looping

Calcium looping is a promising carbon capture technology, circulating CaO-based sorbents between carbonator and calciner units. Natural limestone is a popular choice for the sorbent material due to its abundance and low cost. However, rapid sorbent deactivation is a major challenge to the long-term stable operation of the calcium looping system. Previous studies have mostly focused on the decay of utilization of fully calcined sorbents over multiple cycles of carbonation/calcination. However, lime-based sorbents are rarely fully calcined in practical operations and, hence, the effect of partial calcination on sorbent cyclic performance is of interest.

This work investigates the effect of the extent of limestone calcination on sorbent utilization decay over multiple cycles. Greymont limestone particles (510-700 μm in diameter) were calcined and carbonated for 20 successive cycles in a thermogravimetric analyzer (TGA). The extent of calcination was varied by changing the calcination time, with the calcination time being fixed for each cyclic test. It was observed that partial calcination results in enhanced sorbent utilization over the testing cycles. The experimental results give better understanding of the mechanism of sorbent utilization decay, while also providing more experimental support for prospective calcination kinetics modelling. Future study will verify the observed effects of partial calcination for prolonged numbers of cycles and different calcination operating conditions.

583416 Continuous Viral Purification Using an Aqueous Two-Phase System

Current methods and ordering of bioproduct separation processes are driven by purity, yield, and activity requirements. After inoculation, bioseparation processes are long, sequential unit operations including filtration, extraction, and chromatography. The downstream process of viral products currently accounts for 70% of production overhead with a 30% acceptance yield. Our goal is to improve viral purification techniques with a continuous Aqueous Two-Phase System (ATPS) that produces a higher yield, lower cost, and faster production. ATPS is a liquid-liquid partitioning formed by mixing two immiscible aqueous solutions. Our methods involves the continuous processes of a parvovirus with ATPS, followed by membrane separation. Based on previous experiments, the critical concentration of ATPS components, polyethylene glycol (PEG) and citrate, were pumped continuously and spiked with porcine parvovirus (PPV). Virus hydrophobicity and electrostatic interaction are used as driving forces to concentrate viral particle impurities into the bottom phase and virus into the top phase of the ATPS system. To capture and isolate the product, extraction of PEG from 9 log PPV by ultrafiltration (UF) is being explored as a potential final polishing step. Varied psi and PEG concentrations were tested to find optimal recovery to be applied in the design of the continuous process. PPV virulence and structural integrity determined by qualitative (TEM or DLS) and quantitative (MTT assay) measure. The results from our experiments can be used to show that ATPS is successful at purifying viruses continuously with potential to incorporate into the manufacturing of viral therapeutic products.

585207 Green Solvent Extraction of Soy Biomass

Biofeedstocks contain a variety of components that can be used as fuels or as value-added chemicals and are of particular interest in the energy, pharmaceutical, and nutraceutical industries. Solvent extraction is a useful and well-developed means of extracting many of these components. Because solvents account for a considerable portion of process waste, it is important that green solvents (solvents with fewer environmental, health, and safety impacts), are used to extract these products. In this work, a variety of green solvents were used to extract triglycerides and phospholipids from two biofeedstocks, soy lecithin and soybean flour.The components in each extract were analyzed using supercritical fluid chromatography-mass spectrometry in order to determine the extraction efficiency and selectivity of each solvent. Overall, it is seen that certain solvents, such as acetone and ethyl acetate, were more selective towards triglycerides than phospholipids while other solvents, such as 2-methylhudrofuran, were more selective towards phospholipids than triglycerides. These results make it easier to select a particular solvent when carrying out a solvent extraction based on the yields of the component needed in an extract. In addition, this work analyzes the properties of each solvent to allow for prediction of the effectiveness of other solvents with similar properties in order to aid in the selection of solvents for extraction processes.

584863 Non-Enveloped Virus Purification Using Cation Exchange Chromatography

The supply of vaccines is not consistent and can leave countries without access to lifesaving vaccines. Companies are leaving the market due to multifactorial production issues which leaves countries without access to vaccinations. Vaccine manufacturer's ability to make pure and potent vaccines depend directly on the capability to purify viruses without a decrease in yield. Currently, chromatography is the most common method in industry to purify bioproducts because of its versatility. The mode that is most used for viral purification is ion exchange chromatography. Currently, anion exchange chromatography (AEX) is used more often in virus purification. AEX allows negatively charged viruses to bind to a positively charged chromatography surface at neutral pH, where viruses are the most stable. However, cation exchange chromatography (CEX) has the potential for improved purity. All forms of chromatography share the same basic steps: equilibration, loading, elution, washing, and regeneration. CEX uses two forms of elution, increasing salt gradients or increasing pH gradients. The more common of the two is the salt gradient because of its simplicity and reliability. However, virus elution yields are often low with salt gradients due to the variability of the viral surface. Gradients using a change in pH exploit the biomolecule's isoelectric point (pI). pH gradients are more complex to design but have the potential to provide the necessary separation to rapidly purify vaccine products. The non-enveloped virus porcine parvovirus (PPV), with a reported pI of 5, was purified with CEX. PPV was loaded at pH 3.5 and eluted with a pH step gradient. High recovery of virus particles was found during the elution step of pH 6 with high concentration of the virus eluting over the duration of the pH 6 step. Purity analysis demonstrated a removal of 92% contaminating protein and 98% DNA. Future work will comprise of purification of different viruses and increasing loading volumes to concentrate products. This research will result in improved methods of purification that will bring new vaccines and gene therapy vectors to market and keep costs low.

586509 Low-Cost Microfluidic Devices for Bio-Separation Applications

Single-use vessels have gained significant popularity in the biological processing industry. In this work, we develop software and process equipment for manufacturing a single use or recyclable separation device for the extraction of DNA from solution. Additional software is developed to control the extraction process. Separation units consist of submillimeter expansion-contraction channels, fabricated from machined thermoplastics using a low-cost (<$200) micro-mill. The mill is controlled by a Raspberry Pi, running a homebrew Python code which generates and streams commands to the mill. The separation of DNA from solution is achieved by Electrohydrodynamic Migration, in which DNA molecules are stretched into cigar-shaped conformations by the shearing forces generated by the Poiseuille flow, with the long axis tilted towards the center of the confining channel. An electric field, applied counter to the flow, then has the effect of selectively driving the DNA towards the channel walls. Near the walls, the motion of the DNA is dominated by electrophoresis, which returns the molecules to the inlet of the device. The process leads to a selective concentration of DNA in a small region around the beginning of the contraction.

Separation and concentration require precise control of the applied voltage and pressure differences which are achieved by Arduino microcontrollers controlled via serial connections by a Python interface. Using Python for experimental control allows for reproducible experiments, by saving the controlling instructions as ladder logic in text files. Additionally, the Python interface allows for real-time analysis of the images gathered by the microscope camera, which can be used to fine tune the experiment via application of feedback controls.

AUTHOR INDEX

AUTHOR INDEX

AUTHOR INDEX

AUTHOR INDEX

AUTHOR INDEX

AUTHOR INDEX

AUTHOR INDEX

AUTHOR INDEX

AUTHOR INDEX

AUTHOR INDEX

AUTHOR INDEX

AUTHOR INDEX